T0186533

Gynecologic Oncology

Paola A. Gehrig, MD
Department of Obstetrics and Gynecology
Division of Gynecologic Oncology
University of North Carolina at Chapel Hill
Chapel Hill, North Carolina

Angeles Alvarez Secord, MD
Department of Obstetrics and Gynecology
Division of Gynecologic Oncology
Duke University Medical Center
Durham, North Carolina

LANDES
BIOSCIENCE
AUSTIN, TEXAS
USA

VADEMECUM
Gynecologic Oncology
LANDES BIOSCIENCE
Austin, Texas USA

Please address all inquiries to the Publisher:
Landes Bioscience, 1002 West Avenue, Austin, Texas 78701, USA
Phone: 512/ 637 6050; FAX: 512/ 637 6079

ISBN: 978-1-57059-705-3

Library of Congress Cataloging-in-Publication Data

Library of Congress Cataloging-in-Publication Data

Gynecologic oncology / [edited by] Paola A. Gehrig, Angeles Alvarez Secord.
    p. ; cm. -- (Vademecum)
 Includes bibliographical references and index.
 ISBN 978-1-57059-705-3
 1. Generative organs, Female--Cancer--Handbooks, manuals, etc. I. Gehrig, Paola A.
 II. Secord, Angeles Alvarez. III. Series: Vademecum.
 [DNLM: 1. Genital Neoplasms, Female--Handbooks. WP 39 G9972 2009]
 RC280.G5G8873 2009
 616.99'465--dc22
                              2009021383

While the authors, editors, sponsor and publisher believe that drug selection and dosage and
the specifications and usage of equipment and devices, as set forth in this book, are in accord
with current recommendations and practice at the time of publication, they make no warranty,
expressed or implied, with respect to material described in this book. In view of the ongoing
research, equipment development, changes in governmental regulations and the rapid accumulation of information relating to the biomedical sciences, the reader is urged to carefully review and
evaluate the information provided herein.

Dedication ═══════════════════════

To our husbands, children, and parents, without whose love, tolerance and understanding our careers would not be possible.

About the Editors...

PAOLA A. GEHRIG is an Associate Professor in Gynecologic Oncology in the Department of Obstetrics and Gynecology at the University of North Carolina at Chapel Hill. She is a Fellow of the American Board of Obstetrics and Gynecology and of the American College of Surgeons. She has published over 30 peer-reviewed articles and is a member of the Society of Gynecologic Oncologists, the American Society of Clinical Oncology, the American Association for Cancer Research and the Gynecologic Oncology Group. Her primary research focus is on identifying novel therapeutic regimens for the treatment of women with endometrial cancer. She lives in Chapel Hill with her husband, Tom, and children, Lauren and Joseph.

About the Editors...

ANGELES ALVAREZ SECORD is an Associate Professor in gynecologic oncology in the Department of Obstetrics and Gynecology at Duke University Medical Center in Durham, North Carolina. She is a Fellow of the American Board of Obstetrics and Gynecology. She has published over 40 peer-reviewed articles and is a member of the Society of Gynecologic Oncologists, the American Society of Clinical Oncology, the American Association for Cancer Research, and the Gynecologic Oncology Group. Her research interests include clinical trials for women with gynecologic cancers, the regulation of angiogenesis in ovarian cancer, and the use of genomic expression arrays for directing biologic therapy in ovarian cancer. She lives in Chapel Hill with her husband, Steve, and children, Connor, Joshua and Lena.

Contents

Editors

Paola A. Gehrig, MD
Department of Obstetrics and Gynecology
Division of Gynecologic Oncology
University of North Carolina at Chapel Hill
Chapel Hill, North Carolina

Angeles Alvarez Secord, MD
Department of Obstetrics and Gynecology
Division of Gynecologic Oncology
Duke University Medical Center
Durham, North Carolina
Chapter 5

Contributors

Lisa N. Abaid, MD, MPH
Gynecologic Oncology Associates
Newport Beach, California
Chapter 20B

Ron R. Allison, MD
Department of Radiation Oncology
The Brody School of Medicine
East Carolina University
Greenville, North Carolina
Chapter 7

Victoria Bae-Jump, MD, PhD
Department of Obstetrics
   and Gynecology
Division of Gynecologic Oncology
University of North Carolina
   at Chapel Hill
Chapel, Hill, North Carolina
Chapter 16

Andrew Berchuck, MD
Department of Obstetrics
   and Gynecology
Division of Gynecologic Oncology
Duke University Medical Center
Durham, North Carolina
Chapter 1

Michael A. Bidus, MD
Department of Obstetrics
   and Gynecology
Division of Gynecologic Oncology
Naval Medical Center Portsmouth
Portsmouth, Virginia
Chapter 6

Cecelia H. Boardman, MD
Department of Obstetrics
   and Gynecology
Dianne Harris Wright Professor
   of Obstetrics and Gynecologic
   Oncology Research
Virginia Commonwealth University
   Health System
Richmond, Virginia
Chapters 9, 21

John F. Boggess, MD
Department of Obstetrics
   and Gynecology
Division of Gynecologic Oncology
University of North Carolina
   at Chapel Hill
Chapel Hill, North Carolina
Chapter 20B

Michael E. Carney, MD
Department of Gynecologic Oncology
Women's Cancer Center
Cancer Research Center of Hawaii
University of Hawaii
Honolulu, Hawaii
Chapters 16, 17

David E. Cohn, MD
Department of Obstetrics
  and Gynecology
Division of Gynecologic Oncology
The Ohio State University College
  of Medicine
Columbus, Ohio
Chapter 3

Christopher Darus, MD
Maine Women's Surgery
  and Cancer Center
Scarborough, Maine
Chapter 15

John C. Elkas, MD, JD
Gynecologic Oncologist
Northern Virginia Pelvic
  Surgery Associates
Annandale, Virginia
Chapter 6

Hiram A. Gay, MD
Department of Radiation Oncology
Washington University in St. Louis
St. Louis, Missouri
Chapter 7

Robert L. Giuntoli, II, MD
Department of Gynecology
  and Obstretrics
The Kelly Gynecologic Oncology
  Service
Johns Hopkins Medical Institutions
Baltimore, Maryland
Chapter 13

Warner Huh, MD
University of Alabama at Birmingham
Birmingham, Alabama
Chapter 8

William P. Irvin, Jr, MD
Riverside Cancer Care Center
Newport News, Virginia
Chapter 15

Amy L. Jonson, MD
Department of Obstetrics,
  Gynecology and Women's Health
Division of Gynecologic Oncology
University of Minnesota
Minneapolis, Minnesota
Chapter 11

Patricia L. Judson, MD
Department of Obstetrics,
  Gynecology and Women's Health
Division of Gynecologic Oncology
University of Minnesota
Minneapolis, Minnesota
Chapter 11

Lynne M. Knowles, MD
Department of Obstetrics
  and Gynecology
Division of Gynecologic Oncology
UT Southwestern Medical Center
Dallas, Texas
Chapter 14

Jason A. Lachance, MD
Programin Women's Oncology
Brown Alpert Medical School
Providence, Rhode Island
Chapter 12

Paula S. Lee, MD
Department of Obstetrics
  and Gynecology
Division of Gynecologic Oncology
Duke University Medical Center
Durham, North Carolina
Chapter 1

Brigitte E. Miller, MD
Department of Gynecologic Oncology
Wake Forest University School
   of Medicine
Winston-Salem, North Carolina
Chapter 4

Susan C. Modesitt, MD
Department of Obstetrics
   and Gynecology
Gynecologic Oncology Division
University of Virginia
Charlottesville, Virginia
Chapter 2

T. Michael Numnum, MD
Department of Obstetrics
   and Gynecology
Division of Gynecologic Oncology
University of Alabama
Birmingham, Alabama
Chapter 8

Matthew A. Powell, MD
Department of Obstetrics
   and Gynecology
Division of Gynecologic Oncology
Washington University School
   of Medicine
St. Louis, Missouri
Chapter 10

Heather S. Pulaski, MD
Department of Obstetrics
   and Gynecology
University of Michigan
Ann Arbor, Michigan
Chapter 4

Marcus E. Randall, MD, FACR
Department of Radiation Medicine
University of Kentucky
Lexington, Kentucky
Chapter 7

Laurel W. Rice, MD
Department of Obstetrics
   and Gynecology
School of Medicine and Public Health
University of Wisconsin
Madison, Wisconsin
Chapter 12

Mildred Ridgway, MD
Women's Specialty Center
Jackson, Mississippi
Chapter 19

Jennifer Rubatt, MD
Department of Obstetrics, Gynecology
   and Reproductive Sciences
Division of Gynecologic Oncology
Magee Women's Hospital
Pittsburgh, Pennsylvania
Chapter 5

Teresa Rutledge, MD
Department of Obstetrics
   and Gynecology
Division of Obstetrics
   and Gynecology
University of New Mexico
Albuquerque, New Mexico
Chapter 20C

Bradley Sakaguchi, MD
University of Hawaii
Honolulu, Hawaii
Chapter 16

John O. Schorge, MD
Division of Gynecologic Oncology
Vincent Obstetrics and Gynecology
   Services
Massachusetts General Hospital
Boston, Massachusetts
Chapter 14

Elizabeth N. Skinner, MD
Piedmont Hematology-Oncology
    Associates
Winston-Salem, North Carolina
Chapter 16

John T. Soper, MD
Department of Obstetrics
    and Gynecology
Division of Gynecologic Oncology
University of North Carolina
    at Chapel Hill
Chapel Hill, North Carolina
Chapters 18, 20A

Thanasak Sueblinvong, MD
University of Hawaii
Honolulu, Hawaii
Chapter 17

Edward Tanner, MD
Johns Hopkins Medical Institutions
Baltimore, Maryland
Chapter 13

Sue Valmadre, MD
Department of Obstetrics
    and Gynecology
Division of Gynecologic Oncology
The Ohio State University College
    of Medicine
Columbus, Ohio
Chapter 3

Russell S. Vang, MD
Johns Hopkins Medical Institutions
Baltimore, Maryland
Chapter 13

Tarrik Zaid, MD
Department of Obstetrics
    and Gynecology
University of Mississippi
Jackson, Mississippi
Chapter 19

Israel Zighelboim, MD
Department of Obstetrics
    and Gynecology
Division of Gynecologic Oncology
Washington University School
    of Medicine
St. Louis, Missouri
Chapter 10

Preface

When we first considered this project, we did not want it to be yet another textbook on gynecologic oncology. There are many wonderful, comprehensive books already available and we were not sure what we would be able to add. The publisher had a unique proposition that quickly made us reconsider. This book will be available online to physicians, nurses, allied health care professionals, teachers, and students (to name a few) in countries where access to gynecologic oncologists may be limited. Now this was something that we could get excited about and we felt that our co-authors would be enthusiastic to participate in such a project. In this era of a global economy and tremendous technological advances such as immediate transmission of satellite images in "real-time" and instantaneous cell phone communication from one side of the world to the other, it is hard to believe that global access to health care and knowledge remains limited in many countries. This book is not meant to be a comprehensive guide to the practice of gynecologic oncology, but a reference from which one can be guided in the right direction in the quest for information. We hope that this may help health care professionals to care for the women with gynecologic malignancies or at least help them to make a more timely diagnosis and referral.

This book would not have been possible without the help of our co-authors. We, of course, contacted many of our friends in the field of gynecologic oncology, and we appreciate the time that they took from their already busy schedules and families to make this possible. We would also like to thank our mentors throughout the years. They have motivated, counseled, and driven us to succeed both professionally and personally. Our strongest mentors remain our parents, whose work ethic and life examples have molded us in every way. Our partners at the University of North Carolina at Chapel Hill and at Duke University have been integral to our successes with this project and to our fulfillment with our careers. We appreciate Landes Bioscience and applaud their efforts to narrow the current global medical gap that exists. Finally, we thank our patients. If it were not for these women and for their examples of bravery, we would not feel compelled to try to improve the care of those women who we will never meet.

Paola A. Gehrig, MD
Angeles Alvarez Secord, MD

CHAPTER 1

Biology and Genetics

Paula S. Lee and Andrew Berchuck

Introduction

Cancers arise from genetic alterations that disrupt normal cellular functions such as cell proliferation (growth), senescence (aging) and apoptosis (cell death). The mechanisms of genetic damage have a diverse etiology, with loss of DNA repair mechanisms allowing mutations to accumulate. Cancers are also characterized by their ability to metastasize into surrounding normal tissue and to promote angiogenesis (the formation of new blood vessels). This chapter will review basic molecular mechanisms involved in the development of cancers and molecular alterations that characterize specific gynecologic malignancies.

Mechanisms of Transformation

Malignant tumors are characterized by alterations in genes that control proliferation. The final common pathway for cell division involves distinct molecular switches that control cell cycle progression from G1 to the S phase of DNA synthesis, as well as from G2 to mitosis. Cell cycle progression is regulated by growth factor signals, cyclins, cyclin-dependent kinases (cdk) and cdk inhibitors. The cyclin/cdk complex is critical for phosphorylation and activation of protein substrates involved in DNA replication. The cyclin/cdk complexes of D, E and A phosphorylate the retinoblastoma gene product (pRb), which functions as a tumor suppressor protein. Rb protein forms complexes with the transcription factor E2F and thus inhibits E2F activity. However, phosphorylated Rb protein releases E2F which then is critical in facilitating G1/S transition and in DNA replication. Growth factors trigger the enzymatic cascade of signal transduction. There are several classes of genes involved in growth stimulatory pathways (Table 1.1). However, increased proliferation is only one of several factors that contribute to malignant growth.

Apoptosis is a genetically determined and normal physiologic process of cell self-destruction that eliminates DNA-damaged or unwanted cells. Apoptosis is characterized by condensation of chromatin and cellular shrinkage in contrast to necrosis which is the death of cells through injury or disease and is characterized by loss of osmoregulation and cellular fragmentation. Regulation of apoptosis is dependent upon the balance of oncogene and tumor suppressor gene products. Overexpression of tumor suppressor gene *p53* arrests cell cycle progression. If DNA repair is not sufficient, the *p53* gene promotes apoptosis by downregulating the *bcl-2* gene. The *bcl-2* gene is an oncogene that inhibits apoptosis. Members of the *bcl-2* family have been identified as both apoptotic inhibitors and promoters. Many of the intracellular events involved in the apoptotic pathway occur in the mitochondria. Apoptosis can be triggered by withdrawal of survival factors, by DNA damaging agents such as chemotherapy and radiation, or by activation of natural killer cells.[1]

Gynecologic Oncology, edited by Paola Gehrig and Angeles Secord.
©2009 Landes Bioscience.

Table 1.1. Classes of genes involved in growth stimulatory pathways

Peptide Growth Factors	Corresponding Receptors
• Epidermal growth factor (EGF) and transforming growth factor (TGF-α)	• EGF receptor (ERBB1)
• Heregulin	• ERBB2 (HER-2/neu), ERBB3, ERBB4
• Insulin-like growth factors (IGF-I, IGF-II)	• IGF-I and II receptors
• Platelet-derived growth factor (PDGF)	• PDGF receptors
• Fibroblast growth factor (FGFs)	• FGF receptors
• Macrophage-colony stimulating factor (M-CSF)	• M-CSF receptor (FMS)

Cytoplasmic Factors	Examples
• Nonreceptor tyrosine kinases	• ABL, SRC
• G proteins	• K-RAS, H-RAS
• Serine-threonine kinases	• AKT2

Nuclear Factors	Examples
• Transcription factors	• MYC, JUN, FOS
• Cell-cycle progression factors	• Cyclins, E2F

Reprinted with permission from Boyd J, Hamilton T, Berchuck A. Oncogenes and Tumor Suppressor Genes. In: WJH, CP, RY, eds. Principles and Practices of Gynecologic Oncology. 3rd ed. Philadelphia: Lippincott Williams and Wilkins, 2000:113.

Cellular senescence occurs by shortening of repetitive DNA sequences (TTAGGG) at the ends of each chromosome, called telomeres. Malignant cells avoid senescence and become immortalized by promoting expression of telomerase to prevent telomere shortening. Telomerase is a ribonucleoprotein complex that synthesizes new telomeric DNA. The catalytic subunit of telomerase is human telomerase reverse transcriptase (hTERT). Telomerase expression is elevated in many human cancers, including ovarian, cervical and endometrial cancers.[2-5] Therapeutic approaches targeting telomerase are under development.[6-8] A telomerase-independent form of senescence exists that is referred to as oncogene-induced senescence that may be protective against cancer.[9,10]

Tumor suppressor gene products inhibit cellular division, promote DNA repair and initiate apoptosis (Fig. 1.1). Loss of tumor suppressor gene function occurs when both alleles are inactivated and is often described as the "two hit" hypothesis. Various types of mutations occur in tumor suppressor genes that include missense (a change in a single amino acid in the encoded protein), nonsense (a change in a single base substitution that changes the sequence to a stop codon) and frameshifts from deletions or insertions of nucleotides. In addition, loss of heterozygosity from mitotic nondisjunction, recombination, or large deletions of the remaining wild-type allele is an indication of loss of tumor suppressor gene function. In contrast, oncogenes increase the malignant transformation of cells. Proto-oncogene products are involved in cell growth, cell differentiation and signal transduction. Proto-oncogenes become oncogenes through overexpression by gene duplication, point mutations, or chromosomal translocation.

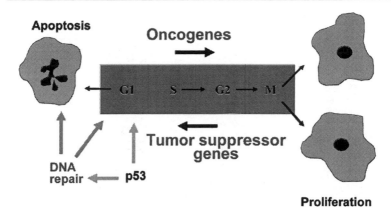

Figure 1.1. Role of oncogenes and tumor suppressor genes in cancer.

Invasion and Metastasis

Metastasis is characterized by loss of adhesion, degradation of surrounding stroma and angiogenesis.[11] Transmembrane proteins involved in cellular adhesion include the families of integrins and cadherins. Integrins interact with extracellular matrix proteins and generate intracellular signals to maintain cytoskeleton structure. E-cadherins are a subgroup of cadherins that are predominantly found in epithelial cells and mediate cell-to-cell adhesion. Cadherin dysfunction results in loss of cell-to-cell cohesion, altered cellular motility and increased invasiveness and metastatic potential. For example in endometrial cancers negative E-cadherin expression was associated with nonendometrioid carcinomas and Grade 3 histology. Combined positive E-cadherin expression with alpha-catenin and beta-catenin was an independent positive prognostic factor for survival in patients with well to moderately differentiated endometrial carcinomas.[12]

Invasion by malignant cells is associated with basement membrane and matrix degradation. Breakdown of the extracellular matrix is mediated by a family of metalloproteinases (MMPs) that are characterized by a zinc atom at their active site. MMPs facilitate tumor invasion and metastasis by degradation of extracellular matrix macromolecules and by modulating cell adhesion. A family of proteinases, called tissue inhibitors of metalloproteinases (TIMPs) inhibits MMP by forming noncovalent associations with the active site of MMPs. Elevated levels of both MMPs and TIMPs have been detected in endometrial, cervical and ovarian malignancies.[13-16]

Expansion of solid tumor beyond 1 mm^3 requires angiogenesis.[17] These new blood vessels provide nutrients and a route by which cancer cells can metastasize. The regulation of angiogenesis is complex and is controlled by both tumor and host tissue factors, which produce angiogenic factors that influence endothelial cell development and migration (Fig. 1.2). Tumor angiogenesis is stimulated by cytokines including basic fibroblast growth factor (bFGF), vascular endothelial growth factor (VEGF) and platelet derived growth factor (PDGF). Angiogenic inhibitors have also been identified and include transforming growth factor beta (TGF-β), angiostatin, thrombospondin-1 (TSBP-1) and maspin. Angiogenesis has been shown to be an independent prognostic predictor of survival in ovarian, endometrial and cervical cancer.[18-21] Specifically, patients whose ovarian cancers demonstrated a high degree of angiogenesis had a significantly

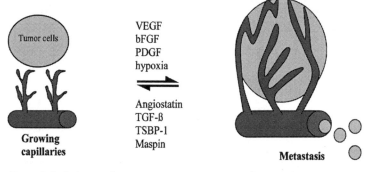

Figure 1.2. Balance of tumor angiogenesis. Factors that stimulate neovascularization include vascular endothelial growth factor (VEGF), basic fibroblast growth factor (bFGF), platelet derived growth factor (PDGF) and hypoxic conditions. Angiogenic inhibitors include angiostatin, transforming growth factor beta (TGF-β), thrombospondin-1 (TSBP-1) and maspin. New blood vessels provide a route for metastasis.

worse survival compared to those patients whose tumors had a low degree of angiogenesis.[18] Recent studies have demonstrated that anti-angiogenic therapies, such as bevacizumab, a monoclonal antibody that binds to VEGF, have significant activity in recurrent ovarian cancer.[22-24] Bevacizumab and other cancer therapies that target the epidermal growth factor receptors (EGFR and HER2/*neu*) for gynecologic malignancies are currently under investigation in clinical trials.[25]

Hereditary Syndromes

Although most cancers arise sporadically from acquired genetic damage, inherited mutations in cancer susceptibility genes are responsible for some cases. In hereditary syndromes, a predisposing mutation is inherited at birth or in the germ line. In contrast, with sporadic cancers, all the mutations are acquired somatically. The most common forms of hereditary cancer predispose to breast/ovarian cancer (*BRCA1* and *BRCA2* genes) and colon/endometrial cancer (*MSH2* and *MLH1* genes). Tumor suppressor genes have been implicated most frequently in hereditary cancer syndromes, followed by DNA repair genes. The penetrance of cancer susceptibility genes is incomplete, as all individuals who inherit a mutation do not develop cancer. The emergence of cancers in carriers is dependent on the occurrence of additional oncogenic mutations. Conversely, genetic polymorphisms may exist that confer slightly increased cancer risks and are associated with sporadic cases of cancer.

Endometrial Cancer

About 5% of endometrial cancers arise due to germline mutations in DNA repair genes in the context of hereditary nonpolyposis colon cancer (HNPCC) syndrome. In addition to early onset of colon cancer, there is increased risk of several other types of cancers including endometrial, ovarian, gastric and biliary tract cancers. Most HNPCC cases are due to alterations in *MSH2* and *MLH1* genes (Table 1.2). Loss of mismatch repair leads to accumulation of genetic mutations throughout the genome, especially in repetitive DNA sequences called microsatellites. Microsatellite instability (MSI) can

Table 1.2. Genetic alterations in endometrial adenocarcinomas

	Class	Activation	Approximate Frequency	Type I/II*
Hereditary				
MSH2	DNA repair	Mutation	Rare	I
MLH1	DNA repair	Mutation	Rare	I
Sporadic				
Oncogenes				
HER-2/neu	Tyrosine kinase	Overexpression	10%	II
K-ras	G protein	Mutation	10-30%	I/II
B-catenin	Transcription factor	Mutation	10%	I
Tumor Suppressor Genes				
MLH1	DNA repair	Promoter methylation	20%	I
p53	Transcription factor	Mutation/ overexpression	20%	II
PTEN	Tyrosine phosphatase	mutation	30-50%	I

*Type I = well-differentiated, endometrioid, estrogen associated cancers, Type II = poorly differentiated, nonendometrioid cancers.

involve both noncoding and coding regions of the genome. Genetic testing is recommended in all families suspected of having HNPCC based on family history. Mutational analysis of *MSH2* and *MLH1* remains the gold standard for diagnosis of HNPCC.

Endometrial cancer is the most common extracolonic malignancy in women with HNPCC. The risk of developing endometrial cancer ranges from 20-60% and is characterized by an earlier age of onset. The risk of ovarian cancer is increased to about 5-12%. The clinical features of HNPCC-associated endometrial cancers are similar to those of most sporadic cases (well-differentiated, endometrioid histology, early stage) and survival is about 90%. The mean age of onset of ovarian cancer in HNPCC families is in the early 40s, usually are early stage, well to moderately differentiated and about 20% occur in the setting of synchronous endometrial cancers.[26]

The majority of endometrial cancers are sporadic cases with two distinct types suggested from epidemiologic studies. Type I cases are associated with unopposed estrogen stimulation and often develop in the background of endometrial hyperplasia. Type I lesions are characterized by well-differentiated and endometrioid histologies, early stage and favorable outcome. Type II lesions are poorly differentiated, nonendometrioid and more aggressive. These lesions often present at an advanced stage and are associated with poor survival.

Some of the genetic alterations involved in the development of sporadic endometrial cancer have been elucidated. Inactivation of tumor suppressor genes is among the most frequent genetic events in endometrial cancers. Overexpression of mutant *p53* occurs in about 20% of endometrial adenocarcinomas and is associated with advanced stage, poor grade and nonendometrioid histology.[27] Endometrial cancers that overexpress *p53* protein usually harbor missense mutations. Mutations in the *PTEN* tumor suppressor gene on chromosome 10q occur in about 30-50% of endometrial cancers and represent the most frequent genetic alteration in endometrial cancer. The majority of these mutations are deletions, insertions and nonsense mutations; whereas only 15% are missense mutations. Unlike p53 mutations, mutations in the *PTEN* gene are associated with well-differentiated, endometrioid histology, early stage and favorable clinical outcome.[28] In addition, *PTEN* mutations have been observed in 20% of endometrial hyperplasias, suggesting that these represent an early event in the development of type I cancers. Microsatellite instability has also been identified in 20% of sporadic endometrial cancers. Loss of mismatch repair in these cases appears to be most often due to silencing of the *MLH1* gene by promoter methylation.[29]

Increased expression of oncogenes occurs less frequently in endometrial cancers. Overexpression of the *HER-2/neu* receptor tyrosine kinase has been noted in 10-15% of endometrial cancers and is associated with advanced stage and poor prognosis. Papillary serous endometrial cancers most frequently overexpress *HER-2/neu* and this may represent a possible therapeutic target.[30] K-ras mutations occur in about 10% of endometrial adenocarcinomas and have been identified in some endometrial hyperplasias. Other oncogenes implicated in endometrial cancers include β-catenin, fms and members of the myc family.[31-33]

Ovarian Cancer

Approximately 10% of ovarian cancers arise in women who carry germline mutations in high penetrance cancer susceptibility genes, predominantly *BRCA1* or *BRCA2* (Table 1.3). *BRCA1* and *BRCA2* are tumor suppressor genes and are located on chromosome 17q and 13q, respectively. In addition, *BRCA* mutations increase the risk of Fallopian tube and primary peritoneal cancers. *BRCA1* and *BRCA2* are associated with 60-90% lifetime risk of breast cancer and this begins to manifest prior to age 30.

Table 1.3. Genetic alterations in invasive epithelial ovarian cancers

	Class	Activation	Approximate Frequency
Hereditary			
BRCA1	DS DNA repair	Mutation/deletion	5%
BRCA2	DS DNA repair	Mutation/deletion	3%
MSH2	DNA mismatch repair	Mutation	1%
Sporadic			
Oncogenes			
HER-2/neu	Tyrosine kinase	Overexpression	10%
K-ras	G protein	Mutation	5%
AKT2	Serine/threonine kinase	Amplification	10%
c-myc	Transcription factor	Overexpression	20%
Tumor Suppressor Genes			
p53	Tumor suppressor Transcription factor	Mutation/deletion Overexpression	60%
p16	Tumor suppressor cdk inhibitor	Homozygous deletion	15%

BRCA2 increases the risk of breast cancer in men. The lifetime risk of ovarian cancer is up to 62% in *BRCA1* carriers and 10-20% in *BRCA2* in carriers, but this risk is not seen until about age 40. This provides the opportunity to perform prophylactic oophorectomy after the completion of childbearing prior to the development of ovarian cancer. Ashkenazi Jewish heritage poses an increased risk due to the presence of founder mutations. Since about 1 in 40 Ashkenazi individuals carries a *BRCA* founder mutation and testing for this panel of specific mutations is less expensive, the threshold for genetic testing is much lower in this population.

The most reliable method of detecting mutations is complete gene sequencing for *BRCA1* and 2. Testing has been advocated when the family history suggests at least a 10% probability of finding a mutation. This correlates into 2 first or second-degree relatives with either ovarian cancer at any age or breast cancer prior to age 50. It is preferable to begin by testing affected individuals. When a specific mutation is identified in an affected individual, others in the family can be tested much more rapidly and inexpensively for that specific mutation.

The vast majority of ovarian cancers are sporadic and occur from accumulation of genetic damage. Epithelial ovarian carcinoma is generally a monoclonal disease involving clonal expansion of a single transformed cell. Alteration of the *p53* tumor suppressor gene is the most frequent genetic event in ovarian cancers and occurs in about 70% of advanced stage cases. Tumor suppressor genes generally are inactivated by mutation of one gene copy and deletion of the second copy resulting in complete loss of gene function. In contrast, most *p53* mutations alter a single amino acid in the DNA binding domain and result in loss of transcriptional regulatory activity. Because these missense mutants are resistant to degradation, they overaccumulate in the nucleus and lead to p53 protein overexpression. *p53* mutation and overexpression are most common in serous and endometrioid histology and rare in clear cell histology. Cyclin-dependent kinase (cdk) inhibitors act as tumor suppressor genes by their role in inhibiting cell cycle progression. Several cdk inhibitors show decreased expression in a significant fraction of ovarian cancers. These include *p16*, *p21* and *p27*.

Activation of oncogenes occurs less commonly in ovarian carcinogenesis. *HER-2/neu* oncogene overexpression was detected in only 11% of ovarian cancers from patients enrolled on a clinical trial and the anti-HER-2/*neu* antibody Herceptin has not proven therapeutically useful in ovarian cancer.[34] Amplification of the related PIK3CA and AKT2 genes also occurs in some high grade serous ovarian cancers. In some cases, mutations of these genes also may occur.[35] Activating K-ras mutations in codons 12 or 13 are present in about 50% of mucinous ovarian cancers but are rare in serous ovarian cancers. However, 20-50% of serous borderline ovarian tumors contain K-ras mutations. In addition, some serous borderline tumors lacking K-ras mutations have mutations in *BRAF*, a downstream mediator.[36] Similar to endometrioid adenocarcinomas of the endometrium, β-catenin and *PTEN* mutations are seen in 30% of endometrioid ovarian cancers.

Cervical Cancer

Molecular and epidemiologic studies have demonstrated that sexually transmitted human papilloma virus (HPV) infections play a role in almost all cervical dysplasias and cancers. The peak incidence of HPV infection is during the second and third decade of life and the incidence of cervical cancer increases from the 20s to a plateau between ages 40-50. Although HPV plays a critical role in the development of most cervical cancers, only a small minority of women who are infected develop invasive cervical cancer. Thus, other genetic and/or environmental factors are involved in cervical carcinogenesis.

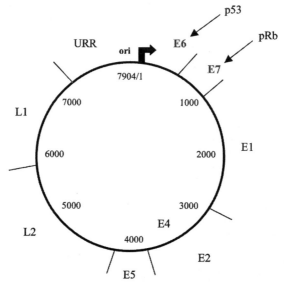

Figure 1.3. HPV 16 genome. HPV is a double stranded DNA, approximately 8,000 bp in size that contains 7 early (E1-E7) and 2 late (L1-L2) open reading frames. E1 is involved in viral DNA replication, E2 regulates viral mRNA synthesis, E4 interacts with cytokeratin, E5 is involved in membrane protein signaling with EGF and PDGF, L1 and L2 are structural capsid proteins, URR regulates viral gene expression and replication. E6 transforms the host cell by binding to p53 tumor suppressor protein and induces telomerase activation. E7 binds to Rb protein, liberating E2F, which results in S phase entry of the cell cycle. Ori: origin of DNA replication.

There are over 80 HPV subtypes and not all infect the lower genital tract. HPV 16 and 18 are the most common types associated with cervical cancer. Other high risk types include 31, 33, 35, 39, 45, 51, 56, 58, 59, 68, 73 and 82.[37] Low risk types that may cause dysplasia or condylomas include types 6, 11, 40, 42, 43, 44, 54, 61, 70, 72, 81. HPV is a circular double stranded DNA virus that contains 7 early (E1-E7) and 2 late (L1-L2) open read frames. E6 and E7 oncoproteins maintain the transformed phenotype and interact with tumor suppressor gene products. E6 proteins inactivate the *p53* gene product, while E7 activates the retinoblastoma (*Rb*) gene product (Fig. 1.3).

In conclusion, understanding the molecular basis of tumor carcinogenesis allows for exciting development of cancer prevention and targeted therapies. Tumor "signatures" to predict responses to therapy and to identify novel tumor markers are under current investigation using microarray and proteomic technologies. The HPV vaccine is one example of how advances in basic science and clinical research may shift our current paradigm of cancer screening, prevention and possible future disease eradication.

References

1. Gasser S, Orsulic S, Brown EJ et al. The DNA damage pathway regulates innate immune system ligands of the NKG2D receptor. Nature 2005; 436(7054):1186-90.
2. Sakamoto M, Toyoizumi T, Kikuchi Y et al. Telomerase activity in gynecological tumors. Oncol Rep 2000; 7(5):1003-9.
3. Braunstein I, Cohen-Barak O, Shachaf C et al. Human telomerase reverse transcriptase promoter regulation in normal and malignant human ovarian epithelial cells. Cancer Res 2001; 61(14):5529-36.
4. Wang PH, Ko JL. Implication of human telomerase reverse transcriptase in cervical carcinogenesis and cancer recurrence. Int J Gynecol Cancer 2006; 16(5):1873-9.
5. Lehner R, Enomoto T, McGregor JA et al. Quantitative analysis of telomerase hTERT mRNA and telomerase activity in endometrioid adenocarcinoma and in normal endometrium. Gynecol Oncol 2002; 84(1):120-5.
6. Ferreira CG, Epping M, Kruyt FA et al. Apoptosis: target of cancer therapy. Clin Cancer Res 2002; 8(7):2024-34.
7. Hu W, Kavanagh JJ. Anticancer therapy targeting the apoptotic pathway. Lancet Oncol 2003; 4(12):721-9.
8. Nakamura M, Masutomi K, Kyo S et al. Efficient inhibition of human telomerase reverse transcriptase expression by RNA interference sensitizes cancer cells to ionizing radiation and chemotherapy. Hum Gene Ther 2005; 16(7):859-68.
9. Braig M, Schmitt CA. Oncogene-induced senescence: putting the brakes on tumor development. Cancer Res 2006; 66(6):2881-4.
10. Mooi WJ, Peeper DS. Oncogene-induced cell senescence—halting on the road to cancer. N Engl J Med 2006; 355(10):1037-46.
11. Bogenrieder T, Herlyn M. Axis of evil: molecular mechanisms of cancer metastasis. Oncogene 2003; 22(42):6524-36.
12. Scholten AN, Aliredjo R, Creutzberg CL et al. Combined E-cadherin, alpha-catenin and beta-catenin expression is a favorable prognostic factor in endometrial carcinoma. Int J Gynecol Cancer 2006; 16(3):1379-85.
13. Davidson B, Goldberg I, Kopolovic J et al. MMP-2 and TIMP-2 expression correlates with poor prognosis in cervical carcinoma—a clinicopathologic study using immunohistochemistry and mRNA in situ hybridization. Gynecol Oncol 1999; 73(3):372-82.
14. Di Nezza LA, Misajon A, Zhang J et al. Presence of active gelatinases in endometrial carcinoma and correlation of matrix metalloproteinase expression with increasing tumor grade and invasion. Cancer 2002; 94(5):1466-75.
15. Lengyel E, Schmalfeldt B, Konik E et al. Expression of latent matrix metalloproteinase 9 (MMP-9) predicts survival in advanced ovarian cancer. Gynecol Oncol 2001; 82(2):291-8.
16. Schmalfeldt B, Prechtel D, Harting K et al. Increased expression of matrix metalloproteinases (MMP)-2, MMP-9 and the urokinase-type plasminogen activator is associated with progression from benign to advanced ovarian cancer. Clin Cancer Res 2001; 7(8):2396-404.
17. Folkman J. Fundamental concepts of the angiogenic process. Curr Mol Med 2003; 3(7):643-51.
18. Alvarez AA, Krigman HR, Whitaker RS et al. The prognostic significance of angiogenesis in epithelial ovarian carcinoma. Clin Cancer Res 1999; 5(3):587-91.
19. Abulafia O, Triest WE, Sherer DM. Angiogenesis in primary and metastatic epithelial ovarian carcinoma. Am J Obstet Gynecol 1997; 177(3):541-7.
20. Obermair A, Wanner C, Bilgi S et al. Tumor angiogenesis in stage IB cervical cancer: correlation of microvessel density with survival. Am J Obstet Gynecol 1998; 178(2):314-9.

21. Kirschner CV, Alanis-Amezcua JM, Martin VG et al. Angiogenesis factor in endometrial carcinoma: a new prognostic indicator? Am J Obstet Gynecol 1996; 174(6):1879-82; discussion 1882-4.

22. Burger R, Sill M, Monk BJ et al. Phase II trial of bevacizumab in persistent or recurrent epithelial ovarian cancer or primary peritoneal cancer: a Gynecologic Oncology Group Study. J Clin Oncol 2007; 25(33):5165-71.

23. Garcia AA, Hirte H, Fleming G et al. Phase II clinical trial of bevacizumab and low-dose metronomic oral cyclophosphamide in recurrent ovarian cancer: a trial of the California, Chicago, and Princess Margaret Hospital phase II consortia. J Clin Oncol 2008; 26(1):76-82.

24. Wright JD, Numnum TM, Rocconi RP et al. A multi-institutional evaluation of factors predictive of toxicity and efficacy of bevacizumab for recurrent ovarian cancer. Int J Gynecol Cancer 2008; 18(3):400-6.

25. Chon HS, Hu W, Kavanagh JJ. Targeted therapies in gynecologic cancers. Curr Cancer Drug Targets 2006; 6(4):333-63.

26. Watson P, Butzow R, Lynch HT et al. The clinical features of ovarian cancer in hereditary nonpolyposis colorectal cancer. Gynecol Oncol 2001; 82(2):223-8.

27. Kohler MF, Berchuck A, Davidoff AM et al. Overexpression and mutation of p53 endometrial carcinoma. Cancer Res 1992; 52(6):1622-7.

28. Risinger JI, Hayes K, Maxwell GL et al. PTEN mutation in endometrial cancers is associated with favorable clinical and pathologic characteristics. Clin Cancer Res 1998; 4(12):3005-10.

29. Simpkins SB, Bocker T, Swisher EM et al. MLH1 promoter methylation and gene silencing is the primary cause of microsatellite instability in sporadic endometrial cancers. Hum Mol Genet 1999; 8(4):661-6.

30. Slomovitz BM, Broaddus RR, Burke TW et al. Her-2/neu overexpression and amplification in uterine papillary serous carcinoma. J Clin Oncol 2004; 22(15):3126-32.

31. Monk BJ, Chapman JA, Johnson GA et al. Correlation of C-myc and HER-2/neu amplification and expression with histopathologic variables in uterine corpus cancer. Am J Obstet Gynecol 1994; 171(5):1193-8.

32. Moreno-Bueno G, Hardisson D, Sanchez C et al. Abnormalities of the APC/beta-catenin pathway in endometrial cancer. Oncogene 2002; 21(52):7981-90.

33. Leiserowitz GS, Harris SA, Subramaniam M et al. The proto-oncogene c-fms is overexpressed in endometrial cancer. Gynecol Oncol 1993; 49(2):190-6.

34. Bookman MA, Darcy KM, Clarke-Pearson D et al. Evaluation of monoclonal humanized anti-HER2 antibody, trastuzumab, in patients with recurrent or refractory ovarian or primary peritoneal carcinoma with overexpression of HER2: a phase II trial of the Gynecologic oncology group. J Clin Oncol 2003; 21(2):283-90.

35. Nakayama K, Nakayama N, Kurman RJ et al. Sequence mutations and amplification of PIK3CA and AKT2 genes in purified ovarian serous neoplasms. Cancer Biol Ther 2006; 5(7):779-85.

36. Singer G, Oldt R 3rd, Cohen Y et al. Mutations in BRAF and KRAS characterize the development of low-grade ovarian serous carcinoma. J Natl Cancer Inst 2003; 95(6):484-6.

37. Munoz N, Bosch FX, de Sanjose S et al. Epidemiologic classification of human papillomavirus types associated with cervical cancer. N Engl J Med 2003; 348(6):518-27.

CHAPTER 2

Cancer Screening in Women

Susan C. Modesitt

Introduction

Cancer remains the second leading cause of death in the developed world behind heart disease. In U.S. women, the most frequent cancers are breast cancers, followed by lung, colorectal and endometrial cancers; however, the leading cause of cancer mortality is lung cancer followed by breast, colon, ovarian and pancreatic cancers (Table 2.1).[1] Worldwide, however, cervical cancer is the cancer responsible for the most deaths in young women. For physicians who take care of women, it is imperative that there is a full and thorough understanding of the cancers that primarily affect women and the recommended cancer screening tests.

Screening tests must fulfill several attributes in order to be effective and merit universal adoption in the preventive care armamentarium of suggested tests (Table 2.2). Essentially, any cancer screening test must be technically feasible, have acceptable detection rates and positively impact the treatment and outcomes of the cancer in question. For women, the only currently accepted gynecologic cancer screening tests are mammograms for breast cancer and Pap tests for cervical cancer. There is no widely accepted screening test for ovarian cancer, endometrial cancer, Fallopian tube, vulvar, or vaginal cancers. The objective of this chapter is to present the data on cancer screening for the gynecologic malignancies.

Breast Cancer Screening

While not often considered a gynecologic malignancy, breast cancer predominantly strikes women and all physicians involved in primary care for women need to know the current recommended screening guidelines. Breast cancer remains the most frequently diagnosed cancer in women with over 180,000 cases expected in the U.S. in 2008. Women are acutely aware of breast cancer and often fear breast cancer more than cardiovascular disease despite the fact that, in the U.S., heart disease will actually kill more women than breast cancer. This fear of breast cancer may be due to the widely publicized statistic that women have a 1 in 8 lifetime chance for developing breast cancer. To try and put this into perspective, the risk is directly impacted by a woman's age. A 25 year-old woman has about a 1 in 20,000 risk of breast cancer. By age 45 this risk has risen to 1 in 93 and for women over the age of 85, it reaches the 1 in 8 lifetime risk. This impact of age, in turn, affects the recommended screening tests for breast cancer based on a woman's age. At various times, three methods for early detection have been recommended for breast cancer screening and they include self breast exam, clinical breast exam and mammography.

Monthly self breast exams were routinely recommended for all women after the age of 18 and were recommended by the American Cancer Society until 2003 at which time

Gynecologic Oncology, edited by Paola Gehrig and Angeles Secord.
©2009 Landes Bioscience.

Table 2.1. Expected cancers in the United States in 2008[1]

Site	Estimated Cases (%)	Estimated Deaths (%)
Breast	182,460 (26%)	40,480 (15%)
Lung	100,330 (14%)	71,030 (26%)
Colorectal	71,560 (10%)	25,700 (9%)
Uterine corpus	40,100 (6%)	7,470 (3%)
Non-Hodgkin's lymphoma	30,670 (4%)	9,370 (4%)
Thyroid	28,410 (4%)	910 (0.3%)
Melanoma	27,530 (4%)	3,020 (1%)
Ovary	21,650 (3%)	15,520 (6%)
Pancreas	18,910 (2%)	16,790 (6%)
Bladder	17,580 (2%)	4,150 (1.5%)
ALL SITES*	692,000 (100%)	271,530 (100%)

*Only top 10 sites for incidence are shown.

the recommendation changed to an optional one. The U.S. Preventative Services Task Force (USPSTF) currently recommends against self breast exam due to the increased false positive rate and a failure of studies to demonstrate any benefit.[2] Currently, at most the recommendation is to continue optional evaluation with any abnormal findings to be reported to a physician.

Clinical breast exams (CBE) are currently recommended for all women as an adjunct to mammography. Specifically, the American Cancer Society (ACS) recommends CBE as part of a routine health maintenance exam every 3 years for women aged 20-39 years and yearly for women 40 and older. The USPSTF recommends CBE in conjunction with mammography (not alone) every 1-2 years for women aged 40 and over.[2] The utility of a CBE in picking up earlier breast cancers has not necessarily been established, yet it is a well-known practice, is acceptable to women and physicians and is easy to implement.

Mammography has been well-established in the breast cancer screening process. Currently, both the ACS and the USPSTF recommend annual mammograms for women over 40 years of age. Based on review of available clinical data, the USPSTF

Table 2.2. Criteria for screening tests

1. Targeted disease must be sufficiently burdensome to the population that a screening program is warranted, minor changes in relative risk (RR) should have substantial impact on the absolute population risk
2. Target disease must have a well-understood natural history with a long preclinical latent period
3. Screening method must have acceptable technical aspects (detecting disease at an earlier stage while minimizing false positive and negative results)
4. Efficacious treatment for the target disease must be available
5. Early detection must improve disease outcome
6. Cost feasibility and acceptability of screening and early treatment should be established

2

Table 2.3. Current cancer screening recommendations for asymptomatic women

Type of Cancer	Recommended Screening Tests
Bladder cancer	None
Breast cancer	Screening mammography with or without clinical breast exam every 1-2 years for women ≥40 years of age. MRI in very high risk women.
Cervical cancer	Periodic Pap test screening beginning 3 years from sexual debut every 1-3 years as indicated
Colon cancer	All testing to begin after age 50
	Yearly fecal occult blood OR
	Sigmoidoscopy every 5 years OR
	Double contrast barium enema every 5-10 years OR
	Colonoscopy every 10 years
Uterine cancer	None
Lung cancer	None
Oral cancer	None
Ovarian cancer	None recommended for general population
	Some experts recommend annual CA125, vaginal ultrasound and pelvic exam for women with a *BRCA1*, *BRCA2*, HNPCC, or a strong family history of ovarian cancer. Enrollment in a screening trial would be ideal if available.
Pancreatic cancer	None
Skin cancer	None
Thyroid cancer	None

found "fair evidence that mammography screening every 12-33 months significantly reduces mortality from breast cancer". The data are most convincing for women between the ages of 50-69 and the benefits may increase with increasing age (due to increased breast cancer prevalence) in the absence of severe comorbidities. There does still appear to be benefit for women in their 40s but the absolute benefit is likely smaller than for older women.[2]

For women with increased breast cancer risk as defined by a strong family history, a genetic mutation (e.g., *BRCA1* or *2* mutation), or a personal history of breast cancer, screening with mammography may be initiated at a younger age, involve more frequent exams, or indicate the need for possible genetic testing. Recent data may also support MRI as a more sensitive supplement to mammography and CBE in women at high risk for breast cancer but currently are not recommended outside of clinical trials.[3-5]

MRI is recommended for the following high risk women:
- Women with a known *BRCA1* or *BRCA2* mutation
- Women who have a first degree relative with a BRCA mutation, but who have not had testing
- Women who have at least a 20% lifetime risk of developing breast cancer based on models that primarily use family history
- Women who have previously received chest radiation

In summary, women aged 40 and over should be strongly encouraged to undergo mammography with or without CBE every 1-2 years.

Cervical Cancer Screening

Few screening tests have had the positive impact as the introduction of the Pap test in the cancer health of women in the United States. Prior to 1940, cervical cancer remained the leading cause of cancer death for U.S. women. Since the introduction of the Pap test in the 1950s, the overall rate of cervical cancer has decreased 100% to an all time low of only 11,000 cases per year in the United States. Most women who are diagnosed with cervical cancer in the United States have not been compliant with cervical screening recommendations.[6] Unfortunately, barriers to screening still exist and women of low socioeconomic status and women of color are still disproportionately represented in women with cervical cancer; additionally worldwide half a million women die from cervical cancer every year.

While Pap tests have been universally endorsed for over 40 years, the specific recommendations continue to change. Currently, the ACS, the American College of Obstetricians and Gynecologists (ACOG) and the USPSTF recommend Pap testing to start 3 years after sexual debut but no later than age 21. Further, screening is recommended annually for conventional cytology or every 2 years for liquid based cytology. For women over 30 with no risk factors and at least 3 prior normal tests, the Pap testing interval can be spaced out to every 2-3 years. Risk factors for cervical cancer include multiple sexual partners, known high risk human papillomavirus (HPV) infection, prior cervical intra-epithelial neoplasia, or immunocompromised states. After age 70, if the woman has had normal Pap tests, screening may be discontinued in the absence of risk factors if desired. Additionally, if women have undergone hysterectomy with removal of the cervix for benign disease (not including cervical dysplasia), they can forgo further vaginal cytology.

Conventional Pap tests suffered from a lack of sensitivity (60-80%[2]) which remains part of the reason that annual exams were recommended. The amazing success of the Pap test, despite these drawbacks, stemmed from the fact that cervical cancer usually has a long precancerous phase that can be detected prior to the development of an invasive cancer. The introduction of the liquid based Pap test has improved the sensitivity and the specificity of the test such that a negative test (in a woman with prior normal screens and a lack of significant risk factors) can allow for the screening interval to be lengthened to every 2-3 years.

The human papillomavirus (HPV) is necessary but not sufficient for development of cervical neoplasia and subsequent invasive cancer and the introduction of the HPV test was hailed as a major breakthrough in women's health. The only commercially available test evaluates the presence of any of the 13 high risk HPV types and can be performed as a stand alone test or as an adjunct to a liquid based cytology. HPV testing is more sensitive for the detection of high grade cervical lesions (almost 100%) than either conventional or liquid based cytology but specificity is still fairly low at 85%.[7] As the prevalence of HPV can be as high as 60-80% in some adolescent populations and the majority of those women will never develop cervical cytologic abnormalities, the HPV test has not been adopted in the United States as a primary screening mechanism. It can be useful in two clinical scenarios. First, in women with a Pap test that demonstrates atypical squamous cells of uncertain significance (ASC-US), the HPV test can be performed. In that situation, if the HPV test is normal, the woman can simply return in one year for a repeat Pap test, if the HPV test is positive, the woman would be triaged to further evaluation with colposcopy and potential biopsies. The second clinical situation in which the HPV test can be useful is in women over 30 years of age without cervical cancer risk factors. For these women, a negative Pap test and simultaneously negative HPV can serve to further space out the interval needed for cervical screening to every 3 years. Currently, in the United States, the HPV test is not recommended as a stand alone screening test; however there are ongoing large prospective trials to further elucidate a role in screening. For example, 44,000 women in the Netherlands have already enrolled in POBASCAM (population-based randomized controlled trial for implementation of high risk HPV testing in cervical cancer screening) that will help determine whether HPV testing can supplant cervical cytology.[8]

In the near future, perhaps the advent of the new HPV vaccines may change the screening recommendations but further data will be required before any alterations are endorsed. Two companies have demonstrated exciting advances in the ability of HPV vaccines (one targeted against HPV 16 and 18, another against HPV Types 16, 18, 6 and 11) to prevent HPV infection and subsequent cervical intra-epithelial neoplasia. Multiple questions remain including, whether universal vaccinations will be recommended and who will pay for them, the length of protection that is afforded by the vaccinations and how Pap tests will be performed in the vaccinated population.

In summary, cervical cancer screening with a Pap test is still recommended every 1-3 years in all women starting at 3 years after initiation of sexual activity but no later than age 21. Screening can be discontinued following hysterectomy for benign reasons, not including cervical dysplasia, or for women over 70 without risk factors.

Ovarian Cancer Screening

An effective screening test is needed more in ovarian cancer than perhaps in any other gynecologic malignancy. Ovarian cancer will strike about 22,000 U.S. women in 2008 and an estimated 15,000 will succumb to the disease.[1] Unfortunately, the high mortality due to the disease can be directly attributed to the fact that the 75-80% of women who are diagnosed with ovarian cancer will have Stage III and IV disease. Overall 5-year survival for women with Stage III and IV disease are 30% and 17% respectively compared to 80-90% five-year survival for women with Stage I disease. To date, there is not an accepted or recommended screening regimen for women with ovarian cancer although several modalities have been in screening trials for years including transvaginal ultrasonography, CA125 testing and, more recently, serum proteonomics. Currently, neither the American Cancer Society, nor the USPSTF, nor ACOG recommend routine screening for ovarian cancer. Part of the difficulty in proving the efficacy of a screening method stems from the low prevalence of the disease in the general population.

Ultrasound was one of the first modalities evaluated for ovarian cancer screening. Ultrasound is very sensitive (almost 100%) in detecting ovarian cancer but results in a false positive rate of 1-5% and many women with benign ovarian pathology receive further evaluation and potential surgery. Furthermore, women who undergo surgery as a result of a positive screen can suffer a significant complications about 0.5-1% of the time and the majority will have benign gynecologic disease.[9] One of the largest ongoing transvaginal ultrasound screening trials has been running at the University of Kentucky since 1987 and has enrolled 25,000 women as of 2007. The last published report of 25,000 women concluded that annual transvaginal ultrasound screening was associated with a decrease in stage at detection and a decrease in case-specific ovarian cancer mortality, but did not detect cancers in normal size ovaries.[10] In order to improve the positive predictive value of vaginal ultrasound (0.6-3%[9]), this institution developed a morphology index (MI) for ovarian masses that may help to discriminate benign lesions (for example simple ovarian cysts) from those with a higher chance of malignant behavior. This scoring system gives points (0-5) based on two parameters, ovarian volume and morphologic characteristics of the mass (simple cyst, presence of septations, wall papillations, or solid components). An MI of <5 is almost always benign (negative predictive value of 99.7%) whereas as MI ≥5 has a higher likelihood of being malignant (positive predictive value 40%).[11] Further refinements of a morphologic screen or improved ultrasound technology may someday improve the specificity and positive predictive value of ultrasound, but currently ultrasound alone is not recommended as a routine ovarian cancer screening test for asymptomatic women.

The CA125 test has been in clinical practice for over 20 years and has been evaluated as a potential screening mechanism. At the time of this antigen's identification, elevations in the serum level were found in 82% of women with ovarian cancer, 28.5% of women with nongynecologic cancers, 6% of women with benign disease and 1% of the general population.[12] To date, the CA125 test alone isn't specific or sensitive enough to be recommended for routine screening. Women with early stage ovarian cancer may have a normal CA125 about 50% of the time and, conversely, women with benign conditions (e.g., endometriosis, pregnancy and fibroids) can have an elevated CA125. Reviews of the efficacy of this single test reveal that CA125 has about an 80% sensitivity and a low positive predictive value (1-15%) and both of these factors limit the effectiveness of the test. Recent studies have focused on the serial use of the test to improve the sensitivity. Skates et al have developed the Risk of Ovarian Cancer Algorithm (ROCA) currently in large trials; using this method, the test is very specific (specificity of 99.7%) with an improved sensitivity (83%) and slightly improved positive predictive value (16%), but further evaluation is ongoing.[13]

The combination of CA125 and ultrasound have demonstrated potential survival benefit[14] in an initial randomized pilot screening trial involving 20,000 women and currently, two large randomized controlled trials are nearing completion and will definitively answer the question for CA125 and ultrasound screening in the general population. The Prostate, Lung, Colorectal and Ovarian Cancer (PLCO) Screening trial accrued 74,000 women and the preliminary results were reported in 2005.[15] This study found that vaginal ultrasound alone was abnormal in 4.7% of the population with a positive predictive value of only 1% for ovarian cancer. CA125 levels were abnormal in 1.4% of the population with a positive predictive value of 3.7%; however, when both tests were abnormal, the positive predictive value improved to 23.5%. Of 570 women who underwent surgery during this trial, only 29 neoplasms were found with 20 malignancies. Cost analysis, morbidity, mortality and survival data are not yet available and the final analysis is still pending. The second randomized trial is the United Kingdom Collaborative Trial of Ovarian cancer screening (UKCTOS) and is nearing accrual completion of 200,000 women in 2005, but results are not yet known. Unless the results of these trials demonstrate a reduction in ovarian cancer mortality due to screening with acceptable morbidity and cost, routine ovarian cancer screening with ultrasound and CA125 should not be implemented in the general population.[16-18]

While ovarian screening is not recommended for the general population, women who are at high risk for the disease have been considered a group that might benefit from screening given their increased disease prevalence. Approximately 10% of ovarian cancers are familial and the most common genetic alterations involve *BRCA1*, *BRCA2*, or hereditary nonpolyposis colorectal carcinoma syndromes (HNPCC). For women with increased risk due to these known mutations and/or a strong family history, some experts have recommended screening with annual pelvic exams, vaginal ultrasound and CA125, yet caution that cancers may still be missed (especially primary peritoneal cancers) and the ability of screening to reduce mortality in this group is still truly unknown and ongoing enrollment into screening trials should be encouraged in this group.[19,20]

The most recent entry into the field as a potential ovarian cancer screening method is proteonomics. Proteonomics refers to the evaluation of multiple serum protein markers to determine differences in protein patterns that can differentiate benign from malignant disease. This was first reported in 2002 by Petricoin[21] and colleagues and demonstrated the ability to discriminate benign from malignant ovarian masses (sensitivity 100%, specificity 95% and positive predictive value of 94%) but the data have not be replicated by others and the application in a screening program is still unknown. Similarly, Mor et al reported on a profile of four proteins (leptin, prolactin, osteopontin and insulin like growth factor 2) that demonstrated similar test values (sensitivity 95%, specificity 95%, positive predictive value of 95% and negative predictive value of 94%) in discriminating benign and malignant ovarian tumors in a small study.[22] Both of these studies represent important and exciting advances but require replication and validation in larger trials before widespread introduction into clinical use.

In summary, further evaluation of current screening modalities are necessary to document reduction in mortality for application in the general population but women at high risk for ovarian cancer should likely be offered screening with available modalities (CA125 and vaginal ultrasound) even knowing the potential limitations of current technology.

Uterine Cancer

Uterine cancer will affect over 40,000 women in the United States in 2008 and over 95% of these will involve the endometrium (endometrial cancer) rather than the myometrium (sarcomas).[1] In contrast to ovarian cancer, women with endometrial cancer are most often diagnosed in early stages due to the fact that they will have abnormal bleeding and seek medical attention. Currently, there is no recommended screening test for endometrial cancer in the general population, but if abnormal bleeding is reported, prompt evaluation with an endometrial biopsy and/or pelvic ultrasound is of paramount importance. Endometrial cancer risk factors include obesity, unopposed estrogen, tamoxifen treatment, diabetes, hypertension, along with genetic alterations; the presence of any of these factors should alert clinicians to potential endometrial pathology.

Hereditary nonpolyposis colorectal carcinoma (HNPCC) is a genetic disorder that predisposes affected individuals to colon cancer, but endometrial cancer is the second most common malignancy in families with this disorder. Yet, even in women with a genetic predisposition to the disease, the recommendations are unclear for screening as most women will still present in early stages with abnormal bleeding. Currently, trials are ongoing in this high risk population regarding the use of screening endometrial biopsies and/or vaginal ultrasound but using these to screen women with HNPCC might merit consideration despite the lack of formal recommendations.

Vaginal Cancer

Currently, vaginal cancer remains so rare (760 expected deaths in the United States in 2008[1]) that routine screening is not recommended outside the routine cervical cancer screening. For women with prior cervical intra-epithelial neoplasia treated with hysterectomy, vaginal cytology should still be performed.

Vulvar Cancer

With only 3,460 cases of vulvar cancers expected in the United States in 2008,[1] routine screening is not recommended. The majority of these cancers will be squamous cell cancers; however, melanoma remains the second leading cause of vulvar cancer. Women should be educated regarding risk factors for vulvar cancer, encouraged to perform self exams and to report any symptoms promptly to their physician.

Conclusion

As cancer is expected to strike almost 700,000 women in the United States in 2006, it is imperative that all physicians who care for women understand the importance of cancer screening. No matter how effective the screening test, if women don't receive the test, cancer cannot be detected in an early stage. Medical research continues to advance in cancer screening paradigms and, in the future, should be able to find effective screening tests for ovarian cancer as well as to define optimal screening mechanisms and risk reduction strategies for high risk women.

Suggested Reading

American Cancer Society www.cancer.org
Women's Cancer Network www.wcn.org
U.S Preventative Services Taskforce www.ahcpr.gov/clinic/uspstfix.htm

References

1. Jemal A, Siegel R, Ward E et al. Cancer statistics. CA Cancer J Clin 2008; 58(2):71-96.
2. U.S. Preventative Services Task Force Guide to Clinical Preventive Services. Agency for Healthcare Research and Quality, U.S. Department of Health and Human Services Available at: http://www.ahcpr.gov/clinic/uspstfix.htm, 2005.
3. Lehman CD, Blume JD, Thickman D et al. Added cancer yield of MRI in screening the contralateral breast of women recently diagnosed with breast cancer: results from the International Breast Magnetic Resonance Consortium (IBMC) trial. J Surg Oncol 2005; 92(1):9-15; discussion 15-6.
4. Lehman CD, Blume JD, Weatherall P et al. Screening women at high risk for breast cancer with mammography and magnetic resonance imaging. Cancer 2005; 103(9):1898-1905.
5. Schnall MD, Blume J, Bluemke DA et al. MRI detection of distinct incidental cancer in women with primary breast cancer studied in IBMC 6883. J Surg Oncol 2005; 92(1):32-8.
6. Janerich DT, Hadjimichael O, Schwartz PE et al. The screening histories of women with invasive cervical cancer, connecticut. Am J Public Health 1995; 85(6):791-4.
7. Clavel C, Masure M, Bory JP et al. Human papillomavirus testing in primary screening for the detection of high-grade cervical lesions: a study of 7932 women. Br J Cancer 2001; 84(12):1616-23.
8. Bulkmans NW, Rozendaal L, Snijders PJ et al. POBASCAM, a population-based randomized controlled trial for implementation of high-risk HPV testing in cervical screening: design, methods and baseline data of 44,102 women. Int J Cancer 2004; 110(1):94-101.
9. Bell R, Petticrew M, Luengo S et al. Screening for ovarian cancer: a systematic review. Health Technol Assess 1998; 2(2):i-iv, 1-84.
10. van Nagell JR Jr, DePriest PD, Ueland FR et al. Ovarian cancer screening with annual transvaginal sonography: Finding of 25,000 women screened. Cancer 2007; 109:1887-96.

11. Ueland FR, DePriest PD, Pavlik EJ et al. Preoperative differentiation of malignant from benign ovarian tumors: the efficacy of morphology indexing and doppler flow sonography. Gynecol Oncol 2003; 91(1):46-50.

12. Bast RC Jr, Klug TL, St John E et al. A radioimmunoassay using a monoclonal antibody to monitor the course of epithelial ovarian cancer. N Engl J Med 1983; 309(15):883-7.

13. Skates SJ, Xu FJ, Yu YH et al. Toward an optimal algorithm for ovarian cancer screening with longitudinal tumor markers. Cancer 1995; 76(10 Suppl):2004-10.

14. Jacobs IJ, Skates SJ, MacDonald N et al. Screening for ovarian cancer: a pilot randomised controlled trial. Lancet 1999; 353(9160):1207-10.

15. Buys SS, Partridge E, Greene MH et al. Ovarian cancer screening in the Prostate, Lung, Colorectal and Ovarian (PLCO) cancer screening trial: findings from the initial screen of a randomized trial. Am J Obstet Gynecol 2005; 193(5):1630-9.

16. Rufford B, Jacobs IJ. Screening and diagnosis of ovarian cancer in the general Population. In: Gershenson DM, McGuire WP, Gore M et al, eds. Gynecologic Malignancies: Controversies in Management. First ed. Philadelphia: Elsevier 2004:355-68.

17. Jacobs I. Screening for familial ovarian cancer: the need for well-designed prospective studies. J Clin Oncol 2005; 23(24):5443-5.

18. Jacobs IJ, Menon U. Progress and challenges in screening for early detection of ovarian cancer. Mol Cell Proteomics 2004; 3(4):355-66.

19. NIH consensus conference. Ovarian cancer. Screening, treatment and follow-up. NIH consensus development panel on ovarian cancer. Jama 1995; 273(6):491-7.

20. Antill Y, Phillips K. Screening and diagnosis of ovarian cancer-high risk. In: Gershenson DM, McGuire WP, Gore M et al, eds. Gynecologic Malignancies: Controversies in Management. First ed. Philadelphia: Elsevier 2004:341-54.

21. Petricoin EF, Ardekani AM, Hitt BA et al. Use of proteomic patterns in serum to identify ovarian cancer. Lancet 2002; 359(9306):572-7.

22. Mor G, Visintin I, Lai Y et al. Serum protein markers for early detection of ovarian cancer. Proc Natl Acad Sci USA 2005; 102(21):7677-82.

CHAPTER 3

Tumor Markers in Gynecological Oncology

Sue Valmadre and David E. Cohn

Introduction

A tumor marker is an entity that occurs in the body either as a direct tumor product or as a reaction to the presence of a tumor. Tumor markers may be detected in the blood, urine or body tissues and may be used for the following:

a. To screen a healthy or high risk population for a specific disease,
b. To assist with the confirmation of a diagnosis of cancer,
c. To assist in assessing a patient's response to treatment and
d. To monitor a patient's disease course.

An ideal tumor marker is sensitive (detects disease if it is present), specific (is negative if the disease is absent), inexpensive and easy to perform. Unfortunately, few of the tumor markers used in malignant gynecological conditions fulfill all of these criteria. This chapter covers the most commonly used tumor markers for gynecological malignancies.

Tumor Markers for Epithelial Ovarian Cancer

CA125 is at present the best available marker for epithelial ovarian cancer. Unfortunately, CA125 lacks specificity for Stage 1 disease and therefore limits its usefulness in the detection of early stage ovarian cancer. In the postmenopausal women a CA125 may assist in determining whether a pelvic mass is benign or malignant. The other important clinical application is in predicting response to treatment i.e., if the CA125 is trending downwards during a course of chemotherapy, it is generally continued whereas an increasing CA125 during chemotherapy would result in cessation of or change of treatment.

Distribution of CA125 in Normal Tissues

CA125 has a widespread distribution in normal human tissue and therefore lacks specificity for the ovary. Earliest immunohistochemical studies using the OC125 antibody demonstrated that the CA125 antigen was not only present in a number of normal adult tissues that were derived from the coelomic epithelium including the endometrium, endocervix and Fallopian tube but was also in tissues of mesothelial origin including the pleura, pericardium and peritoneum.[1,2] Other studies have also shown that the epithelia of kidney, lung, stomach, gall bladder, pancreas and colon also express the CA125 antigen.[3-5]

Structure of CA125

CA125 was originally identified as an antigenic determinant found on a high molecular weight glycoprotein and recognized by the OC 125 monoclonal antibody.[6]

Gynecologic Oncology, edited by Paola Gehrig and Angeles Secord.
©2009 Landes Bioscience.

The process of producing this antibody involved immunizing mice with a cell line designated OVCA433, which was derived from a human ovarian serous cystadenocarcinoma. CA125 was originally characterized as a high molecular weight glycoprotein with an estimated molecular weight of between 200 and 2000 kD.[7] More recently CA125 has been thought to have a number of features characteristic of mucin and has been designated as MUC16 by Yin and Lloyd.[8] Studies have also shown the CA125 antigen to be a highly glycosylated molecule with a protein content consisting largely of serine, threonine and proline.[8,9]

Serum Levels of CA125 in Healthy Subjects

It is important to realize that basal or low levels of CA125 are found in serum from apparently healthy males and females. The value of 35 U/mL is widely regarded as the upper limit for a CA125 to be regarded as within the normal range. This "normal range" was defined by the work published by Bast et al where the CA125 serum levels in 888 apparently healthy subjects (537 males and 351 females) were tested using a immuno-radiometric assay with OC125 as both the capture and labeled antibody. They found that 1% of the 888 subjects had serum levels of CA125 greater than 35 U/mL.[10,11] As a result 35 U/mL became the arbitrary, but widely accepted, upper limit for a normal CA125 serum level. Other investigators have suggested that in some circumstances, e.g., in a postmenopausal woman or women after a hysterectomy, a lower CA125 level may be more appropriately regarded as "normal".[11-13]

CA125 is known to carry two major antigenic domains classified as A, the domain that binds the monoclonal antibody OC125 and B, the domain that binds the monoclonal antibody M11.[14] The current immunoassay which is universally used for the quantitation of serum levels of CA125 is a heterologous assay, CA125 II, which uses both OC125 and M11 monoclonal antibodies. This process has replaced the homologous assay which only used the OC125 monoclonal antibody.[14]

Factors That Influence CA125 Levels

Age

Several studies have documented consistently higher CA125 serum levels in healthy premenopausal women as compared to their postmenopausal counterparts.[11,12] Bon et al studied CA125 serum levels in 1,026 apparently healthy women and found mean levels of 18 U/mL (range 2-98) and 12 U/mL (range 2-37) in the premenopausal and postmenopausal groups respectively.[11] Similarly, Bonfrer et al found the 95th percentile values of CA125 serum levels were 36 U/mL for those women aged 40-44 years, 30 U/mL for those aged 45-55 years and 25 U/mL for those women over 55 years.[12]

Race

Pauler et al investigated that effect of race on serum CA125 levels in 18,748 postmenopausal women who participated in an ovarian cancer screening program but who were not diagnosed with ovarian malignancy during the 12-year follow-up period. The authors found significantly higher levels of CA125 in Caucasian women with mean levels of 14.2 U/mL, compared to the Asian or African women involved in the study with their mean levels being 13 U/mL and 9 U/mL respectively.[15]

Menstrual Cycle

In some women, serum levels of CA125 have been shown to fluctuate throughout the menstrual cycle. Grover et al measured serum CA125 levels in 1,478 apparently healthy women and found levels greater than 35 U/mL in 77 subjects. Forty of these women had weekly serum CA125 measurements. In 29 subjects, higher values were found at the time of menstruation compared with other times of the menstrual cycle.[16]

Pregnancy

CA125 may increase in pregnancy, especially during the first trimester. Gocze et al measured the CA125 levels in sera from 20 apparently healthy women in the first trimester of pregnancy. They found 4 cases of elevated CA125 ranging from 65 to >500 U/mL.[17] It is thought that the increased CA125 level is most likely to be derived from the decidualized endometrium.

CA125 in Benign and Malignant Disease

Serum CA125 Levels in Benign Disease

Multiple benign gynecological and nongynecological conditions can elevate CA125 serum levels. Benign gynecological conditions that may be associated with an elevated CA125 serum level include endometriosis, uterine leiomyomata, acute and chronic salpingitis, and pelvic inflammatory disease.[18-20] Nongynecological conditions which can elevate CA125 serum levels include congestive heart failure, liver cirrhosis, chronic active hepatitis, acute and chronic pancreatitis, other malignancies (pancreas, breast, lung), lung and pleural disease and ascites of benign or malignant origin. From this list, it is generally accepted that any disease process that inflames the peritoneum, pericardium or pleura can elevate CA125 serum levels.

Serum CA125 Levels in Epithelial Ovarian Cancer

Currently CA125 is the best tumor marker available for epithelial ovarian cancer.

In the original paper published by Bast et al, it was reported that the serum level of CA125 was elevated in 83 of 101 (82%) women with ovarian cancer.[21] Subsequent studies demonstrated that both the proportion of patients with elevated levels and the extent to which the levels were elevated depended largely on the stage and the histological type of disease. Jacobs and Bast in their review combined the data from 15 different studies and showed that CA125 levels were elevated in 49 of 96 women (50%) with FIGO Stage 1, 55 of 61 women (90%) with Stage II, 199 of 216 women with Stage III (92%) and 77 of 82 women (94%) with Stage IV disease.[18] Analysing the data from 12 studies, Jacobs and Bast reported that elevated CA125 levels were detected in 254 of 317 women (80%) with serous ovarian carcinomas, 35 of 51 women (69%) with mucinous type, 39 of 52 women (75%) with endometrioid type, 28 of 36 women (78%) with clear cell type and 56 of 64 women (88%) with undifferentiated type.[1] In summary, although elevated serum levels of CA125 occur in 80-90% of women with ovarian cancer the chances of an elevated level increases with nonmucinous histologies and in more advanced disease.

CA125 as a Screening Test for Ovarian Cancer

Approximately 75% of ovarian cancers are diagnosed at an advanced stage of disease (Stage III and IV) Which is largely due to the lack of specific symptoms associated with this disease. A 5-year survival rate of 90% or greater can be achieved in women who are diagnosed and treated when their disease is confined to their ovary. Survival significantly decreases to about 30% and 10% for those women diagnosed with Stage III and IV disease, respectively. So it follows that early detection of ovarian cancer improves survival. Various modalities including transabdominal and transvaginal ultrasound and serum CA125 measurements have been suggested as worthwhile screening tools. As discussed previously, serum CA125 levels have a low sensitivity for Stage I disease and a low specificity when used in premenopausal women. This low sensitivity and specificity combined with the low prevalence of ovarian cancer means that serum CA125 is limited as a screening tool as it has a low positive predictive value.

In order to enhance the effectiveness of CA125 as a screening tool it has been used in conjunction with ultrasound in so-called multimodal screening. In a systematic review of the literature, Bell et al analyzed four prospective, nonrandomized studies using multimodal screening for ovarian cancer in the general population.[22] Over 27,000 women were screened with 14 ovarian cancers being detected. Seven of the 14 ovarian cancers diagnosed were Stage 1 disease. In the largest of these studies, the positive predictive value of serum CA125 followed by ultrasound when the CA125 was elevated was 26%, which translates into approximately four surgeries for every one case of ovarian cancer.[23]

In 1995, a National Institutes of Health (NIH) Consensus Conference concluded that there was no evidence that screening the general population with serum CA125 measurements and transvaginal ultrasound could reduce the mortality from ovarian cancer, and therefore screening the general population could not be supported. However, in those patients with an increased genetic risk due to family history or a known inherited predisposition to ovarian cancer, rectovaginal pelvic examinations, transvaginal ultrasound, and serum CA125 measurements were recommended.[24]

Currently there are large prospective population, based screening studies being conducted in the United States (Prostate, Lung, Colorectal and Ovarian (PLCO) Cancer Screening Trial) and in the United Kingdom (United Kingdom Collaborative Trial of Ovarian Cancer Screening (UKCTOCS)). The results of these studies will not be known for many years.

CA125 to Evaluate a Pelvic Mass

Several retrospective studies have demonstrated that CA125 levels can assist in predicting whether a pelvic mass is benign or malignant.[25,26] Schutter et al reported on the results of a prospective multicenter study involving 228 postmenopausal women with pelvic masses evaluated by CA125, transvaginal ultrasound and pelvic examination.[27] The accuracy of an elevated CA125 (>35 U/mL) in differentiating between a benign and malignant process was 77%. This approximated the accuracy using pelvic examination (76%) or transvaginal ultrasound alone (74%). Importantly, no malignancy was detected in any of the women in whom all three tests were considered either negative or not indicative of a malignant process.

Therefore CA125 contributes important preoperative information to guide the management of a pelvic mass, particularly in the postmenopausal setting, as an elevated value alerts the physician to a higher likelihood of the presence of a malignancy. The American College of Obstetricians and Gynecologists (ACOG) recently suggested that if a premenopausal woman with a pelvic mass has a CA125 of >200 U/L consideration should be given to consultation with or referral to a gynecological oncologist.[28]

In 1993, Davies et al suggest the Risk of Malignancy Index (RMI) which consisted of an algorithm combining menopausal status, ultrasound appearance of the pelvic mass and the absolute CA125 level as a triage tool predict whether a pelvic mass would be malignant or not. They reported that the RMI carried a high sensitivity but had a lower specificity; however it remained a useful tool in the evaluation of a pelvic mass.[29] These sentiments were shared by Obeidat et al who reported that the RMI was more accurate than any single criterion when predicting the nature of a pelvic mass. In this study involving 100 women, the sensitivity, specificity, positive predictive value and negative predictive value of the RMI were reported as 90%, 89%, 96% and 78% respectively.[30]

CA125 to Assess Prognosis

There have been a number of small retrospective studies reporting on the rate of fall of CA125 or the absolute levels of CA125 during chemotherapy for ovarian cancer giving independent prognostic information.[31-36] In an attempt to further clarify this issue, Fayers et al evaluated the use of CA125 as a prognostic indicator in 248 women with ovarian cancer and found that the absolute CA125 value after two courses of chemotherapy was the single most important factor for predicting disease progression at 12 months.[37] In fact the authors reported that attainment of a CA125 value less than or equal to 70 IU/mL prior to the 3rd cycle of chemotherapy was the optimal predictive value in the study. However, the authors also cautioned that the absolute value of CA125 gave a false positive rate of 19% and was therefore deemed not reliable enough to direct individual patient management.

Similarly Colakovic et al performed a retrospective study of 222 women and found that treatment response may be predicted by the CA125 half life during treatment. Whilst the prechemotherapy CA125 level was found to have no prognostic value, the normalization of CA125 prior to the third cycle of chemotherapy was associated with a median survival time of 101 months compared to 21 months in those patients who did not have normalization of their CA125.[38]

Markman et al recently evaluated previously reported survival and sequential CA125 data from a Southwest Oncology Group (SWOG) randomized Phase 3 trial comparing the combination of cisplatin and cyclophosphamide to a regimen of carboplatin and cyclophosphamide.[39,40] The authors noted that the 101 women involved in the more recent study were representative of the original study population. Several conclusions were drawn including that the baseline CA125 level did not predict outcome, particularly in those patients with suboptimally debulked Stage III and IV disease. The authors inferred that the majority of cancers were heterogenous in the amount of CA125 they produced, and therefore it was not able to predict the tumor's sensitivity to chemotherapy. As in the papers by Fayers and Colakovic, Markman et al agreed that a favorable change in the CA125 levels after the first cycle of chemotherapy was associated with a modest impact on survival.[39]

Nonepithelial Ovarian Cancer Tumor Markers

The nonepithelial ovarian cancers include germ cell and sex-cord/stromal tumors; most of these tumors produce very characteristic tumor markers. Alpha-fetoprotein (AFP), human chorionic gonadotrophin (hCG) and lactate dehydrogenase (LDH) are the most commonly utilized ovarian germ cell tumor markers (Table 3.1). The particular marker expressed in an ovarian germ cell tumor depends on the differentiation of the tumor as well as the histological type.[41] Ovarian germ cell tumors are commonly of mixed histologies, so they can have a number of positive markers. Immunohistochemistry using these ovarian germ cell tumor markers can be used to aid in the pathological diagnosis. In fact, placental alkaline phosphatase (PLAP) which can be elevated in dysgerminoma is a much more valuable histochemical rather than serological marker.[42] As with epithelial ovarian cancer, the serologic tumor markers are important when assessing treatment response or disease progression.[42]

Table 3.1. Characteristic tumor marker profiles for ovarian germ cell tumors

	LDH	AFP	hCG
Dysgerminoma	+		+/–
Endodermal sinus tumor		+	
Choriocarcinoma			+
Immature teratoma		+/–	
Embryonal carcinoma			+
Mixed germ cell tumor	+/–	+/–	+/–

Abbreviations: LDH: lactate dehydrogenase; AFP: alpha-fetoprotein; hCG: human chorionic gonadotropin.

Lactate Dehydrogenase (LDH)

LDH is produced by liver cells under physiological conditions. Apart from being produced by ovarian germ cell tumors it can also be produced by cutaneous melanoma, pleural mesothelioma and lung cancer cells.

Alpha-Fetoprotein (AFP)

AFP is a glycoprotein that is structurally related to albumin with a molecular weight of 69 kD. In normal physiology, AFP is made by human yolk sac cells and in later embryonic life, by the fetal liver. As the fetal liver matures it transitions to synthesize albumin. AFP is commonly raised in pregnancy and in various liver diseases.

Human Chorionic Gonadotrophin (hCG)

HCG is a sialoglycoprotein with a molecular weight of about 36.5 kD. The physiological source of hCG are trophoblastic cells of the human placenta. Elevated levels are found in normal or ectopic pregnancies as well as molar pregnancy and gestational choriocarcinoma. For this reason, it is a very sensitive marker for nongestational (ovarian) choriocarcinoma and gestational trophoblastic disease.

Tumor markers in Ovarian Germ Cell Tumors

Similar to that in epithelial ovarian cancer, tumor markers in ovarian germ cell tumors are used to aid in diagnosis, to arrange referral to or consultation with a gynecologic oncologist, or to follow the response to therapy in patients diagnosed with these diseases. Given the difference in epidemiology of germ cell and epithelial ovarian tumors, a young woman or girl with a pelvic mass should be evaluated with the appropriate tumor markers for ovarian germ cell tumors.

Sex Cord/Stromal Ovarian Tumors

Ovarian sex cord and stromal tumors are hormonally active; as such, tumor markers for these diseases relate to the hormones that they produce. Ovarian granulosa cell tumors, which account for approximately 2% of ovarian malignancies, have been demonstrated to produce both estradiol and inhibin. Approximately 30% of granulosa cell tumors and most extraovarian recurrences do not produce estradiol. However, inhibin is considered to be a particularly sensitive marker of granulosa cell tumor activity and is elevated often before these tumors become symptomatic.[43] Inhibin is a polypeptide hormone consisting of two subunits (alpha and beta) produced by ovarian granulosa cells and inhibits follicle-stimulating hormone secretion by the anterior pituitary.[43]

The Sertoli-Leydig tumors, another of the sex cord/stromal ovarian tumors, contains components resembling the Sertoli and Leydig testicular cells. It most commonly presents in women of reproductive age and is commonly associated with amenorrhea and signs of virilization as it classically secretes androgens.[44] The cells of the Sertoli-Leydig tumors secrete steroids similar to those of ovarian theca cells including dehydroepiandrosterone, androstenedione, testosterone and 17-hydroxyprogesterone. The hormonal profile is dominated by testosterone; however there does not seem to be a correlation between the degree of tumor differentiation and the amount of testosterone produced.[45] Some Sertoli-Leydig tumors can produce excessive estrogen and progesterone; however this is thought to be due to secondary to hypertrophied thecal tissue associated with the tumor rather than from the Sertoli-Leydig cells themselves.[45] Other authors have also described production of inhibin and alpha-fetoprotein from a Sertoli-Leydig tumor.[46,47]

Diagnosis prior to surgery is suggested by a combination of one or more of the following factors: a history of rapid androgenization in a woman of reproductive age, a pelvic mass and elevated serum testosterone. Recently there have been reports of direct laparoscopic venous ovarian blood sampling for androgen levels, which have been used to aid in the diagnosis.[48]

The therapy and prognosis of Sertoli-Leydig tumors depends on age, stage of tumor and degree of differentiation. The mainstay of treatment is surgical resection with the plasma testosterone and other hormonal levels rapidly returning to normal after this has been achieved. Once treatment is completed the patient can be monitored with periodic plasma testosterone measurements.[49,50]

Gestational Trophoblastic Neoplasia (GTN)

Human chorionic gonadotropin (hCG) is produced in all gestational trophoblastic tumors. It is used as an important marker for detection and monitoring response to treatment in these tumors. A logarithmic regression curve is used based on the serum half-life of the hCG molecule to monitor the coarse of primary treatment.[44] If an abnormal regression of the hCG is documented then commencing chemotherapy or changing chemotherapy is considered.[45] The hCG level corresponds to the tumor mass and is therefore one of the prognostic variables used in evaluating this disease (please refer to Chapter 18, *Gestational Trophoblastic Disease*).

Placental site trophoblastic tumors arise from intermediate trophoblastic cells and, in contrast to other forms of GTN, these tumors tend to produce small amounts of hCG relative to their mass.[46] Placental site trophoblastic tumors however produce human placental lactogen (hPL) which can be helpful in confirming the diagnosis of this disease.[47]

Tumor Markers in Other Gynecologic Cancers

Endometrial Cancer

The marker most frequently elevated in endometrial cancer is CA125, and it normally indicates extrauterine disease. When results from five studies were combined, Jacobs and Bast calculated that high levels of CA125 (>35 U/mL) were present in 22% of women with Stage I and II disease and in 82% of women with Stage III and IV disease.[3]

Cervical Cancer

In squamous cell carcinoma of the cervix the squamous cell carcinoma antigen (SCCA) tends to increase with tumor volume, stage and lymph node involvement.[48] Duk et al suggested that elevated pretreatment SCCA levels in early stage squamous cell carcinoma of the cervix may be an independent predictor of lymph node metastases.[49] Despite this there are no serum markers commonly used in the management of cervix cancer.

Vulvar Cancer

There are no clinically valuable serum markers for the use in vulvar cancer.

References

1. Kabawat SE, Bast RC, Bhan AK et al. Tissue distribution of a coelomic-epithelium related antigen recognized by the monoclonal antibody OC125. Int J Gynecol Pathol 1983; 2:275-85.
2. Nouwen EJ, Hendrix PG, Dauwe S et al. Tumor markers in the human ovary and its neoplasms. Am J Pathol 1987; 126:230-42.
3. Jacobs I, Bast RC. The CA125 tumor-associate antigen: a review of the literature. Hum Reprod 1989; 4:1-12.
4. Tuxen MK, Soletormos G, Dombernowsky P. Tumor markers in the management of patients with ovarian cancer. Cancer Treat Rev 1995; 21:215-43.
5. Tuxen MK. Tumor marker CA125 in ovarian cancer. J Tumor Markers Oncol 2001; 16:49-68.
6. Bast RC, Feeney M, Lazarus H et al. Reactivity of a monoclonal antibody with a human ovarian carcinoma. J Clin Invest 1981; 68:1331-7.
7. Davis HM, Zurawski VR, Bast RC, Klug TL. Characterization of the CA125 antigen associated with human epithelial ovarian carcinomas. Cancer Res 1986; 46:6143-8.
8. Yin BWT, Lloyd KO. Molecular cloning of the CA125 ovarian cancer antigen: identification as a new mucin, muc16. J Biol Chem 2001; 276:27371-5.
9. O'Brien TJ, Beard JB, Underwood LJ et al. The CA125 gene: an extracellular superstructure dominated by repeat sequences. Tumor Biol 2001; 22:348-66.
10. Bast RC, Klug TL, St John E et al. A radioimmunoassay using a monoclonal antibody to monitor the course of epithelial ovarian cancer. N Engl J Med 1983; 309:883-7.
11. Bon GG, Kenemans P, Verstraeten R et al. Serum tumor marker immunoassays in gynecologic oncology: establishment of reference values. Am J Obstet Gynecol 1996; 174:107-14.
12. Bonfrer JMG, Korse CM, Verstraeten RA et al. Clinical evaluation of the Byk LIA-mat CA125 assay: discussion of a reference value. Clin Chem 1997; 43:491-7.
13. Alagoz T, Buller RE, Berman M et al. What is a normal CA125 level? Gynecol Oncol 1994; 53:93-7.
14. Nustad K, Bast RC Jr, Brien TJ et al. Specificity and affinity of 26 monoclonal antibodies against the CA125 antigen: first report from the ISOBM TD-1 workshop, International Society for Oncodevelopmental Biology and Medicine. Tumour Biol 1996; 17:196-219.
15. Pauler DK, Menon U, McIntosh M et al. Factors influencing serum CA125 II levels in healthy postmenopausal women. Cancer Epidemiol Biomarkers Prev 2001; 10:489-93.
16. Grover S, Koh H, Weiderman P, Quinn MA. The effect of menstrual cycle on serum CA125 levels: a population study. Br J Obstet Gynecol 1990; 97:930-3.

17. Gocze PM, Szabo DG, Than GN et al. Occurrence of CA125 and CA19-9 tumor associated antigens in sera of patients with gynaecologic, trophoblastic and colorectal tumors. Gynecol Obstet Invest 1988; 25:268-72.

18. Jacobs I, Bast RC. The CA125 tumor-associate antigen: a review of the literature. Hum Reprod 1989; 4:1-12

19. Tuxen MK, Soletormos G, Dombernowsky P. Tumor markers in the management of ovarian cancer. Cancer Treat Rev 1995; 21:215-43.

20. Tuxen MK. Tumor marker CA125 in ovarian cancer. J Tumor Markers Oncol 2001; 16:49-68.

21. Bast RC, Klug TL, St John E et al. A radioimmunoassay using a monoclonal antibody to monitor the course of epithelial cancer. N Engl J Med 1983; 309:883-7.

22. Bell R, Petticrew M, Sheldon T. The performance of screening tests for ovarian cancer: results of a systematic review. Br J Obstet Gynecol 1998; 105:1136-47.

23. Jacobs I, Davies AP, Bridges J et al. Prevalence screening for ovarian cancer in postmenopausal women by CA125 measurements and ultrasonography. Br Med J 1993; 306:1030-4.

24. NIH Consensus Development Panel on Ovarian Cancer. NIH Consensus Conference. Ovarian Cancer: screening, treatment and follow-up. JAMA 1995; 273:491-7.

25. Mogensen O, Mogensen B, Jacobsen A. CA125 in the diagnosis of pelvic masses. Eur J Cancer Clin Oncol 1989; 25:1187-90.

26. Jacobs IJ, Rivera H, Oram DH, Bast RC. Differential diagnosis of ovarian cancer with tumour markers CA125, CA 15-3 and TAG-72-3. Br J Obstet Gynaecol 1993; 100:1120-4.

27. Shutter EMJ, Kenemans P, Sohn C et al. Diagnostic value of pelvic examination, ultrasound and serum CA125 in postmenopausal women with a pelvic mass. Cancer 1994; 74:1398-406.

28. ACOG Committee Opinion. The role of the generalist obstetrician-gynaecologist in the early detection of ovarian cancer. Obstet Gynecol 2002; 100:1413-6.

29. Davies AP, Jacobs I, Woolas R et al. The adnexal mass: benign or malignant? Evidence of a risk of malignancy index. Br J Obstet Gynaecol 1993; 100(10):927-31.

30. Obeidat BR, Amarin ZO, Latimer JA, Crawford RA. Risk of Malignancy Index in the preoperative evaluation of pelvic masses. Int J Gynaecol Obstet 2004; 85(3):255-8.

31. van der Burg ME, Lammes FB, van Putten WL, Stoter G. Ovarian cancer: the prognostic value of the serum half-life of CA125 during induction chemotherapy. Gynecol Oncol 1988; 30:307-12.

32. Mogensen O. Prognostic value of CA125 in advanced ovarian cancer. Gynecol Oncol 1992; 44:207-12.

33. Fisken J, Leonard RCF, Stewart M et al. The prognostic value of early CA125 serum assay in epithelial ovarian cancer. Br J Cancer 1993; 68:140-5.

34. Ron IG, Inbar M, Gelernter I et al. The use of CA125 response to predict survival parameters of patients with advanced ovarian carcinoma. Acta Obstet Gynecol Scand 1994; 73:658-62.

35. Gadduchi A, Zola P, Landoni F et al. Serum half-life of CA125 during chemotherapy as an independent prognostic variable for patients with advanced epithelial ovarian cancer: results of a multicentric Italian study. Gynecol Oncol 1995; 58:42-7.

36. Paramasivam S, Tripcony L, Crandon A et al. Prognostic importance of preoperative CA125 in International Federation of Gynecology and Obstetrics Stage 1 Epithelial Ovarian Cancer: An Australian Multicenter Study. J Clin Oncol 2005; 23(25):5938-42.

37. Fayers PM, Rustin GJS, Wood R et al. The prognostic value of serum CA125 in patients with advanced ovarian carcinoma: an analysis of 573 patients by the Medical Research Council Working Party on Gynecological Cancer. Int J Gynecol Cancer 1993; 3:285-92.

38. Colakovic S, Lukic V, Mitrovic L et al. Prognostic value of CA125 kinetics and half-life in advanced ovarian cancer. Int J Biol Markers 2000; 15(2):147-52.

39. Markman M, Federico M, Liu PY et al. Significance of early changes in the serum CA125 antigen level on overall survival in advanced ovarian cancer. Gynecol Oncol 2006; 103:195-8.

40. Alberts DS, Green S, Hannigan EV et al. Improved therapeutic index of carboplatin plus cyclophosphamide versus cisplatin plus cyclophosphamide: final report by the Southwest Oncology Group of a Phase III randomized trial in stages III and IV ovarian cancer. J Clin Oncol 1992; 10:706-17.

41. Bower M. The value of tumor markers in ovarian germ cell tumors. Tumor Marker Update 1996; 8:1-7.

42. Patterson DM, Rustin GJS. Controversies in the management of germ cell tumours of the ovary. Curr Opin Oncol 2006; 18:500-6.

43. Lappohn RE, Burger HG, Bouma J et al. Inhibin as a marker for granulosa cell tumors. N Engl J Med 1989; 321:790-3.

44. Sawetawan C, Rainey WE, Word RA et al. Immunohistochemical and biochemical analysis of a human Sertoli-Leydig cell tumor: autonomous steroid production characteristic of ovarian theca cells. J Soc Gynecol Investig 1995; 2(1):30-7.

45. Stegner H-E, Lisboa BP. Steroid metabolism in an androblastoma (Sertoli-Leydig cell tumor). A histopathological and biochemical study. Int J Gynecol Pathol 1985; 2:410-25.

46. Kommoss F, Oliva E, Bahn AK et al. Inhibin expression in ovarian tumors and tumor-like lesions: an immunohistochemical study. Mol Pathol 1998; 11:656-64.

47. Singh ZN, Singh MK, Chopra P. Sertoli Leydig cell tumor with malignant heterologous elements and raised alpha-fetoprotein: a case report. J Obstet Gynaecol Res 1996; 22:595-8.

48. White LC, Buchanan KD, O'Leary TD et al. Direct laparoscopic venous sampling to diagnose a small Sertoli-Leydig tumor. Gynecol Oncol 2003; 91(1):254-7.

49. Lantzsch T, Stoerer S, Lawrenz K et al. Sertoli-Leydig cell tumor. Arch Gynecol Obstet 2001; 264:206-8.

50. Gershenson DM, Morris M, Burke TW et al. Treatment of poor-prognosis sex cord stromal tumors of the ovary with the combination of bleomycin, etoposide and cisplatin. Obstet Gynecol 1996; 87:527-31.

51. Morrow CP, Kletzky OA, DiSaia PJ et al. Clinical and laboratory correlates of molar pregnancy and trophoblastic disease. Am J Obstet Gynecol 1977; 128:424-30.

52. Feltmate CM, Batorfi J, Fulop V et al. Human Chorionic Gonadotrophin follow-up in patients with molar pregnancy: a time for reevaluation. Obstet Gynecol 2003; 101:732-6.

53. Feltmate CM, Genest DR, Wise L et al. Placental site trophoblastic tumor: a 17 year experience at the New England Trophoblastic Disease Center. Gynecol Oncol 2001; 82:415-9.

54. Papadopoulos AJ, Foskett M, Seckl MJ et al. Twenty-five years' clinical experience with placental site trophoblastic tumors. J Reprod Med 2002; 47:460-4.

55. Scambia G, Panici BP, Foti E, Amoroso M et al. Squamous cell carcinoma antigen: prognostic significance and role in the monitoring of neo-adjuvant chemotherapy response in cervical cancer. J. Clin. Oncol. 1994 12:2309-16.

56. Duk JM, de Bruijn H, Klass KH et al. Pre-treatment serum squamous cell carcinoma antigen: a newly identified prognostic factor in early stage cervical cancer. J Clin Oncol 1996; 14:111-8.

CHAPTER 4

Epidemiology

Heather S. Pulaski and Brigitte E. Miller

Introduction

Although breast cancer is the most frequent cancer in women, gynecologic malignancies involving the upper and lower reproductive tract likewise pose a significant threat. In the United States, 78,490 new cases and 28,490 deaths from these cancers will occur in the year 2008 (Fig. 4.1).[1] The incidence of new gynecologic cancers and deaths worldwide in 2002 were 896,525 and 448,692 respectively (Fig. 4.2) with cervical cancer posing the largest threat.[2] The study of epidemiology examines these cases to determine incidence, risk factors, protective and preventive factors and survival projections.

In order to understand the study of epidemiology it is important to be familiar with several terms. Incidence refers to the number of new cases of a disease in a population during a specified time period. Age-specific incidence is defined by the number of new cases in a particular age group divided by the total population in that age group. The proportion of people who have a condition or disease at a specific time is prevalence. Relative survival is the ratio of observed survival rate for a patient group compared to the survival rate expected for a population with similar demographics. Relative risk (RR) is the risk of disease or death in a population exposed to the factor of interest divided by the risk of those who were not exposed. An increased risk of disease is indicated by a relative risk greater than one, while a relative risk less than one shows a decreased risk from the exposure. Studies of epidemiology need to be evaluated closely and the findings weighed objectively since most are observational in nature and as such are subject to bias. Nonetheless, these studies are useful and can provide information regarding the underlying etiology of disease states.

Ovarian Cancer

Worldwide, carcinoma of the ovary is the sixth most common cancer in women accounting for 4% of all female cancer cases worldwide.[2] Ovarian cancer (OC) is the fourth leading cause of cancer deaths in the United States and the most common cause of death from a gynecological malignancy. Similar findings are reported from all well-developed countries of the northern hemisphere. The American Cancer Society estimates there will be 21,650 new cases with 15,520 deaths from the disease expected in 2008.[1] Over the last 20 years, incidence rates have remained basically stable with an age-adjusted incidence rate for all cases of approximately 13.3 per 100,000 women per year of which the majority (95%) will be epithelial tumors (Fig. 4.3).[3,4] Based on data from 2000-2003, 1 in 72 women born today will be diagnosed with ovarian cancer during their lifetime.[3]

Gynecologic Oncology, edited by Paola Gehrig and Angeles Secord.
©2009 Landes Bioscience.

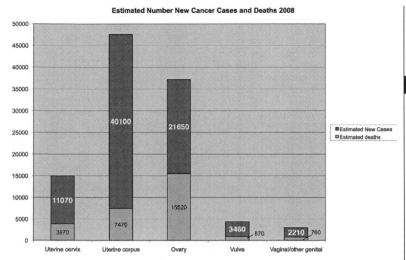

Figure 4.1. Estimated new gynecologic cancer cases and deaths in the United States in 2008.[1]

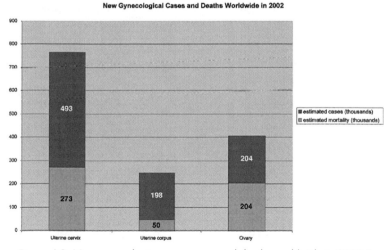

Figure 4.2. New gynecologic cancer cases and deaths worldwide in 2002.[2]

Risk Factors

Many theories exist regarding the etiologies of OC. The development of a conclusive system is hampered by the heterogeneity of epithelial OC and the rarity of other ovarian malignancies. The best developed concept at this time is the ovulation model, which theorizes that the trauma to the ovarian epithelium at time of ovulation and subsequent contact with fluids containing high estrogen concentrations may increase proliferation and inclusion cyst formation. The regenerating epithelium may also be

4

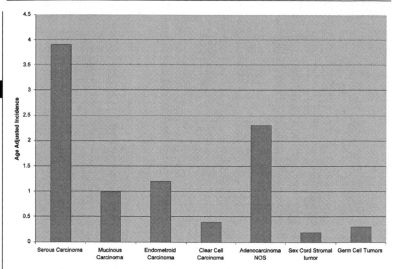

Figure 4.3. Age adjusted incidence rates of ovarian cancer by histological type.[5]

more susceptible to external factors such as infectious agents and chemicals.[6] Another hypothesis identifies high gonadotropin levels as a cause for ovarian cancer; however recently conflicting results have been published.[7] The estimated relative risk of ovarian cancer is increased for women who are nulliparous (RR 2-3), have an early menarche (RR 1.5), late menopause (RR 1.5-2), or a history of infertility (RR 2-5).[8]

Genetic predisposition is the strongest risk factor but only affects 10% of patients with ovarian cancer. Having one first degree relative with OC confers a relative risk of 1.9-18 while having two or more first degree relatives with OC has a relative risk of 7.18-16.[10,11] The association of nulliparity with OC is stronger if a family history of disease is also present. The relative risk of nulliparity alone is 1.4 but rises to 2.7 if there is a positive family history of the disease.[11] The most frequently seen mutation in 60-65% of cases involves the *BRCA1* and *2* genes. Women of Ashkenazi heritage have a significantly increased risk for a *BRCA* mutation over 1%; however, the penetrance of the mutation is lower.[12] By age 70, the average cumulative risk of having OC is 39% for *BRCA1* carriers and 11% for *BRCA2* carriers. Probabilities published in the literature vary due to the populations examined.

Increasing age has been shown to increase the risk as it allows more time for noxious stimuli to initiate the development of a malignancy. However, in women with a genetic predisposition an initial genetic insult has already occurred and therefore malignancies tend to develop at an earlier age. Data collected between 1998-2002 including all patients with ovarian cancer reveal a median age at diagnosis of 63 years with 41.8% of cases diagnosed in patients between the ages of 55 and 74 (Fig. 4.4).[3] A history of endometriosis has also been associated with an increased incidence of the rare endometrioid or clear cell carcinoma.[11] An inflammatory reaction in experimental studies has been shown to increase development of ovarian cancer. Epidemiologic data confirms this only by inference as a decreased risk is noted in the absence of infection or after long-term anti-inflammatory medication.[6,13] The contaminant theory arose since the use of talc in genital hygiene has

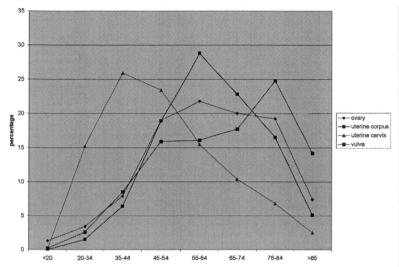

4

Figure 4.4. Age at diagnosis.[3]

been linked to an increased risk of OC, possibly due to increased inflammatory reaction. This effect may be related to small admixtures of asbestos, which is related to mesotheliomas. Overall this impact is minor.[14] Regardless, both endometriosis and inflammatory reaction are thought to contribute to the development of OC by causing transformation of the cells on the ovarian surface.

Protective Factors

Most protective factors supposedly decrease the risk for OC by reducing the number of ovulations, decreasing gonadotropin secretion or increasing progesterone levels.[15] Increasing parity would lead to a lower risk through both mechanisms since pregnancy stops ovulation, decreases gonadotropin secretion and increases progesterone. The first pregnancy grants a 40% reduction in risk and each subsequent pregnancy decreases the risk 10-15%. The Nurses Health Study found a relative risk of 0.84 per pregnancy. This benefit exists across all histological subtypes.[11,16] By the same token, oral contraceptive (OCP) use offers a 40% reduction in risk with use and an additional 5-10% reduction with each subsequent year by inducing anovulation.[11] The hormonal effects are similar to pregnancy. Overall, women with a history of OCP use have a relative risk of 0.64 compared to women who have never used OCPs. These effects appear to last 10-20 years after use has been discontinued. The impact may vary depending on histology. An analysis of multiple case controlled studies showed a relative risk of 0.73 for epithelial tumors and 0.56 for serous tumors but 1.96 for mucinous tumors.[11] Breastfeeding also confers a risk reduction of almost 20% with an improved benefit for increasing duration secondary to its anovulatory effects. The relative risk of OC for women who have breastfed is 0.7.[9,11] This factor may also vary depending on histological type. Some studies have shown an effect towards nonmucinous tumors only while a large analysis of case controlled studies showed an inverse association with all types except clear cell tumors.[16] Hysterectomy has also shown to be protective. Multiple studies have shown a 30-40% reduction in risk even after controlling for parity and OCP use. The reason for this is unclear although it could be due to impaired ovarian blood supply following surgery or by the prevention of passage of carcinogens and infections through the vagina and uterus. Other studies have raised the possibility of bias since many patients may be unaware of concomitant oophorectomy at the time of operation, or because possibly diseased ovaries were removed. However, the protective effects appear to be long lasting which seems to indicate some relationship. Tubal ligation is thought to confer a reduced risk of approximately 15% through mechanisms similar to those proposed for hysterectomy. Relative risk ratios have been reported at 0.59-0.87.[9,19,20]

Undetermined Factors

Many other factors have been studied but no firm conclusions have been reached about their effects on ovarian cancer risk. Since most risk factors seem to be time-dependent, patient age at the time of initial reproductive events such as menarche and menopause have been extensively studied. Specifically, the age at menarche and menopause have been heavily examined given the theory of ovulation-induced damage. The majority of studies have shown no clear association, but a large case control study did show a relative risk of 0.8 associated with a late age at menarche as well as a relative risk of 0.7 associated with premature menopause (<45 years-old).[11] Studies investigating the age at first delivery have also had inconsistent results. Most US studies show a trend of decreasing risk with increasing age at first birth, but European studies show a significantly increased risk with increasing age at first child. Many other studies have no associated affect.[9,11]

Another area of focus has been the effect of medical treatments on ovarian function. While radiation generally increases the rate of pelvic cancer, there has been no reported increased risk of ovarian cancer in patients with radiation-induced menopause.[6] Hormone replacement could theoretically increase the rate of ovarian cancer by prolonging ovarian stimulation, but analysis of multiple case control studies did not find any significant alteration in risk.[11] A recent prospective study, however, did rekindle concern with relative risk of 1.7 with greater than 11 years of HRT use prior to study inclusion and a relative risk of 1.7 with current use of 6 years or greater.[11] Patients with a history of infertility have a multitude of compounding variables that may contribute to their ovarian cancer risk such as parity, breast feeding and OCP use. It is also difficult to compare results across multiple studies since the definition of infertility, mode of diagnosis and methods of treatment differ greatly. In general, infertility is considered to be a risk factor for OC, but the role of fertility treatments, underlying ovarian dysfunction and confounding factors has yet to be fully determined.[9,11]

Other studies have examined demographics such as height, weight and socioeconomic factors. Some large prospective studies show a higher risk associated with increasing height especially in regard to endometrioid and serous borderline tumors but definitive studies need to be done.[11] In terms of weight, several large cohort studies have shown a relationship between adolescent obesity and increased risk of ovarian cancer but have not shown an association with obesity later in life. Large case control studies have shown slight elevation in risk with body mass index (BMI) >30 (RR 1.1-1.2) while many others show no associated risk change. This disparity may be due to the different histological subtypes involved since some tumor types such as endometrioid and clear cell may be more sensitive to estrogen and thus more affected by obesity.[11,16,20]

Lifestyle factors have also been thought to play a role. Overall, no clear evidence links diet choices to an alteration in ovarian cancer risk. The Nurses Health Study found no association except for decreased risk among patients that ate at least 2.5 servings of fruit and vegetables per day as adolescents. Other factors that have been examined are the consumption of red meat, milk, fish, eggs, saturated fat, polyunsaturated fatty acids, carbohydrates, fiber and vitamins A, C, D and E.[9,11] Data regarding tobacco use is conflicting. A relative risk of 2.9 for patients that smoke more than one pack a day compared to never smokers has been shown, but others have shown only an increased association with mucinous tumors.[11] There is also a proposed inverse association of current smoking with endometrioid subtype similar to endometrial cancer since smoking alters estrogen metabolism. Results regarding alcohol use are conflicting, but chronic use is associated with increased estrogen levels and so may increase the risk of endometrioid types.[9,11] Ovarian cancer risk may increase with rising socioeconomic status but this may be attributed to confounding factors such as lower fertility rates in this group or a different diet.[9,11]

Primary Prevention

No preventive intervention has been shown to decrease the incidence of ovarian cancer or improve survival in the general population. If not contraindicated, patients may use oral contraceptives and should be counseled to avoid talc. For women with a mutation within the *BRCA* gene, prophylactic removal of ovaries and Fallopian tubes may lead to a risk reduction as high as 90% and surgery should be considered after genetic counseling and completion of childbearing.[17,18] The risk of a peritoneal malignancy although low is still present.

Survival

Since ovarian cancer has a very insidious onset with few specific symptoms, the majority of patients (approximately 70%) present with advanced stage disease (Fig. 4.5). Since the most important prognostic factor is the stage at presentation, ovarian cancer has the highest mortality to case rate of all the gynecological malignancies. It is the cause of 4% of cancer deaths in women worldwide. The median age at death from ovarian cancer was 71 years; between 2001-2005 the age adjusted death rate was 8.8 per 100,000 women per year (Fig. 4.6). The overall 5-year survival rate is about 45%, but with disease confined to the primary site the rate approaches 93%. Unfortunately, the rate falls to 31% when distant metastasis are present (Fig. 4.7). A woman's lifetime risk of dying from ovarian cancer is almost 1%.[3]

Endometrial Cancer

Endometrial cancer is the most common gynecologic malignancy. Worldwide it is the seventh most common cancer in women and in the United States it makes up 6% of all cancer cases.[1-3] In America, one out of 40 women will develop endometrial cancer. For 2008, the American Cancer Society projects that there will be 40,100 new cases and 7,470 deaths (Fig. 4.1). The rates of uterine cancer had increased dramatically, peaking in the mid-1970s at which point the increased risk of endometrial cancer secondary to unopposed estrogen was recognized and postmenopausal hormone replacement therapy was adjusted to include progestins to prevent endometrial cancer.[19] Since then the incidence has decreased significantly and eventually stabilized. Between 1997-2005 the incidence of endometrial cancer declined by 0.6%.[3]

Risk Factors

Endometrial lesions can be stratified by type based on whether or not tumors are hormonally sensitive. Type I tumors comprise approximately 80% of cases and are associated with unopposed estrogen exposure. These cancers are usually low grade and are

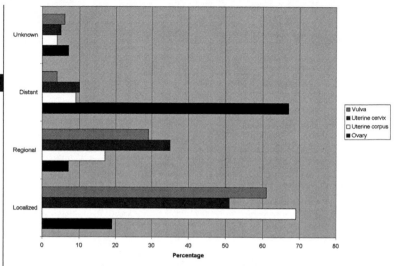

Figure 4.5. Stage at presentation.[3]

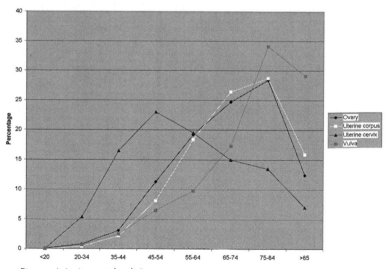

Figure 4.6. Age at death.[3]

associated with endometrial hyperplasia. These cancers result when there is an excess of estrogen over progesterone either through increased production, decreased degradation or exogenous intake. The other 20% of cases are Type II lesions which do not seem to be hormone-dependent and usually occur in older patients with no increased estrogen exposure. These tumors are usually high grade and associated with poor prognostic histologies such as clear cell or papillary serous.[21,22]

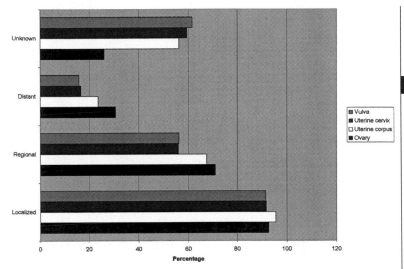

Figure 4.7. Average 5-year survival by stage.[3]

Chronic unopposed estrogen is the strongest risk factor identified in relationship to endometrial cancer. The endometrium is stimulated by estrogen to proliferate and in the absence of progesterone, this leads to increased mitotic activity, inhibition of apoptosis, increased angiogenesis and finally cellular atypia. The estrogen can be internally produced and therefore conditions such as early menarche or late menopause that prolong the duration of estrogen exposure, often without adequate progesterone production, can increase risk. Obesity is another major risk factor and at least 50% of patients with endometrial cancer are overweight. The excess adipose tissue contributes to peripheral conversion of estrogen precursors into estradiol, insulin resistance, ovarian androgen excess, anovulation and chronic progesterone deficiency. A BMI of greater than 25 kg/m² doubles the risk and a BMI of greater than 30 mg/m² triples the risk. For patients more than 50 pounds over their ideal body weight the risk of developing cancer is increased tenfold.[22] Anovulation also contributes to endometrial cancer risk by causing a lack of progesterone control in chronic conditions such as polycystic ovarian syndrome.[22] Exogenous estrogen therapy has been shown to increase the risk for endometrial cancer in a dose- and time-dependent fashion. After 1 year of unopposed estrogen therapy, 20-50% of women will develop endometrial hyperplasia. The relative risk of developing endometrial cancer while on estrogen only therapy ranges from 3.1 to 15.[19,20,22] Tamoxifen, a common therapy for breast cancer, functions as an estrogen antagonist in breast tissue but as an estrogen agonist within the uterus. With 2 to 5 years of use the relative risk of developing endometrial cancer is 2.0 and with at least 5 years of use, it rises to 6.9. Although most are endometrioid lesions, the likelihood of a papillary serous tumor is increased two to six times.[23]

Apart from estrogen effects, other factors do appear to correlate with a higher rate of uterine cancer. Age does appear to influence risk as the usual patient is postmenopausal with a median age of 63 years (Fig. 4.4). The incidence among different ethnic groups is similar, but African-American women are more frequently diagnosed with poorly differentiated or papillary serous high risk endometrial cancers.[3,24] Family history is important since the most common extracolonic manifestation of hereditary nonpolyposis colon cancer syndrome (HNPCC) is endometrial cancer. Women with HNPCC have a lifetime risk of endometrial cancer of 40-60%.[22,25] A history of malignancy also leads to an elevated incidence of uterine cancer. Tumors of the breast, ovary and liver can metastasize to the uterus, they can produce hormones that stimulate secondary tumors or since similar risk factors are present, a separate primary tumor may develop. The best example is the granulosa tumor of the ovary, which increases the risk of endometrial cancer through increased estrogen production. Comorbid conditions such as diabetes and hypertension, which often accompany obesity, have also been shown to have independent effects.[22]

Protective Factors

Factors that decrease the risk of cancer of the uterus apply their effects mostly by decreasing the amount of available estrogen and/or increasing the progesterone effect. The progesterone component of oral contraceptives suppresses estrogen effect and at least 1 year of use of oral contraceptive use decreases the relative risk to 0.6.[26] This effect appears to last at least 15 years after use has been discontinued. Increasing parity also decreases the risk for endometrial cancer due to the increased production of progesterone by the placenta.[19,22] Current and past tobacco use decreased the risk of endometrial cancer by causing weight reduction, altered estrogen metabolism and early menopause. The Nurses Health Study showed a relative risk 0.63 for current smokers and 0.73 for past smokers compared to never smokers.[27] When the results were adjusted for BMI, the relative risk increased to 0.72 for current smokers but did not change for past smokers. Others have found that women with at least a 50 pack year history had approximately a 50% reduction in risk compared to never smokers.[27] Lastly, even among obese women, physical activity exerts a protective effect. The exact mechanism is unknown but is theoretically due to alterations in estrogen metabolism.[20,22]

Primary Prevention

The best ways to prevent endometrial cancer are to maintain an ideal body weight, participate in physical activity and avoid unopposed estrogen therapy. Long periods of oligomenorrhea should be treated with intermittent progestins to prevent hyperplasia and cancer. It is also important to educate patients on the importance of seeking treatment for changes in bleeding patterns and postmenopausal bleeding.[19]

Survival

Unlike ovarian cancer, most cases (70-90%) of endometrial cancer present when still confined to the uterine corpus when it can be easily cured by surgical removal (Fig. 4.5). Besides stage, the histological type and grade and the presence of lymphovascular invasion are the main prognostic factors. The median age at death from endometrial tumors is 73 years with an age-adjusted death rate of 4.1 per 100,000 women per year (Fig. 4.6). The overall 5-year survival is 83% although if regional or distant disease is present the rates are 68% and 24%, respectively (Fig. 4.7).

Cervical Cancer

The incidence of cervical cancer varies by geographical region due to access to screening. Worldwide it is the seventh most common cause of cancer overall and accounts for 9.7% of all cancers in women (Fig. 4.2).[2] Three-quarters of cases occur in developing countries where the lifetime risk of having cervical cancer is 3%. The highest rates occur in Latin America, subsaharan Africa, the Caribbean, Southern and Southeast Asia. In developing countries, cervical cancer is the second most common cause of cancer morbidity. By contrast, in the United States, the lifetime risk is about 1%. It is projected that in 2008, 11,070 new cases of invasive cervical cancer will be diagnosed in the United States and 3,870 deaths will result from the disease (Fig. 4.1). The age adjusted incidence rate is 8.8 per 100,000 women per year. Over a lifetime, 1/138 women will develop cancer of the cervix with a median age of diagnosis of 48 years (Fig. 4.4).[3] The majority of cases are squamous cell tumors but approximately 15% are adenocarcionomas.[28]

Risk Factors

The most significant risk factor for the development of cervical cancer is infection with the human papillomavirus (HPV). There are over 100 types of the HPV virus, some of which infect the genital tract and are sexually transmitted. Among these there are types leading to integration of the HPV DNA into the host genome which interferes with gene function, specifically with p53 gene function, thus promoting the development of dysplasia, a precancerous lesion also known as cervical intra-epithelial neoplasia (CIN). Among the high risk types, HPV 16 and18 are the most frequent.[29] Low risk HPV types remain episomal and do not increase the risk of malignancy, although they may give rise to genital condyloma. Persistent infection with HPV has now been shown to have a causal relationship with severe cervical dysplasia and invasive cervical cancer. The incidence of HPV is hard to predict since many cases are subclinical. An estimated 30 million cases of genital HPV infection are diagnosed each year worldwide, and at least half of the cases are in individuals between the ages of 15 and 24 years. While nearly 80% of newly diagnosed cases are cleared within 8 to 12 months, women with high risk HPV types are more likely to have persistent infections and develop cervical cancer. Risk factors for acquiring HPV are similar to other sexually transmitted diseases (STDs) such as a history of multiple sexual partners, early age at first intercourse, a history of STDs, high parity, drug use, alcohol consumption and a higher frequency of vaginal intercourse.[21,28-32] Even after treatment for CIN, the risk of invasive cervical cancer remains approximately 56/100,000.[33]

Nearly all other risk factors for cervical cancer are important mostly due to their influence on HPV infections. Tobacco use confers an increased risk of cervical cancer even when adjustments are made for the other confounders of age and number of sexual partners.[34] The cervical mucous of smokers has been found to contain potentially mutagenic factors such as nicotine, hydrocarbons and tars; in addition smoking alters the immune system which promotes HPV infection. Also related to immune modulation, chronic immunosuppression secondary to HIV infection, chronic steroid medication or chronic immunosuppressive therapy after organ transplantation, can lead to an increased risk of dysplasia and cervical cancer, a higher recurrence risk after treatment for cervical dysplasia and more rapidly progressing disease. A history of vulvar or vaginal dysplasia is also linked to higher rates of cervical neoplasm most likely secondary to higher rates of HPV infection in these patients and the multifocal field effect of this infectious disease. Oral contraceptive use has also been implicated in cancer of the cervix. While use of birth control pills could indicate promiscuity, the association persists once other factors are controlled and the risk increases with the duration of use. While the incidence of CIN doubles after 5 years of oral contraceptive use, squamous cell tumors have less of a correlation with oral contraceptives than adenocarcinomas. Finally, since cervical lesions usually progress very slowly, lack of access to screening is an important risk factor in the development of invasive cervical disease. Patients of low socioeconomic status have been shown to have higher rates of cervical cancer due to decreased screening and lack of health care resources.[28,30] In the US most patients diagnosed with advanced invasive cervical cancer have not participated in screening programs. The lack of adequate screening is also the reason for the high rate of cervical cancer in countries without such programs.

Protective Factors

Protection from the development of cervical cancer occurs mainly through the prevention of HPV transmission or by improving the body's natural immunity. Delay of initiation of sexual activity and reducing the number of partners is an important factor in this respect. Barrier contraception works by protecting against sexually transmitted disease and male circumcision seems to decrease the transmission of the disease. Diets rich in folate, vitamin A and beta carotene have been shown to be protective against cervical dysplasia possibly by increasing the level of micronutrients at the level of the cervix.[35]

Undetermined Factors

Other points that have been examined with no clear conclusion regarding their effect on cervical cancer are obesity, age at menopause, hormone replacement therapy and douching. The most extensively studied are the effects of weight and douching. There is no data to support a relationship between body weight and squamous cell carcinoma. Glandular tumors may be more hormonally sensitive and therefore obese women may have an increased risk. It has been shown that increasing weight does decrease both compliance with screening and the effectiveness of screening which could also explain the increased incidence among overweight women.[20] In the early 20th century coal tar douches were often used and were associated with an increased incidence of cervical cancer indicating that coal may be involved in carcinogenesis. In modern times, douching may increase the risk by increasing HPV infection. In one study douching within the last 90 days was associated with a two-fold increase in risk of detection of HPV.[11]

Primary Prevention

Patients should be counseled on safe sex practices including the use of barrier contraceptives. Avoiding tobacco and eating a diet rich in folate may also be preventative. The HPV vaccine holds great promise for preventing HPV infection and thus preventing cervical cancer.

Survival

The worldwide ratio of mortality to incidence of disease is 55%.[2] The age-adjusted death rate is 2.6 per 100,000 women per year in the United States. For all stages taken together the 5-year survival rate is 72% (Fig. 4.7).[3] The most important prognostic factor is lymph node status. Approximately 52% of cases present with localized disease, and these patients have an average 5-year survival rate of 92% (Figs. 4.5 and 4.7). For the 9% of patients who present with distant disease, the survival rate decreases to about 15%.

Vulvar Cancer

Since the vulva undergoes less constant transformation, the disease progression from in situ lesion is less clear for the vulva than for the cervix. Vulvar cancer makes up approximately 4% of gynecologic malignancies. The American Cancer Society estimates that 3,460 women will be diagnosed with and 870 women will die of cancer of the vulva in 2008 (Fig. 4.1).[1] The overall age adjusted incidence rate was 2.2 per 100,000 women per year and the incidence has increased slightly (0.6%) between 1975 and 2003.[3] The median age at diagnosis is 69 years old, but the trend is turning towards younger patients (Fig. 4.4). The increased incidence of vulvar cancer in younger patients may be due to increasing awareness of the disease; however others speculate that the rising incidence of vulvar intra-epithelial neoplasia in these patients could be the source.[41]

Risk Factors

If patients are younger than 55 years old they tend to have the same risk factors as are related to other genital cancers such as low socioeconomic status, high risk sexual behavior and presence of HPV. In patients over 55 years, there typically is no history or STD or tobacco use. In these patients, HPV DNA is found only in about 15%.[19,37,38]

While vulvar intra-epithelial neoplasia (VIN) has traditionally been thought to rarely progress to invasive disease, recent studies have shown that approximately 9% of untreated patients with VIN III will progress to invasion over 12-96 months and occult invasion was found in 20% of patients at the time of treatment for presumed VIN III. Of the patients who were treated, only 3% subsequently developed invasive disease. There was no difference in the rates of progression for multifocal or unifocal VIN III. Many of these patients have a history of HPV. As with cervical cancer, tobacco use and immunosuppression increase the risk of vulvar cancer through similar mechanisms of immune modulation and carcinogenesis. Likewise, patients with a history of other anogenital cancers or tobacco related neoplasms such as lung and oropharynx cancers have an increased rate of invasive vulvar disease.[39-41]

In older patients the most common risk factor is a history of vulvar dystrophy without associated HPV infection or tobacco use. Lichen sclerosis, a disease characterized by vulvar inflammation and epithelial thinning, has a 5% chance of developing into a vulvar carcinoma.[38]

Primary Prevention

Patients should be counseled to avoid tobacco and perform self exams once a month. This is especially important for immunocompromised patients and those with a history of VIN or vulvar dystrophies.

Survival

The age-adjusted death rate for vulvar cancer is 0.4 per 100,000 women per year.[3] The overall survival rate is 78% for all stages. If distant metastasis are present, the 5-year survival falls to 18% (Fig. 4.7). The main prognostic factor is lymph node involvement and patients with positive inguinal nodes have a 31% risk of recurrence in the first 2 years after treatment versus 5% in patients with negative nodes.[42]

Vaginal Cancer

As with vulvar lesions, the progression from in situ dysplastic lesions to invasive cancer is less clear than with cervical disease. Primary malignancies of the vagina are quite rare with an incidence of 1/100,000 women and make up only 3% of all female genital neoplasms.[1,41] Of these tumors, approximately 90% are squamous lesions and 25% of these tumors are in situ only. In children, sarcomas are the most common tumor type.[31] Metastatic tumors of the vagina are much more common than primary tumors. These result either by direct extension from the uterus, cervix or vulva or by spread through the hematogenous or lymphatic systems from the ovary, breast or kidney.[19]

Risk Factors

Since the incidence of vaginal cancer is so low, knowledge about risk factors and etiology is sparse. In general, similar risk factors to cervical dysplasia/neoplasia such as early sexual intercoure, increasing number of sexual partners, history of other ano-genital cancers and HPV subtypes are present. Squamous cell carcinoma occurs mostly in postmenopausal women with a mean age at diagnosis of 60 years. Adenocarcinoma occurs mostly in younger women with history of diethylstilbestrol (DES) exposure prior to 12th week of gestation. A family history of vaginal cancer in a first degree relative is associated with a three-fold increase in the risk for invasive vaginal lesions raising the possibility of a genetic susceptibility to HPV. Other factors that appear to increase the risk of vaginal carcinoma are hysterectomy, a history of frequent vaginitis and endometriosis.[19,43,44]

Primary Prevention

As with cervical and vaginal tumors, careful sexual practices that decrease the risk of HPV and the avoidance of tobacco will decrease the risk of vaginal cancer. It is also important to carefully screen patients with a history of immunosuppression or a history of anogenital cancers with cytology and colposcopy.

Survival

The prognosis of vaginal cancer is mainly related to the stage at presentation and the type of histology. For squamous tumors the relative survival at 5 years by stage ranges from 96% for in situ lesions, 73% for Stage I lesions, 58% for Stage II lesions to 36% for Stage III-IV disease. Melanoma is associated with a very poor prognosis since the 5-year survival rate is only 14%. Children under 12 who are diagnosed with sarcoma have a 5-year survival rate of 90% in contrast to patients with a sarcoma over 50 who have a 5-year survival rate of only 48%.[44]

Fallopian Tube Cancer

Primary carcinoma of the Fallopian tube is reported to be the rarest of all gyneco-logic malignancies (<1%) with a yearly incidence of 3.6 per million women. As with vaginal cancer metastatic disease seems more common; however in advanced stages it is impossible to differentiate from primary ovarian cancer. The mean age at presenta-tion is 57 years old. The only risk factors that have been consistently identified are the BRCA1 and 2 mutations where a significant percentage of lesions develop in the distal aspect of the Fallopian tube.[45,46]

Survival

More patients with Fallopian tube cancer present at early stages when compared to ovarian cancer. Approximately 57% of patients have Stage I/II disease and 41% present with Stage III/IV disease. Because of the earlier stage at presentation, the 5-year survival rate is 56% overall with 84% of Stage I patients surviving compared to only 36% of patients with Stage III disease.[47]

Gestational Trophoblastic Disease

The incidence of choriocarcinoma (CC) is closely related to the incidence of gestational trophoblastic disease (GTD) since this type of malignancy usually arises after an antecedent molar pregnancy. In the United States where the incidence of GTD is 1/1000 deliveries, the incidence of choriocarcinoma is approximately 1 out of every 24,096 pregnancies and 1 out of every 19,920 live births. Compare this to Asia where the rate of GTD approaches 1 of every 100 deliveries which leads to an incidence of choriocarcinoma as high as 1 out of every 500 to 3000 pregnancies. In the American population, African-Americans have a rate of CC two times higher than Caucasians and the overall age adjusted incidence was 1.78 per million women.[48]

Risk Factors

Since choriocarcinoma usually follows GTD, it stands to reason that the risk factors for molar pregnancy would also contribute to the risk for trophoblastic malignancies. Risk for GTD has been shown to be age related. Higher rates occur in women under 20 and over 40 possibly secondary to defects in premature or postmature ovarian function. Smoking has also been positively correlated with an increased risk for trophoblastic disease with a relative risk of 2.6 for patients who smoke more than 15 cigarettes per day and a relative risk of 4.3 if they have smoked for at least 10 years when compared to non-smokers. Maternal blood type has also been implicated: patients with Type B blood have a higher incidence of recurrent GTD. A personal history of GTD is probably the largest risk factor with future pregnancies resulting in GTD 0.6-2.6% of the time, a rate 20-40 times higher than that seen in the general population. Other risk factors that have been examined are the age at first pregnancy, a history of induced abortions or spontaneous miscarriages, spacing between pregnancies, infertility, paternal age, a history of twins, diet and oral contraceptive use. For choriocarcinoma specifically, the strongest risk factor is a history of a complete molar pregnancy. With this condition, the risk is 1000 times higher than the risk with a normal pregnancy. Even when compared to partial moles which have an incidence of 0.1-0.5%, the incidence of CC following a complete mole is much higher (up to 20%). Choriocarcinoma is also more common in certain ethnic groups such as those of Asian descent, American Indians and African-Americans. Since many of these groups may have a higher level of inbreeding, this seems to point to a possible influence of consanguinity. The use of oral contraceptives has also been examined with one study showing a six-fold increased risk in women who received OCPs for more than 5 years. Given this association some have hypothesized that the below normal levels of estrogen driven by OCPs can predispose to choriocarcinoma. Blood type has also been investigated, but in contrast to GTD, patients with Type A blood or a partner with an incompatible blood type seem to have a higher risk of CC.[48]

Survival

Prior to the advent of combined chemotherapy, mortality rates for CC used to approach 80%. Now with chemotherapy and surgical interventions, survival rates are greater than 95%. Lower survival still exists for nonwhites likely secondary to poorer post mole surveillance in minority groups.[47,48]

Conclusions

Understanding the epidemiology of gynecologic malignancies can impact both patient care by providing information regarding patient screening, diagnosis and treatment selection as well as future research by elucidating information regarding the carcinogenic process and the effects of screening and lastly will lead to trials of chemoprevention. For instance, the Gynecologic Oncology Group (GOG) trial #199 is a prospective evaluation of a cohort of women at increased genetic risk for ovarian cancer some of whom elect to undergo a risk-reducing prophylactic bilateral salpingo-oophorectomy whereas others decide to be followed closely with regular clinical screening including physical exam, CA125 level determination and pelvic ultrasound. This study addresses the incidence of critical cancer endpoints and quality of life in high risk patients by offering information regarding the risk and time span for the development of ovarian cancer in high risk women who retain their ovaries as well as important factors such as the long-term risks of estrogen deficiency and the psychological impact of surgery or continued screening among others.[50] Future epidemiological research will also focus on chemoprevention which is the use of a nutrient or pharmaceutical agent to prevent, delay or reverse cancer formation. It requires an agent that can be used for long periods of time with low toxicity. To increase the feasibility of these studies they are most often done in high risk populations such as those with *BRCA* and *HNPCC* mutations. In addition, the effectiveness must be measured by an endpoint prior to cancer development such as precursor lesions or indicators of cell proliferation or apoptosis.[51-53] For example, knowledge of the impact of HPV infection on the development of cervical carcinoma has led to the development of preventative and therapeutic vaccines. Other studies have looked at chemopreventative agents for cervical dysplasia such as retinoic acid, folate, vitamin C, beta carotene, Difluoromethylornithine (DMFO) and cox-2 inhibitors and their effects on HPV persistence and progression.[51,54] To examine the possibility of chemoprevention for ovarian cancer, the GOG trial #190 is evaluating the use of Fenretinide (4-HPR), a vitamin A derivative, in patients at high risk for developing OC by randomizing patients to either an immediate prophylactic oophorectomy or 6 to 8 weeks of therapy followed by surgery. Histological data from this study will provide information regarding possible precursor lesions of the ovary and changes in markers for cell proliferation and apoptosis in cells exposed to 4-HPR.[50] Since many cases of endometrial cancer are related to estrogen exposure, the focus on prevention for these tumors relies mostly on hormonal management with medications such as progestins, aromatase inhibitors and oral contraceptives. Recent studies focus on hyperinsulinemia and the reliability of complex atypical endometrial hyperplasia as a predictable histological endpoint.[53]

These studies and others like them will help to define risk groups and molecular biomarkers which will hopefully provide ways to identify patients at higher risk and at earlier cancer stages. They may also help elucidate the process of carcinogenesis which will lead to the development of new and complementary therapies.

Suggested Reading

1. Banks E, Beral V, Reeves G. The epidemiology of epithelial ovarian cancer: a review. Int J Gynecol Cancer 1997; 7:425-38.
2. Duarte-Franco E, Franco E. Other gynecological cancers: endometrial, ovarian, vulvar and vaginal cancers. BMC Women's Health 2004; 4:S14.
3. Amant F, Moerman P, Neven P et al. Endometrial cancer. Lancet 2005; 366:491-505.
4. Fracno E, Schlecht N, Saslow D. The epidemiology of cervical cancer. Cancer J 2003; 9(5):348-59.
5. Hoskins W, Perez C et al. Principles and practice of gynecologic oncology. Chapter One: Epidemiology of gynecologic cancers. Lippincott Williams and Wilkins; 4th Illus edition 2004.

References

1. Jemal A, Siegel R, Ward E et al. Cancer statistics, 2008. CA Cancer J Clin 2008; 58:71-96.
2. Parkin D, Bray F, Ferlay J et al. Global cancer statistics. CA Cancer J Clin 2005; 55:74-108.
3. http://www.seer.cancer.gov
4. Quirk J, Natarajan N. Ovarian cancer incidence in the united states 1992-1999. Gyn Onc 2005; 97:519.
5. Goodman M, Correa C, Tung K et al. Stage at diagnosis of ovarian cancer in the United States, 1992-1997. Cancer 2003; 97(10):2648-59.
6. Fleming JS, Beaugie CR, Haviv I et al. Incessant ovulation, inflammation and epithelial ovarian carcinogenesis: revisiting old hypothesis. Mol Cell Endocrinol 2006; 247:4-21.
7. Arslan A, Zeleniuch-Jacquotte A, Lundin E et al. Serum follicle-stimulating hormone and risk of epithelial ovarian cancer in postmenopausal women. Cancer Epid Biomarkers Prev 2003; 12:1531-5.
8. Brinton LA, Lamb EJ, Moghissi KS et al. Ovarian cancer risk associated with varying causes of infertility. Fertil Steril 2004; 82:405-14.
9. Banks E, Beral V, Reeves G. The epidemiology of epithelial ovarian cancer: a review. Int J Gynecol Cancer 1997; 7:425-38.
10. Carlson K, Skates S, Singer D. Screening for ovarian cancer. Ann Int Med 1994; 121(2):124-32.
11. Zografos G, Panou M, Panou N. Common risk factors of breast and ovarian cancer: recent view. Int J Gynecol Cancer 2004; 14:721-40.
12. Steuwing J, Hartge P, Wacholder S et al. The risk of cancer associated with specific mutations of BRCA1 and BRCA2 among ashkenazi jews. N Eng J Med 1997; 336(20):1401-8.
13. Harris RE, Beebe-Donk J, Doss H et al. Aspirin, ibuprofen and other nonsteroidal anti-inflammatory drugs in cancer prevention: a critical review of nonselective COX-2 blockade. Oncol Rep 2005; 13:559-83.
14. Whittemore A, Wu M, paffenbarger R et al. Personal and environmental characteristics realted to epithelial ovarian cancer. II. Exposures to talcum powder, tobacco, alcohol and coffee. Am J Epidemiol 1988; 128(6):1228-40.
15. Rodriguez G. New insights regarding pharmacologic approaches for ovarian cancer prevention. Hematol Oncol Clin North Am 2003; 17:1007-20.
16. Kurian A, Balise R, McGuire V. Histologic types of epithelial ovarian cancer: have they different risk factors. Gyn Onc 2005; 96:520-30.
17. Rosen B, Kwon J, Fung M et al. Systemic review of management options for women with a hereditary predisposition to ovarian cancer. Gyn Oncol 2004; 93(2):280-6.
18. Domchek SM, Friebel TM, Neuhausen SL et al. Mortality after bilateral salpingo-oophorectomy in BRCA1 and BRCA2 mutation carriers: a prospective cohort study. Lancet Oncol 2006; 7:223-9.
19. Duarte-Franco E, Franco E. Other gynecological cancers: endometrial, ovarian, vulvar and vaginal cancers. BMC Women's Health 2004; 4:S14.
20. Modesitt S, van Nagall J. The impact of obesity on the incidence and treatment of gynecologic cancers: A review. Obstet Gyn Surv 2005; 60:683-92.
21. Berek J, Hacker N. Practical gynecologic oncology. 4th ed. Lippincott Williams Wilkins, Philadelphia, 2005.
22. Amant F, Moerman P, Neven P et al. Endometrial cancer. Lancet 2005; 366:491-505.
23. Bergman L, Beelen M et al. Risk and prognosis of endometrial cancer after tamoxifen for breast cancer. Lancet 2000; 356:881-7.

24. Plaxe S, Saltzstein S. Impact of ethnicity on the incidence of high-risk endometrial carcinoma. Gyn Oncol 1997; 65(1):8-12.
25. Hemminki K, Bermejo J, Granstrom C. Endometrial cancer: Population attributable risks from reproductive, familial and socioeconomic factors. Eur J Cancer 2005; 41:2155-9.
26. Combination oral contraceptive use and the risk of endometrial cancer. The cancer and steroid hormone study of the centers for disease control and the national institute of child health and human development. JAMA 1987; 257(6):796-800.
27. Viswanathan A, Feskanich D, De Vivo I et al. Smoking and the risk of endometrial cancer: results form the nurses' health study. Int J Cancer 2005; 114:996-1001.
28. Gonzalez A, Sweetland S, Green J. Comparison of risk factors for squamous cell and adenocarcinomas of the cervix: a meta-analysis. Br J Cancer 2004; 90:1781-91.
29. Clifford G, Franceschi S, Diaz M et al. HPV type-distribution in women with and without cervical neoplasic diseases. Vaccine 2006; S3:26-34.
30. Fracno E, Schlecht N, Saslow D. The epidemiology of cervical cancer. Cancer J 2003; 9(5):348-59.
31. Tarkowski T, Koumans E, Sawyer M et al. Epidemiology of human papillomavirus infection and abnormal cytologic test results in an urban adolescent population. J Inf Disease 2004; 189:46-50.
32. Ho G, Beirman R, Beardsley L et al. Natural history of cervicovaginal papillomavirous infection in young women. NEJM 1998; 338(7):423-8.
33. Soutter W, Sasieni P et al. Long term risk of invasive cervical cancer after treatment of squamous cervical intraepithelial neoplasia. Int J Cancer 2006; 188(8):2048-55.
34. Winkelstein W. Smoking and cervical cancer—current status: a review. Am J Epidemiol 1990; 131(6):945-57.
35. Garcia-Closas R, Castellsague X, Bosch X et al. The role of diet and nutrition in cervical carcinogenesis: A review of recent evidence. Int J Cancer 2005; 117:629-37.
36. Bell M, Alvarez R. Chemoprevention and vaccines: a review of the nonsurgical options for the treatment of cervical dysplasia. Int J Gynecol Cacner 2005; 15:4-12.23.
37. van der Avoort I, Shirango H, Hoervenaars B et al. Vulvar squamous cell carcinoma is a multifactorial disease following two separate and independent pathways. Int J Gynecol Pathol 2006; 25(1):22-9.
38. Carlson J, Ambros R, Malfetano J et al. Vulvar lichen sclerosus and squamous cell carcinoma: a cohort, case control and investigational study with historical perspective; implications for chronic inflammation and sclerosis in the development of neoplasia. Human Pathol 1998; 29(9):932-48.
39. Messing M, Gallup D. Carcinoma of the vulva in young women. Obstet Gyn 1995; 86:51-4.
40. van Seters M, van Beurden M, de Craen A. Is the assumed natural history of vulvar intraepithelial neoplasia III based on enough evidence? A systemic review on 3322 published patients. Gyn Onc 2005; 97:645-51.
41. Jones R, Baranyai J, Stables S. Trends in squamous cell carcinoma of the vulva: the influence of vulvar intraepithelial neoplasia. Obstet Gynecol 1997; 90(3):448-52.
42. Bosquet J, Magrina J, Gaffey T et al. Long term survival and disease recurrence in patients with primary squamous cell carcinoma of the vulva. Gyn Onc 2005; 97:828-33.
43. Daling P, Madeleine M, Schwartz S. A population based study of squamous cell vaginal cancer: HPV and cofactors. Gyn Onc 2002; 84:263-70.
44. Creasman W, Phillips J, Menck H. The national cancer data base report on cancer of the vagina. Cancer 1998; 83(5):1033-40.

4

45. Aziz S, Kupersteink GK, Rosen B et al. A genetic epidemiological study of carcinoma of the Fallopian tube. Gyn Onc 2001; 80:341-5.
46. Nordin A. Primary carcinoma of the Fallopian tube: a 20-year literature review. Obstet Gynecol Surv 1994; 49(5):349-61.
47. Pecorelli S, Odicino F, Maisonneuve P et al. Carcinoma of the Fallopian tube: FIGO annual report on the results of treatment in gynaecological cancer. J Epidem Biostat 1998; 3:363-74.
48. Smith H. Gestational trophoblastic disease epidemiology and trends. Clin Obstet Gynecol 2003; 46(3):541-56.
49. Goto S, Ino K, Mitsui T. Survival rates of patients with choriocarcinoma treated with chemotherapy without hysterectomy: effects of anticancer agents on subsequent births. Gyn Onc 2004; 93:529-35.
50. Greene M, Skates S et al. Prospective study of risk reducing salpingo-oophorectomy and longitudinal CA-125 screening among women at increased genetic risk. Protocol GOG-0199.
51. Follen M, Meyskens F, Alvarez R et al. Cervical cancer chemoprevention, vaccines and surrogate endpoint biomarkers. Cancer 2003; 98(9 Suppl):2044-2051.
52. Daly M, Ozols R et al. An exploratory evaluation of fenretinide (4-HPR) (IND#39,812) as a chemopreventative agent for ovarian cancer. Protocol GOG-0190.
53. Kelloff G, Lippmann S et al. Progress in chemoprevention drug development: The promise of molecular biomarkers for prevention of intraepithelial neoplasia and cancer—a plan to move forward. Clin Cancer Res 2006; 12:3661-97.
54. Bell M, Alvarez R. Chemoprevention and vaccines: a review of the nonsurgical options for the treatment of cervical dysplasia. Int J Gynecol Cancer 2005; 15:4-12.

4

CHAPTER 5

Biological Therapy for Gynecologic Cancers

Jennifer Rubatt and Angeles Alvarez Secord

Introduction

Traditional chemotherapy drugs used to treat gynecologic cancers are cytotoxic in nature. DNA alkylaters (cyclophosphamide), cytoskeletal destabilizers (taxanes), DNA cross-linkers (platins), antimetabolites (methotrexate), antitumor antibiotics (dactinomycin) and topoisomerase inhibitors (topotecan) are all classes of drugs whose main action is to halt cellular growth. These agents do not discriminate between cancer and normal cells. Because of this, significant side effects exist with traditional cytotoxic drugs extending from their effects on normal tissues. These side effects may be dose limiting in clinical practice and adversely affect patients' quality of life. Targeting cancer cells more effectively with therapeutic agents that target the intracellular molecular mechanisms that have gone awry is a novel and interesting field of cancer therapy research.

Cells gain malignant potential through stepwise alterations in molecular pathways that are responsible for cell proliferation, death and senescence. A better understanding of the molecular pathways that become altered in cancer cells when compared with normal cells provides us with a unique opportunity to devise therapeutic agents that differentiate between cancer and normal cells. These novel agents have been designed to interfere specifically with the molecular pathways that the cancer cells use to proliferate, invade normal tissue and metastasize to new locations throughout the body. Because targeted therapies center on molecular changes that are particular to malignant cells they will hopefully spare normal cells and be associated with less side effects.

Optimal molecular targets for drug development should have (1) differential expression between malignant and normal cells, (2) target cancer cell-dependent pathways for proliferation, invasion, metastasis and apoptosis and (3) a function that is not essential for normal cell function. Several molecular targeted cancer therapies already have received Food and Drug Administration (FDA) approval and many others are undergoing testing in clinical trials.

Currently, there are several approaches to targeting aberrant molecular pathways in cancer cells. One way to interfere with pathways is to develop monoclonal antibodies to bind to either receptors or to the growth factors themselves to selectively turn off pathways that lead to uncontrolled cells growth, invasion and metastasis. Another design employed is to produce small molecule inhibitors that are able to cross cell membranes and inhibit a receptor driven pathway by binding to the inner portion of the receptor (known as the tyrosine kinase domain) or its downstream substrates. These small molecule inhibitors are referred to as tyrosine kinase inhibitors (TKIs). Both monoclonal antibodies and TKIs are used to target many pathways, including those involved in angiogenesis and cellular proliferation.

Gynecologic Oncology, edited by Paola Gehrig and Angeles Secord.
©2009 Landes Bioscience.

Anti-Angiogenic Therapies

Targeting tumor blood supply to limit tumor size and ability to metastasize has shown promising results in initial clinical studies. Angiogenesis (new blood vessel development) (Fig. 5.1) is a complex process with many pro- and anti-angiogenic factors influencing endothelial cell growth and migration. In malignant tumors these factors are primarily pro-angiogenic, in fact for a tumor to grow more than 2 mm in diameter new blood vessels must grow to feed the tumor.[1,2]

Vascular endothelial growth factor (VEGF) along with its receptors, VEGFR-1 (FLT-1) and VEGFR-2 (KDR) are part of a potent pro-angiogenic pathway involved in physiological as well as pathological angiogenesis.[3,4] Tumor angiogenesis seems to be regulated primarily by VEGF and VEGFR-2. Bevacizumab (Avastin®), a monoclonal antibody that targets VEGF (Fig. 5.2), was approved to treat colon cancer in 2005. It starves tumors by inhibiting new blood vessel formation therefore blocking nutrient delivery. It has been shown to extend overall survival by 5 months when added to standard first-line therapy for metastatic colorectal cancer.[5] Bevacizumab has also shown an increase in survival time in breast and lung cancer patients when combined with traditional cytotoxic therapy.[6,7]

Even though the use of bevacizumab in gynecologic cancers has not been FDA approved, some exciting results have been reported in recent trials in ovarian cancer. The Gynecologic Oncology Group (GOG) investigated using bevacizumab in women

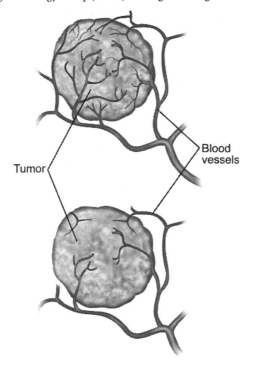

Figure 5.1. Angiogenesis (new blood vessel formation) is a process that must develop to provide nutrients for a tumor to grow beyond 2 mm.

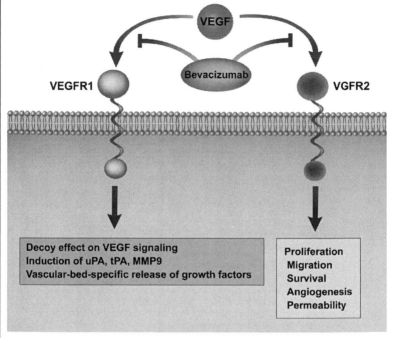

Figure 5.2. VEGF is a potent pro-angiogenic growth factor. After treatment with bevacizumab, which binds and neutralizes VEGF, downstream signal transduction (cell migration and proliferation) is prevented halting angiogenesis.

who have previously failed traditional cytotoxic chemotherapy in a Phase II trial and found an 18% response rate with three complete responders.[8] Another Phase II study combined low-dose cyclophosphamide with bevacizumab and reported a 28% response rate with a 57% rate of 6-month progression-free survival.[9] Proteinuria, hypertension and arterial thrombosis were the most common adverse events experienced. These significant results provoked excitement for targeting anti-angiogenesis in ovarian cancer research. The excitement was dampened when a third Phase II study investigating bevacizumab in platinum-refractory disease reported an alarmingly high (11%) incidence of bowel perforation. Although the cause of this has yet to be determined, it has been suggested that extensive intraperitoneal disease involving the large and small bowel may be responsible. It also appears that patients with heavily pretreated, refractory disease are at higher risk for bowel perforation compared to those patients treated earlier in the course of their disease. Despite these findings the GOG has commenced a Phase III trial evaluating the addition of bevacizumab to traditional first line agents in patients with advanced ovarian cancer. The initial results of this study show that the addition of bevacizumab to carboplatin and paclitaxel is well tolerated and active in treating ovarian cancer.

Targeting VEGF and its Receptors

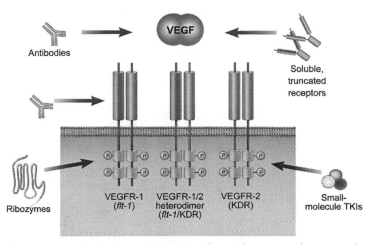

Figure 5.3. Multiple steps in VEGF signal transduction can be targeted. Anti-VEGF antibodies and faulty soluble receptors can bind VEGF preventing VEGF/VEGFR activation. Antibodies can also bind the VEGF receptor directly to block VEGF binding. Small molecule tyrosine kinase inhibitors (TKIs) and ribozymes can bind the intracellular portion of the receptor preventing activation.

Bevacizumab is also undergoing investigation for advanced cervical cancer and persistent/recurrent endometrial cancer. At the 2006 annual meeting of the American Society of Clinical Oncology (ASCO), scientists demonstrated that distinct molecular subtypes of VEGF and VEGF-R are differentially expressed in pre-invasive versus invasive cervical cancer.[10] This information can potentially be used to stratify patients into treatment groups.

VEGF and its receptors are also being targeted using TKIs (Fig. 5.3). Two of these, sorafenib (Nexavar®) and sunitinib (Sutent®), have shown activity in renal cell carcinoma and are currently FDA approved for this indication.[11,12] A Phase II trial coordinated by the GOG has demonstrated a 33% response rate when sorafenib was used in women with recurrent/persistent ovarian or peritoneal cancer.

Thalidomide, originally developed as a sedative to treat hyperemesis and subsequently found to cause limb defects (phocomelia), has now been found to have potent anti-angiogenic properties.[13] The role of thalidomide in the treatment of gynecologic cancers is currently under study.

Epidermal Growth Factor Receptor Family

The epidermal growth factor receptor (EGFR) family is comprised of four similar receptor tyrosine kinases including EGFR, ErbB-2 (HER2/*neu*), ErbB-3 and ErbB-4.[14] The binding of growth factors to the receptor results in the activation of a tyrosine kinase pathway which facilitates communication between the cell's surface and nucleus. Growth factor receptors can become mutated to be overly active in malignant cells.

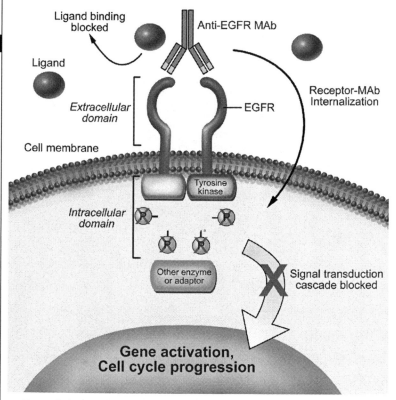

Figure 5.4. Anti-EGFR antibody blocks ligand binding, effectively inactivating the intracellular domain of the EGFR receptor.

These transmembrane receptors can also be overexpressed on the cell surface making them attractive targets for targeted biological therapy. Monoclonal antibodies have been developed that either bind the growth factors or bind to the receptors themselves to inhibit activation of these receptors (Fig. 5.4). Two examples of monoclonal antibodies that target the EGFR pathway are trastuzumab (Herceptin®) and cetuximab (Erbitux®). Imatinib (Gleevec®), gefitinib (Iressa®) and erlotinib (Tarceva®) are small molecule tyrosine kinase inhibitors (TKIs) that have been previously or are currently undergoing investigation in gynecologic cancers (Fig. 5.5).

Trastuzumab, which binds to the HER2/*neu* receptor, can slow the growth and spread of tumors that overexpress the HER2/*neu* receptor (Fig. 5.6). The FDA approved trastuzumab in 1998 for the treatment of metastatic breast cancers that overexpress HER2/*neu* as determined by immunohistochemistry (IHC). It had been

EGFR Inhibition Via Tyrosine Kinase Inhibitors

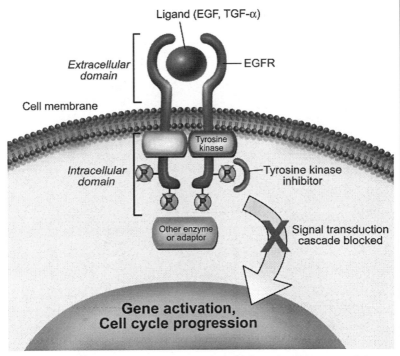

Figure 5.5. Small molecule tyrosine kinase inhibitor bind the intracellular domain of the EGFR receptor preventing further signal transduction.

established that up to 30% of breast cancers have high expression of HER2/*neu*.[15,16] Trastuzumab has shown the most promise for survival and recurrence prevention when combined with traditional cytotoxic chemotherapy in the front line setting. Women with metastatic breast cancer who were given a combination of trastuzumab with front line chemotherapy showed a 5-month survival advantage over those treated with chemotherapy alone.[17,18] This regimen also demonstrated a dramatic 52% decrease in disease recurrence.

Unfortunately in gynecologic malignancies trastuzumab has not demonstrated an overwhelming response. In contrast to breast cancers the overexpression of HER2/*neu* is relatively uncommon in gynecologic malignancies.[19,20] Only 11.4% of over 800 ovarian cancer specimens exhibited HER2/*neu* overexpression by IHC in a Phase II GOG study.[20] Trastuzumab therapy demonstrated a disappointing response rate of only 7.3%. The academic community attributed the poor response rate of single agent trastuzumab in this study to low frequency of HER2/*neu* overexpression. The value of combining trastuzumab with traditional chemotherapy has yet to be researched in a systematic way.

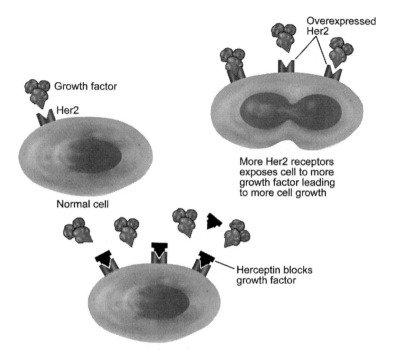

Figure 5.6. Herceptin binds to HER2/*neu* receptors blocking growth factor binding and inhibiting further cell proliferation.

However, there has been a case report of tumor response to combination tras-tuzumab and taxane-based therapy in a patient with metastatic endometrial cancer indicating that perhaps trastuzumab in combination with chemotherapy may be worthy of further investigation in endometrial cancer.[21]

A significant percentage of cancers demonstrate overexpression of EGFR including ovarian, endometrial, cervical, colorectal, breast and head and neck.[22-24] Several agents have been approved by the FDA to target EGFR: cetuximab (colon cancer), gefitinib (lung cancer) and erlotinib (lung and pancreatic cancers).[25,26] The FDA has since withdrawn approval for gefitinib when subsequent clinical trials demonstrated no increase in survival benefit. The GOG is currently studying combination cetux-imab and carboplatin in platinum-sensitive ovarian cancer. Gefitinib and erlotinib have been studied as single agents in ovarian cancer; however the response has been unacceptably low. The only responder in a GOG study of gefitinib for persistent or recurrent ovarian/peritoneal carcinoma possessed an EGFR mutation.[27] The relative success (bevacizumab) and failure (gefitinib) of novel targeted biological therapies bring to light the importance of developing laboratory tests that can accurately predict which patients will respond to specific novel therapeutics.

Monoclonal antibodies and TKIs that have dual or multiple inhibitory functions of the EGFRs have been developed. A dual inhibitor would be capable of inactivating

HER2/*neu*/Erb-3 heterodimers in addition to inactivating EGFR or HER2/*neu* alone. This action broadens the scope of pathways affected by one inhibitor. Heterodimer formation may be a resistance mechanism and thus a dual inhibitor would be a great advantage in treatment regimens. Pertuzumab (Omnitarg®) is a monoclonal antibody that binds to the dimerization domain of the HER2/*neu* receptor and prevents it from binding to activated EGFR as well as Erb-3 receptors. This action interferes with multiple signaling pathways involving not only HER2/*neu* but also its action as a coreceptor.[28] An additional advantage of targeting the dimerization domain on the HER2/*neu* receptor is that pertuzumab theoretically should have activity even when HER2/*neu* expression is not amplified. A Phase II trial is currently underway to evaluate the utility of pertuzumab for ovarian cancer patients who have not responded to platinum-based therapy.[29]

Additional Therapies

A number of other targeted biological cancer therapies have been developed. Dasatinib (Sprycel®) is a TKI that inhibits several molecular pathways including SRC, BCR-ABL and c-KIT.[30,31] SRC is a family of membrane-associated protein kinases that functionally interact with multiple transmembrane tyrosine kinase receptors, including platelet derived growth factor (PDGF), epidermal growth factor receptor (EGFR), basic fibroblast growth factor (bFGF) and colony stimulating growth factor-I (CSF-1).[30] SRC couples the signals from these membrane-bound receptors to cytoplasmic signaling systems to influence cell growth, migration, differentiation and survival. Interrupting multiple pathways by targeting one molecule makes SRC a very attractive anticancer target. Dasatinib is currently approved for use in imatinib-resistant CML. Multiple SRC inhibitors are currently undergoing investigation in Phase I and II trials for gynecologic cancers.

A new biological target for endometrial carcinoma is mammalian target of rapamycin (mTOR). mTOR is a serine-threonine kinase downstream from PTEN. Loss of PTEN function is found in 36-83% of endometrioid-type endometrial cancer making mTOR a highly desirable therapy target.[32] Although the solubility and stability in solution of rapamycin make it an undesirable anticancer agent, several rapamycin analogs have been developed and currently being evaluated in Phase I-III trials.

Imatinib (Gleevec®) was FDA approved in 2001 for treatment of chronic myeloid leukemia (CML) and gastrointestinal stromal tumors (GIST). Imatinib is a TKI that specifically binds and inhibits the BCR-ABL and KIT kinases. A lot of interest for biological agents has been generated by the high numbers of patients who have shown impressive and prolonged responses to imatinib when treated for CML and GIST. The GOG evaluated imatinib in gynecologic cancers and demonstrated minimal activity for ovarian cancer.[33]

Greater than 80% of epithelial ovarian cancers express CA125 on the cell surface. Oregovomab (Ovarex®) is a monoclonal antibody targeted against CA125 and intended to produce an immune response against tumor cells. Even though the first Phase III randomized trial failed to demonstrate a survival benefit when used as maintenance chemotherapy, women who were completely debulked and had an excellent response to first line chemotherapy showed a doubling of time to cancer recurrence (from 11 months in controls to 24 months in those treated with oregovomab).[34] Randomized controlled trials to examine oregovomab in combination with front line agents as well as single drug maintenance therapy in select patients are underway.

5

Table 5.1. Targeted cancer therapies

Agent	Trade Name	Target	Cancer Types	Current Standing
Monoclonal Antibodies				
Bevacizumab	Avastin®	VEGF receptor	Colorectal	FDA approved 2004
			Non squamous, non small cell Lung	FDA approved 2006
			Ovarian	Phase III
			Breast, prostate and renal	Phase III
Sorafenib	Nexavar®	VEGF receptor	Renal	FDA approved 2005
			Ovarian	Phase II
Sunitinib	Sutent®	VEGF receptor	Renal, GI stromal tumors (GIST)	FDA approved 2006
Trastuzumab	Herceptin®	HER-2/neu	Breast-HER-2/neu+	FDA approved 1998
			Ovary and Endometrial	Phase II
Pertuzumab	Omnitarg®	HER-2/neu homo and heterodimers	Ovary	Phase II
			Pancreas	Phase I/II
			Breast, prostate, lung	Phase II
Cetuximab	Erbitux®	EGF receptor	Colorectal	FDA approved 2004
			Ovary	Phase II

continued on next page

Table 5.1. Continued

Agent	Trade Name	Target	Cancer Types	Current Standing
Monoclonal Antibodies				
Oregovomab	OvaRex®	CA125	Ovarian	Phase III
Rituximab	Rituxan®	CD20	Non-Hodgkin's lymphoma	FDA approved 1997
Small Molecules				
Imatinib	Gleevec®	BCR-ABL and KIT receptors	Chronic myelogenous leukemia	FDA approved 2001
			GI stromal tumors	FDA approved 2002
Gefitinib	Iressa®	EGF receptor	Non small cell lung cancer	FDA approved 2003*
Erlotinib	Tarceva®	EGF-1 receptor	Non small cell lung cancer	FDA approved 2004
			Pancreatic cancer	FDA approved 2005
Dasatinib	Sprycel®	SRC kinase inhibitor	Chronic myelogenous leukemia (CML)	FDA approved 2006
			GI stromal tumors	Phase III
			Ovary	Phase I

*The FDA has since removed approval for gefitinib given no increase in survival benefit.
Note: VEGF-vascular endothelial growth factor; FDA-Food and Drug Administration; HER-2/*neu*-specific growth factor receptor; EGF-epidermal growth factor, CA125- a protein that is produced by some ovarian cancers; CD20-a B-cell specific protein; BCR-ABL, KIT and SRC-tyrosine kinase proteins.

Other biologic agents have been developed that target the RAS pathway, matrix metalloproteinases, heat shock proteins, the mitogen-activated protein kinase pathway (MAPK) and cell-cycle regulators.

Conclusion

Targeted biological therapies hold great promise for the future of treating gynecologic malignancies. However, we must continue to improve our understanding of cancer molecular biology to facilitate the development and implementation of targeted therapies and identify patients most likely to benefit from these treatments. Numerous novel biological agents are currently undergoing investigation for the treatment of gynecologic cancers in the Phase I/II arena. The drugs discussed in this chapter are summarized in Table 5.1 and are under investigation as single agents as well as in combination with traditional cytotoxic chemotherapy.

Clinical studies are mirrored in the laboratory to understand the details of the pathways affected and the molecules involved. In the future, a more comprehensive understanding of the molecular basis of carcinogenesis will hopefully lead to diagnostic assays that will be able to predict the most appropriate combination of chemotherapy for each patient. This will usher in a new age of individualized patient specific chemotherapy treatment algorithms.

Suggested Reading

1. Targeted Cancer Therapies: National Cancer Institute www.cancer.gov/cancertopics/factsheet/Therapy/targeted.
2. Chon HS, Hu W, Kavanagh JJ. Targeted therapies in gynecologic cancers. Curr Cancer Drug Targets 2006; 6(4):333-63.

References

1. Folkman J. What is the evidence that tumors are angiogenesis dependent? J Natl Cancer Inst 1990; 82:4-6.
2. Diaz Flores L, Gutierrez R, Varela H. Angiogenesis: an update. Histol Histopathol 1994; 9:807-43.
3. Ferrara N, CarverMoore K, Chen H et al. Heterozygous embryonic lethality induced by targeted inactivation of the VEGF gene. Nature 1996; 380:439-42.
4. Carmeliet P, Ferreira V, Breier G et al. Abnormal blood vessel development and lethality in embryos lacking a single VEGF allele. Nature 1996; 380:435-9.
5. Hurwitz H, Fehrenbacher L, Novotny W et al. Bevacizumab plus irinotecan, fluorouracil and leucovorin for metastatic colorectal cancer. N Engl J Med 2004; 350:2335-42.
6. Sandler AB, Gray R, Brahmer J et al. Randomized phase II/III Trial of paclitaxel (P) plus carboplatin (C) with or without bevacizumab (NSC #704865) in patients with advanced nonsquamous nonsmall cell lung cancer (NSCLC): An Eastern Cooperative Oncology Group (ECOG) Trial—E4599. ASCO 41st Annual Meeting, a #4 2005.
7. Miller KD, Wang M, Gralow J et al. A randomized phase III trial of paclitaxel versus paclitaxel plus bevacizumab as first-line therapy for locally recurrent or metastatic breast cancer. ASCO 41st Annual Meeting 2005.
8. Burger RA, Sill M, Monk BJ et al. Phase II trial of bevacizumab in persistent or recurrent epithelial ovarian cancer (EOC) or primary peritoneal cancer (PPC): A gynecologic oncology group (GOG) study. J Clin Oncol 2005; 23:457S.
9. Garcia AA, Oza AM, Hirte H et al. Interim report of a phase II clinical trial of bevacizumab (Bev) and low dose metronomic oral cyclophosphamide (mCTX) in recurrent ovarian (OC) and primary peritoneal carcinoma: A california cancer consortium trial. J Clin Oncol 2005; 23:455S.

10. Landt S, Thomas A, Fueger A et al. Analysis of the VEGF family and their receptors in serum/plasma of patients with pre-invasive and invasive cervical cancer. ASCO 42nd Annual Meeting 2006.

11. Motzer RJ, Michaelson MD, Redman BG et al. Activity of SU11248, a multitargeted inhibitor of vascular endothelial growth factor receptor and platelet-derived growth factor receptor, in patients with metastatic renal cell carcinoma. J Clin Oncol 2006; 24:16-24.

12. Favaro JP, George DJ. Targeted therapy in renal cell carcinoma. Expert Opin Investig Drugs 2005; 14:1251-8.

13. D'Amato RJ, Loughnan MS, Flynn E et al. Thalidomide is an inhibitor of angiogenesis. Proc Natl Acad Sci USA 1994; 91:4082-5.

14. Franklin WA, Veve R, Hirsch FR et al. Epidermal growth factor receptor family in lung cancer and premalignancy. Semin Oncol 2002; 29:3-14.

15. Harries M, Smith I. The development and clinical use of trastuzumab (Herceptin). Endocrine-Related Cancer 2002; 9:75-85.

16. Press MF, Godolphin W, Slamon D. Expression of the HER-2/neu/c-erbB-2 oncogene in breast cancer. Lab Invest 1989; 60:A73.

17. Slamon DJ, Leyland-Jones B, Shak S et al. Use of chemotherapy plus a monoclonal antibody against HER2 for metastatic breast cancer that overexpresses HER2. N Engl J Med 2001; 344:783-92.

18. Eiermann W. Trastuzumab combined with chemotherapy for the treatment of HER2-positive metastatic breast cancer: Pivotal trial data. Ann Oncol 2001; 12:57-62.

19. Bookman MA, Darcy KM, Clarke-Pearson D et al. Evaluation of monoclonal humanized anti-HER2 antibody, trastuzumab, in patients with recurrent or refractory ovarian or primary peritoneal carcinoma with overexpression of HER2: A phase II trial of the gynecologic oncology group. J Clin Oncol 2003; 21:283-90.

20. Fleming GF, Sill MA, Thigpen JT et al. Phase II evaluation of trastuzumab in patients with advanced or recurrent endometrial carcinoma: A report on GOG 181B. Proc Am Soc Clin Oncol 2003; 22:453 abstract#1821.

21. Jewell E, Secord AA, Brotherton T et al. Use of trastuzumab in the treatment of metastatic endometrial cancer. Int J Gynecol Cancer 2006; 16:1370-3.

22. Scambia G, Panici PB, Battaglia F et al. Significance of epidermal growth factor receptor in advanced ovarian cancer. J Clin Oncol 1992; 10:529-35.

23. Salomon DS, Brandt R, Ciardiello F et al. Epidermal growth factor-related peptides and their receptors in human malignancies. Crit Rev Oncol Hematol 1995; 19:183-232.

24. Scambia G, Panici PB, Ferrandina G et al. Significance of epidermal growth factor receptor expression in primary human endometrial cancer. Int J Gynecol Cancer 1994; 56:26-30.

25. Cohen MH, Williams GA, Sridhara R et al. FDA drug approval summary: Gefitinib (ZD1839) (Iressa (R)) tablets. Oncologist 2003; 8:303-6.

26. Cohen MH, Johnson JR, Chen YF et al. FDA drug approval summary: Erlotinib (Tarceva (R)) tablets. Oncologist 2005; 10:461-6.

27. Gordon AN, Finkler N, Edwards RP et al. Efficacy and safety of erlotinib HCl, an epidermal growth factor receptor (HER1/EGFR) tyrosine kinase inhibitor, in patients with advanced ovarian carcinoma: results from a phase II multicenter study. Int J Gynecol Cancer 2005; 15:785-92.

28. Baselga J, Arteaga CL. Critical update and emerging trends in epidermal growth factor receptor targeting in cancer. J Clin Oncol 2005; 23:2445-59.

29. Gordon MS, Matei D, Aghajanian C et al. Clinical activity of pertuzumab (rhuMab 2C4) in advanced, refractory or recurrent ovarian cancer (OC) and the role of HER2 activation status. 2005 ASCO Annual Meeting 2006; Abstract No:5051.

30. Parsons JT, Parsons SJ. SRC family protein tyrosinekinases: cooperating with growth factor and ashesion signaling pathways. Curr Opin Cell Biol 1997; 9:187-92.
31. Parsons SJ, Parsons JT. SRC family kinases, key regulators of signal transduction. Oncogene 2004; 23:7906-9.
32. Gadducci A, Cosio S, Genazzani AR. Old and new perspectives in the pharmacological treatment of advanced or recurrent endometrial cancer: Hormonal therapy, chemotherapy and molecularly targeted therapies. Crit Rev Oncol Hematol 2006; 58:242-56.
33. Fox Chase Cancer Center 2005 Scientific Report.
34. Berek J, Taylor P, Gordon A et al. Randomized, placebo-controlled study of oregovomab for consolidation of clinical remission in patients with advanced ovarian cancer. J Clin Oncol 2004; 22:3507-16.

5

CHAPTER 6

Chemotherapy in Gynecologic Oncology

Michael A. Bidus and John C. Elkas

Introduction

Since the modern introduction of chemotherapy for the treatment of cancer in the 1940s, considerable progress has been made, affording opportunities for the clinician to cure or extend survival of women afflicted with these diseases. However, the use of chemotherapy for the treatment of gynecologic malignancies requires a broad understanding of the biology of cancer, the pharmacology of chemotherapeutic agents and management of the complications of therapy. This chapter will provide an overview of cell cycle kinetics, individual drug characteristics, current drug regimens used in gynecologic malignancies, major clinical studies in gynecologic cancer chemotherapy, chemotherapy complications and anti-emetic therapy.

Cell Cycle Kinetics and Mechanism of Action

Cell Cycle Kinetics

Knowledge of the cell cycle is important in order to understand the mechanism of action of chemotherapeutic agents. Cells that are quiescent and not actively dividing are in the G0 phase. Quiescent cells enter the cycle in the G1 phase where cellular processes are initiated in preparation for DNA synthesis. G1 cells then progress into the S phase where DNA synthesis is undertaken, followed by the short G2 phase where the cell prepares for mitosis. Mitosis takes place in the M phase, after which cells may re-enter the G1 or G0 phase. Chemotherapy may kill cells in specific phases of the cell cycle (cell cycle specific agents) or independent of where the cell is in the cycle (cell cycle nonspecific).

Chemotherapy exploits a fundamental difference between cancer cells and normal tissue to effect cell kill: cellular growth. With respect to cell growth, there are three populations of cells in the human body. The static population is fairly constant, rarely undergoes cellular division and is not actively proliferating. Neural and muscular tissues are examples of static populations of cells. Expanding populations such as the liver, are usually not actively proliferating but retain this capacity in response to stress or injury. Finally, the renewing population such as intestinal mucosa or bone marrow, are continuously proliferating and moving through the cell cycle. It is the renewing population of cells that are affected the most by chemotherapy. Cancer cells, by virtue of their unregulated cell growth, can be thought of as a renewing population and can be exquisitely sensitive to chemotherapy.

In general, tumor cells display Gompertzian growth characteristics. As the tumor mass increases, the time required for the tumor to double in size increases as well. This pattern results in rapid initial growth of the tumor but slower growth when the tumor burden is large. However, a slower doubling time can still result in dramatic increases

---

Gynecologic Oncology, edited by Paola Gehrig and Angeles Secord.
©2009 Landes Bioscience.

in tumor burden as only a few doublings are required in late stage disease to overwhelm the patient. Gompertzian growth implies that not all tumor cells are actively dividing. In fact, the smaller the tumor burden the greater the percentage of cells actively in the cell cycle. Conversely, the larger the tumor burden, the greater the percentage of cells are quiescent.

Chemotherapy kills cells that are actively in the cell cycle and does so according to first order kinetics. That is, chemotherapy kills a constant fraction of cells exposed to the drug as opposed to a constant number. With each additional cycle of chemotherapy, a similar fraction of cells is killed. This fact has led to the applications of repetitive cycles of chemotherapy in order to maximally effect tumor cell kill.

Classes of Drugs and Mechanisms of Action (Table 6.1, see fold-out poster)

Antimetabolites

Antimetabolites are either structural or chemical analogues of naturally occurring compounds and substrates that are required for the synthesis of DNA and RNA. They are typically involved in the synthesis of nucleic acids, purines and pyrimidines. Because these compounds are involved in the synthesis of DNA, these drugs are typically S phase cell cycle specific agents. Examples of antimetabolites include methotrexate, flourouracil, capecitabine and gemcitabine. Because they preferentially kill cells in the cell cycle, the side effects seen clinically are most associated with renewing tissues such as myelosupression, diarrhea and GI toxicity.

Alkylating Agents

Alkylating agents are highly reactive molecules which bind with great affinity to DNA, forming DNA adducts. This binding may introduce single or double strand DNA breaks or induce conformational changes in the DNA. In addition, these agents may inhibit the normal DNA repair mechanisms. These agents typically are cell cycle nonspecific. Examples of alkylating agents include cyclophosphamide, ifosfamide, cisplatin, carboplatin, melphalan, oxaliplatin, altretamine and chlorambucil.

Antitumor Antibiotics

This class of agents is usually derived from microorganisms and work by a variety of mechanisms such as DNA intercalation (resulting in inhibition of DNA transcription and RNA translation) and free radical formation (resulting in single or double strand DNA breaks). Actinomycin D, bleomycin, adriamycin, liposomal doxorubicin, mitomycin C and mitoxantrone are examples of antitumor antibiotics.

Agents Derived from Plants

Paclitaxel, docetaxel, vinblastine, vincristine, vinorelbine, etoposide, topotecan and irinotecan are examples of plant-derived cytotoxic agents. Paclitaxel and docetaxel stabilize and promote the assembly of intracellular microtubules causing microtubule bundling throughout the cell and are cell cycle specific in the M phase. Vinblastine, vincristine and vinorelbine are vinca alkaloids and inhibit microtubular polymerization resulting in mitotic arrest as a result of failure of formation of the mitotic spindle. These agents are specific for the S, M and G phases of the cell cycle. Etoposide, topotecan and irinotecan are topoisomerase inhibitors, altering the supercoiling of DNA and are S phase specific.

Hormonal Agents

The mechanisms of action of hormonal agents are not well-understood. These agents are translocated to the nucleus of the cell where hormonally mediated transcription and protein synthesis are altered. These agents are cell static agents and not cidal agents. Therefore, prolonged use is recommended.

Small Molecules

Small molecules are the newest agents for use with chemotherapy. This field is rapidly expanding and the mechanisms of action are varied, including inhibiting tyrosine kinases.

Disease Sites: Treatment Regimens for Gynecologic Oncology

Ovarian Cancer: Epithelial

Primary Therapy

Primary therapy for epithelial ovarian cancer, after surgical staging and/or cytoreduction, is based upon stage of disease and risk of recurrence:

- Stage IA or IB, Grade 1 or 2—Low risk disease, no chemotherapy indicated.
- Stage IC, Grade 3, Clear cell histology or Stage II—High risk, early stage disease. Based upon the results of GOG 157 showing no survival benefit of six cycles versus three cycles of carboplatin and paclitaxel, appropriate therapy for this group of patients is three cycles carboplatin (dosed to AUC 5-7.5) and paclitaxel (dosed 175 mg/m^2).[1]
- Stage III and IV disease—standard therapy, based on the results of GOG 158, is six cycles of paclitaxel (dosed 175 mg/m^2 over 3 hours) and carboplatin (dosed AUC 5-7.5 at 1 mg/min).[2] However, recent data from GOG 172 has shown an overall survival benefit for intraperitoneal paclitaxel and cisplatin in well-selected patients with optimally cytoreduced disease. Toxicity however, was high with the intraperitoneal regimen.[3]

Consolidation

Following completion of initial chemotherapy, patients with a complete clinical response may be considered for maintenance chemotherapy, traditionally paclitaxel 40 mg/m^2 monthly for 12 months. The results of GOG 178 have shown a progression free survival benefit of 7 months for this regimen, compared to 3 months of additional therapy. Overall survival was not reported.[4]

Recurrent Disease

There are a myriad of agents available for patients with recurrent disease, to include paclitaxel, docetaxel, etoposide, liposomal doxorubicin, topotecan, gemcitabine, vinorelbine, ifosfamide, oxaliplatin, cisplatin and carboplatin. These patients and those who do not respond or progress on initial therapy are generally selected for additional chemotherapy based on response to initial treatment.

Generally, platinum-sensitive patients are those with a complete clinical response and a treatment-free interval of at least 6 months. Patients with greater than a 12-month treatment free interval should have strong consideration given to retreatment with a platinum combination. The results of ICON4 suggest that retreating this cohort with carboplatin and paclitaxel is superior in terms of progression-free survival and overall survival. Other agents including gemcitabine have also been investigated and in one study there was improved PFS with gemcitabine and carboplatin as compared to carboplatin alone; however there was no difference in OS and there was more associated toxicity.[5] For the patient with a 6-12 month treatment-free interval, controversy exists regarding the optimal treatment strategy.[6] Generally, combination regimens have a slightly higher overall response rate than single agent regimens; however toxicity is higher.

Patients with recurrent disease and a treatment free interval of less than 6 months, who do not achieve a complete clinical response, or who progress on initial therapy are labeled as platinum resistant or refractory. In this cohort, there is little compelling evidence to suggest that combination chemotherapy is better than single agent. In addition, no one agent has been shown to be clearly superior. As these patients are incurable and have a very poor overall prognosis, treatment goals should be palliative and toxicities should be a strong factor in selecting a second line agent for use in this cohort.

Ovarian Cancer: Germ Cell and Stromal Tumors

The standard therapy for ovarian germ cell tumors is BEP (bleomycin 20 units/m^2 weekly for 12 doses; etoposide 100 mg/m^2 daily for 5 days every 3 weeks for four cycles; and cisplatin 20 mg/m^2 daily for 5 days every 3 weeks for four cycles) and response rates of 90% can be expected.[7] Alternatively, some authors have recommended the use of paclitaxel and carboplatin in this cohort. Chemotherapeutic regimens used in germ cell tumors have also been used in ovarian stromal tumors with moderate success.

Uterine Cancers: Endometrial

Patients with advanced, high risk or recurrent endometrial cancer can be treated with either hormonal or cytotoxic chemotherapy. Hormonal therapy in these patients has been traditionally progestin therapy, and response rates of up to 20% can be expected.[8] The current cytotoxic chemotherapy regimen for patients with endometrial cancer consists of cisplatin (50 mg/m^2), doxorubicin (45 mg/m^2) and paclitaxel (3 hour infusion, 160 mg/m^2) repeated every 21 days for six cycles. Response rates approach 70% with this regimen.[9] Alternatively, some clinicians support the use of carboplatin and paclitaxel with nearly equivalent response rates reported.

Uterine Cancers: Sarcomas

Uterine leiomyosarcomas are traditionally not responsive to chemotherapy. The most active single agent for this disease is doxorubicin with a 25% response rate.[10] Recently, gemcitabine plus docetaxel has been reported to be active in 53% of patients and has been associated with a median survival of 17.9 months, the longest reported in Phase II trials to date.[12] Single agent response rates to uterine carcinosarcoma have been reported to be 20-30%. The most active agent in carcinosarcoma is ifosfamide with a 30% response rate. While response rates of 56% have been reported with cisplatin and ifosfamide combination therapy, there has been no overall survival benefit to combination therapy in this cohort, and toxicities are considerably higher.[11] The combination of ifosfamide and paclitaxel has been shown to have a 45% response rate and a 5-month overall survival advantage compared to ifosfamide alone. This trial is the only combination chemotherapy trial for carcinosarcoma to show an overall survival advantage.

Cervical Carcinoma

Cervical carcinoma has been shown to be only moderately responsive to chemotherapy. Most agents have a 10-15% response rates in this disease. Cisplatin appears to be the most active single agent with a 23% response rate as reported by the GOG.[13] Other active agents in this disease are irinotecan, topotecan, paclitaxel and vinorelbine, all with response rates of 17-20%.[13] Recently, several authors have published results of combination chemotherapy in advanced stage or recurrent cervical cancer patients. Combination therapy consisting of cisplatin (50 mg/m^2 every 3 weeks) and topotecan (0.75 mg/m^2 days 1 to 3 every 3 weeks) has been shown by the GOG to have a several month survival advantage compared to cisplatin alone.[14] Similarly, cisplatin and vinorelbine has shown promise as combination therapy in these patients.

In cervical carcinoma patients, cisplatin has been shown to be beneficial as a radiosensitizer in the treatment of patients with locally advanced disease or for those at high risk for recurrence.[15] In this setting, cisplatin is given weekly at a dose of 40 mg/m^2 during radiation treatment.

Chemotherapy Sequelae

Anemia

Anemia is commonly encountered in the cancer patient, the etiology of which may be multifactorial, including bleeding, hemolysis, nutritional deficiencies, bone marrow disease, anemia of chronic disease and the myelosuppressive effects of chemotherapy. All cancer patients with anemia should undergo full evaluation to determine severity and rule out common correctable causes. If clinically indicated, the following studies should be completed: peripheral smear, iron studies, folate, B12, stool guiac, reticulocyte count, LDH, hemoglobin electrophoresis, bilirubin, direct Coombs test and creatinine or creatinine clearance.

In the female cancer patient, mild anemia is defined by National Cancer Institute guidelines to be a hemoglobin level of 10-12 g/dL, with a level of 8-10 being moderate anemia, 6.5-7.9 severe anemia and less than 6.5 life threatening. Using a cutoff of 12 g/dL, up to 80% of cervical cancer patients can be diagnosed with anemia by the completion of therapy. The incidence of anemia however, varies with treatment and with regards to chemotherapy, dose intensity and chemotherapeutic agent.

Published studies comparing epoetin alpha with placebo have demonstrated a statistically significant decrease in transfusion requirement and increased hemoglobin level with epoetin alpha. This benefit is most pronounced for chemotherapy patients with hemoglobin levels less than 10 g/dL. The value of epoetin in patients with hemoglobin levels 10 g/dL or higher is unclear. In addition to decreasing the need for blood transfusions and increasing the hemoglobin levels, epoetin has been shown to decrease cancer-related fatigue in chemotherapy responders and patients with stable disease and improve quality of life.[16] Although the available data shows transfusion reduction benefits to treating to a hemoglobin of 10 g/dL, QOL improvements are seen with treatment until hemoglobin of 12 g/dL is obtained. Similar results have been obtained with darbepoetin alpha.[17,18] Compared to placebo, patients receiving darbepoetin require nearly 50% fewer transfusions.

Low hemoglobin levels have been associated with disease progression in some cancers (particularly cervical cancer); however few studies exist addressing treatment outcome for cancer patients treated with erythropoetic therapy.

Anemia: Treatment Guidelines

Observation for erythropoetic therapy should be considered for asymptomatic patients at risk for developing anemia. For symptomatic patients, transfusion or erythropoetic therapy should be instituted. If the hemoglobin is between 10-11 g/dL, therapy with transfusion, erythropoetic factors, or both should be considered.

Clinical judgment with regard to comorbid diseases should be used. Hemoglobin levels less than 10 g/dL should be treated, especially in patients undergoing radiation therapy. Iron supplementation should be given to deficient patients and treatable causes of anemia should be corrected, especially since erythropoetic therapy may result in measurable deficiencies in patients with borderline nutritional status. There is evidence to suggest that parenteral iron may be more effective than oral iron. Iron dextran, ferric gluconate and iron sucrose are parenterally available iron formulations. All require test dosing (without pretreatment) to evaluate for hypersensitivity reactions and treatment with diphenhydramine and acetaminophen during therapy prior to full dosing. Although all three preparations are associated with adverse events, ferric gluconate and iron sucrose have better safety profiles than iron dextran.

Doses

In patients who respond to therapy, medication should be continued to maintain an optimal hemoglobin of 12 g/dL. If hemoglobin increases greater than 1 g/dL in a 2-week period, the dose should be decreased by 25%. All medications should be discontinued if the hemoglobin is greater than 12 g/dL to prevent thrombotic complications.

Darbepoetin is dosed 100 mcg SQ weekly, 200 mcg every 2 weeks, or 300-500 mcg every 3 weeks. Patients should be evaluated for response at 6 weeks with dose increases to 300 mcg every 2 weeks implemented if there is less than 1 g/dL response. If there is no response by 12 weeks, darbopoetin should be discontinued.

Epoetin alpha 40,000 units weekly SQ. Dose may be increased to 60,000 units TIW (three times per week) after 4 weeks if hemoglobin response is less than 1 g/dL. Epoetin alpha should be discontinued at 8 weeks if there is no response in spite of dose escalation and iron supplementation.

Intravenous iron dextran, ferric gluconate, iron sucrose test doses of 25 mg slow IV push should be given followed by standard doses of prophylactic medications and then planned doses of iron. For iron dextran, the total body deficit of iron can be calculated by one of the following formulas and replaced over several hours (Bwt = body weight; Hbt = target hemoglobin; Hbo = observed hemoglobin):

- Total Dose (mg) = 50 {0.0442 Bwt (Hbt-Hbo) + (0.26 Bwt)}
- Total Dose (mg) = 2.4 Bwt (15-Hbo) + 500

Ferric gluconate should be dosed as 125 mg over 10 minutes (no total dose infusion) and iron sucrose is dosed at 100 mg over 5 minutes.

Neutropenia

Neutropenia is defined as (1) an absolute neutrophil count of less than 500 cells/microliter or (2) an ANC < 1000 cells/microliter and a predicted decline to 500 cells/microliter within the next 48 hours. The ANC can be calculated by the following formula:

- ANC = WBC (cells/microliter) × percent (PMN + bands) / 100

Metamyelocytes, when present, are not included in the calculation although they are a reassuring sign that the ANC has reached the nadir and bone marrow recovery is progressing.

Neutropenia is important as the risk of infection is inversely proportional to the decline in the ANC. The decision to hospitalize febrile neutropenic patients for treatment is based on the presence of high risk factors. Patients should be hospitalized and treated with intravenous therapy if there is significant medical comorbidity, an unstable patient, an anticipated prolonged episode (≥7 days) of severe neutropenia (≤100), hepatic or renal insufficiency, or pneumonia or other complex infections. Patients not meeting these criteria may be considered low risk patients and treated with either inpatient or outpatient regimens.[19]

Neutropenia may be managed and prevented by chemotherapy delay, dose reductions and myeloid growth factors. Myeloid growth factors, when used appropriately, have been associated with a 62% reduction in the risk of febrile neutropenia, and a 50% reduction in documented infection. As the inpatient mortality rate from febrile neutropenia has been reported to be between 7-11%, the reduction in risk associated with the appropriate use of G-CSF provides significant clinical and economic benefit.[19,20]

Guidelines

Prevention of Febrile Neutropenia: Primary Prophylaxis

There is no role for the routine administration of myeloid growth factors (G-CSF or GM-CSF) as primary prophylaxis in the untreated patient unless the incidence of febrile neutropenia is at least 20%.[19,20] Febrile neutropenic risk assessment is based on the chemotherapeutic regimen chosen, as well as the presence of clinical risk factors such as renal or hepatic dysfunction, prior exposure to radiation or chemotherapy, and comorbid diseases. Chemotherapy agents used in gynecologic cancer regimens which carry a risk of febrile neutropenia greater than 20% include topotecan, paclitaxel, and docetaxel for ovarian cancer, and cisplatin + paclitaxel for cervical cancer.[19,20] In general, for gynecologic malignancy the use of G-CSF for primary prophylaxis before febrile neutropenia occurs is not supported in previously untreated patients. Additionally, G-CSF should not be given with concomitant administration of chemotherapy and radiation due to the higher incidence of thrombocytopenia.

Secondary Prophylaxis

Secondary prophylaxis refers to the use of myeloid growth factors to prevent febrile neutropenia in subsequent chemotherapy cycles after febrile neutropenia has occurred in a previous cycle. Unless chemotherapy is being given with curative intent, chemotherapy dose reduction should be the primary management option to prevent febrile neutropenia since the use of G-CSF has not been shown to be associated with improved disease-free or overall survival. In such cases, treatment with G-CSF may permit maintaining chemotherapy dose intensity.

Additionally, although the severely afebrile neutropenic patient (ANC <100) who is treated with G-CSF may have a shorter duration of neutropenia (2 days) there is no effect on the rate of hospitalization or the number of culture positive infections.

G-CSF Treatment of Febrile Neutropenia

There is no proven clinical benefit to the routine use of G-CSF in the febrile neutropenic patient, and its use is not routinely recommended. There may be benefit for G-CSF use in the complicated febrile neutropenic patient who has one or more of the following factors: ANC <100, uncontrolled primary disease, pneumonia, hypotension, multiorgan system dysfunction, invasive fungal infection, sepsis or clinically documented infection, failure of previous management, performance status ≥3, or prior inpatient status. Use in such patients has been associated with a reduction in hospital stay by 2 days, a reduction in the median duration of antibiotic therapy (5 vs 6 days) and a reduction in the mean duration of Grade 4 neutropenia (2 vs 3 days).[21]

Dosage

Filgrastim (Neupogen). Recommended dose is 5 µg/kg/day. Treatment is daily until the ANC is at least 10,000 or until a clinically appropriate response is obtained.

Pegfilgrastim (Neulasta). Recommended dose is a single 6 mg dose SQ administered at least 24 hours after chemotherapy (to prevent chemotherapy-related killing of progenitor cells) and with at least 14 days prior to the next chemotherapy administration. Pegfilgrastim, while easier to administer, is slightly more costly than filgrastim, although the two are equivalent in reducing the incidence and severity of chemotherapy-induced neutropenia and febrile neutropenia.

Antibiotic Use in the Neutropenic and Febrile Neutropenic Patient

Greater than 50% of neutropenic patients who are febrile will have an established infection and at least 20% with an ANC <100 will have bacteremia.[18-21] A wide spectrum of bacteria is responsible for infections, and fungi can cause secondary infections in the neutropenic host. Primary sites of infection are the gastrointestinal tract where chemotherapy related mucosal damage occurs and indwelling catheters. A single oral temperature greater than 101 degrees or a persistent 100.4 degrees for one hour define the febrile state.

Febrile patients should be fully evaluated, with attention given to common sites of infection such as the periodontium, pharynx, esophagus, lung, perineum, anus, eye, skin and nails and indwelling catheter sites. Laboratory evaluation should consist of CBC, serum creatinine, urea nitrogen and transaminases. Blood cultures both from an indwelling catheter and peripheral site, as well as a chest X-ray if clinically indicated, should be obtained. Interestingly, CT scan of the chest will reveal evidence of pneumonia in more than 50% of febrile neutropenic patients with normal findings on chest X-ray. Routine oral, urine, or rectal swabs are of little value.

Initial Antibiotic Therapy

Low-Risk Patients: Low risk patients may be treated either as inpatients or outpatients. Acceptable oral outpatient regimens include the combination of ciprofloxacin with amoxicillin/clavulanate, and ciprofloxacin with clindamycin. Oral monotherapy with ciprofloxacin is probably not adequate therapy. Acceptable intravenous outpatient regimens include cefepime, ceftazidime, imipenem/cilastatin and aztreonam plus clindamycin. Outpatient should have daily evaluations for response. Reevaluation in the hospital setting is indicated if positive cultures, persistent fevers after 3-5 days of therapy, or if the clinical condition of the patient worsens.[19]

High-Risk Patients: All other patients are considered high risk and should be treated with IV monotherapy (cefepime, ceftazidime, or imipenim-cilastin, or meropenem), IV dual therapy (aminoglycoside + antipseudomonal penicillin, or ciprofloxacin + anti-pseudomonal penicillin, or aminoglycoside + cefepime or ceftzidime) or dual therapy with vancomycin. Vancomycin should normally be reserved for specific indications such as clinically suspicious catheter infections, known MRSA (methicillin-resistant *Staphylococcus aureus*), positive blood cultures with Gram-positive organisms, or hypotension. No one combination regimen has been proven superior to another. The advantage of combination therapy is antibiotic synergy against Gram-negative organisms. The disadvantage is toxicity. Linezolid is a promising new agent for the treatment of vancomycin resistant enterococci and other drug resistant Gram-positive organisms. Initial therapy for clinically unstable patients includes combination therapy with either imipenem, meropenem, or piperacillin-tazobactum plus an aminoglycoside and vancomycin. Serious consideration to adding and antifungal should be made. [19]

If Afebrile within 3-5 Days: All patients should be treated until the ANC is >500 cells/µL. If a causative organism has been identified, the antibiotic regimen should be adjusted to the most appropriate antibiotic and the duration of overall therapy should be tailored to the source of infection. If no organism has been identified, low risk patients initially on IV monotherapy can be placed on oral ciprofloxacin and amoxicillin-clavulanate until the ANC is greater than 500 cells/microliter at which point all antibiotics may be discontinued. High risk patients should remain on their current regimen.[19]

Persistent Fever for 3-5 Days: If the patient remains febrile, initial antibiotics may be continued if the patient remains stable. Vancomycin should be discontinued if cultures are negative. If the patients' condition worsens, antibiotics should be changed and vancomycin added if indicated. An alternative is to add an antifungal agent (amphotericin B is the drug of choice although fluconazole is an acceptable alternative), with or without a change in antibiotic regimen. Caspofungin may be used for invasive aspergillosis refractory to amphotericin B. After reassessment of the appropriateness of antibiotics and ensuring antifungal coverage, unstable patients should be considered candidates for G-CSF therapy and an infectious disease consult obtained.[19]

Duration of Antibiotic Therapy

The most important predictor of successful termination of antibiotic therapy is the recovering neutrophil count (ANC). If no infection has been identified, the patient is afebrile by 3-5 days, and the ANC is greater than 500 for 2 days, then antibiotics may be discontinued. If the patient is persistently neutropenic but all other criteria have been met, then antibiotics (oral or IV) should be continued until the neutropenia is resolved. An alternative is to discontinue antibiotics when the patient is stable and has been afebrile for 7-14 days. Patients who remain febrile at 3-5 days should be reassessed to determine the cause of the fever. Unstable patients should have coverage broadened, and treated as above. Stable patients should continue on antibiotics and antifungal coverage initiated. Documented infections should be treated with a regimen tailored to the sensitivity of the organism recovered.[19]

Viral Drugs

There is no indication for the routine use of antiviral medications in the febrile neutropenic patient who is without evidence of a viral infection.

Antibiotic Prophylaxis for the Afebrile Neutropenic Patient

Routine use of antibiotic prophylaxis for the afebrile neutropenic patient is not recommended. Although there is evidence to suggest that the frequency of febrile illnesses and infectious diseases can be reduced with early antibiotic use in the afebrile neutropenic patient, concern regarding emerging drug resistance precludes routine use in this patient population. Severely neutropenic patients (ANC <100) who are afebrile however, may benefit from prophylactic antibiotics as they are at the greatest risk for developing infections. Appropriate antibiotic choices for these patients are TMP-SMZ, ofloxacin or ciprofloxacin. Patients who are expected to have prolonged neutropenia (>7 days, ANC <1000 cells/µL) may be considered candidates for flouroquinolone prophylaxis.[20,21]

Anti-Emetic Therapy (Table 6.2)

The primary goal of anti-emesis treatment is prevention. Therapy is required for patients undergoing treatment with moderate or severely emetogenic chemotherapy regimens. Typically, treatment begins with the recognition and treatment of anticipatory

Table 6.2. Antineoplastic agents, emetogenic potential and treatment recommendations

Level of Risk	Drug	Treatment Recommendations
High, Frequency >90%	Cisplatin >50 mg/m² Cyclophosphamide >1500 mg/m² Adriamycin and cycophosphamide combination	Give prior to initiating each cycle of chemotherapy. Regimen: Aprepitant 125 mg PO day 1, 80 mg PO days 2-3 PLUS Dexamethasone 12 mg PO or IV day 1, 8 mg PO/IV days 2-4 PLUS 5-HT3 antagonist (Ondansetron 16-24 mg PO or 8-12 mg IV day 1; or Granisetron 2 mg PO or 1 mg PO BID or 0.01 mg/kg IV day 1; or Dolasetron 100 mg PO or 1.8 mg/ kg IV or 100 mg IV day 1; or Palono- setron 0.25 mg IV day 1). Consider add ing Lorazepam 0.5-2 mg PO q 6 hours days 1-4.
Moderate Risk, 30-90%	Methotrexate >1000 mg/m² Melphalan >50 mg/m² Doxorubicin Carboplatin Cisplatin <50 mg/m² Cyclophosfamide <1500 mg/m² Amifostine >500 mg/m² Hexamethylmelamine Ifosfamide Irinotecan Oxaliplatin >75 mg/m²	Give prior to initiating each cycle of chemotherapy. Regimen (day 1): Dexamethasone PLUS 5-HT3 antagonist; consider adding Lorazepam and/or Aprepitant. Regimen (day 2-4): Dexamethasone or 5-HT3 antagonist or Metoclopramide 20 mg PO QID with Diphenhydramine 25-50 mg PO q 6 hours or Aprepitant or Lorazepam.
Low Risk, 10-30%	Amifostine <300 mg/m² Cytarabine Capecitabine Docetaxol Liposomal Doxorubicin Etoposide 5-Fluorouracil Gemcitabine Paclitaxel Topotecan	Give prior to initiating each cycle of chemotherapy. Administer daily: Dexamethasone or Prochlorperazine or Metaclopromide with Diphenhydramine. Consider adding Lorazepam if necessary.
Minimal Risk, <10%	Bevacizumab Bleomycin Cetuximab Chlorambucil Gefitinib Hydroxyurea Melphalan >50 mg/m² Methotrexate <50 mg/m² Vinblastine Vincristine Vinorelbine	No routine anti-emetics are required.

nausea and is carried through the administration of chemotherapy and for 4 days thereafter. While there are many options for anti-emetic agents, it should be remembered that most agents are equally efficacious whether given intravenously or orally and the administration of agents should be in accordance with good pharmacologic principles, i.e., the lowest fully effective dose should be given with due consideration given to toxicity and side effects. Finally, it should be remembered that cancer patients undergoing chemotherapy might experience nausea due to other causes such as bowel obstruction, gastroparesis, psychological disturbances, electrolyte imbalances, radiation therapy etc. When clinically appropriate, such conditions warrant careful evaluation and treatment.

There are three types of nausea encountered in the cancer patient undergoing chemotherapy:

Anticipatory Nausea

Anticipatory nausea is best managed through primary prevention. Successful management options include behavioral therapy, hypnosis and chemoprevention. Anti-anxiety agents in combination with anti-emetic medications given the day prior to treatment have been shown to be effective. Benzodiazipines are currently the drug of choice for the management of this condition (please refer to supportive and palliative care chapters). Effective treatment of nausea during the first cycle of chemotherapy diminishes the impact of anticipatory nausea during subsequent cycles.

Acute Nausea

Acute nausea typically occurs during the first 24 hours of after chemotherapy and is managed with prophylaxis before and during chemotherapy. It typically is self-limiting if untreated.

Delayed Nausea

Delayed nausea occurs at least 24 hours after chemotherapy and gradually subsides over the course of the following week. It most commonly occurs after platinum therapy but may occur after any highly emetogenic drug. The most significant predictor of delayed nausea is poor control of acute or anticipatory nausea.

Suggested Reading

1. Chemotherapy of Gynecologic Cancers, Second Edition. Stephen C. Rubin, ed. Society of Gynecologic Oncologist, Lippincott Williams and Wilkins, 2004.
2. Cancer and Treatment-Related Anemia, Version 2. National Comprehensive Cancer Network Clinical Practice Guidelines in Oncology, 2005.
3. Lyman GH. Guidelines of the national comprehensive cancer network on the use of myeloid growth factors with cancer chemotherapy: A review of the evidence. JNCCN 2005; 3(4):557-71.
4. Prevention and treatment of cancer-related infections, Version 1.2008. National Comprehensive Cancer Network Clinical Practice Guidelines in Oncology, 2008.
5. Myeloid Growth Factors, Version 1.2008. National Comprehensive Cancer Network Clinical Practice Guidelines in Oncology, 2008.

References

1. Bell J, Brady MF, Young RC et al. Randomized phase III trial of three versus six cycles of adjuvant carboplatin and paclitaxel in early stage epithelial ovarian cancer: a Gynecologic oncology group study. Gynecol Oncol 2006; 102:432-9.
2. Ozols RF, Bundy BN, Greer BE et al. Phase III trial of carboplatin and paclitaxel compared with cisplatin and paclitaxel in patients with optimally resected stage III ovarian cancer: a Gynecologic oncology group study. J Clin Oncol 2003; 21:3194-200.

3. Armstrong DK, Bundy B, Wenzel L et al. Intraperitoneal cisplatin and paclitaxel in ovarian cancer. N Engl J Med 2006; 354:34-43.

4. Markman M, Liu PY, Wilczynski S et al. Phase III randomized trial of 12 versus 3 months of maintenance paclitaxel in patients with advanced ovarian cancer after complete response to platinum and paclitaxel-based chemotherapy: a southwest oncology group and gynecologic oncology group trial. J Clin Oncol 2003; 21:2460-5.

5. Pfisterer J, Plante M, Vergote I et al. Gemcitabine plus carboplatin compared with carboplatin in patients with platinum-sensitive recurrent ovarian cancer: an intergroup trial of the AGO-OVAR, the NCIC CTG and the EORTC GOG. J Clin Oncol 2006; 24:4699-707.

6. Columbo N, Gore M. Treatment of recurrent ovarian cancer relapsing 6-12 months post platinum-based chemotherapy. Crit Rev Oncol Hematol 2007; 64:129-38.

7. de Wit R. Refining the optimal chemotherapy regimen in good prognosis germ cell cancer: interpretation of the current body of knowledge. J Clin Oncol 2007; 25:4346-9.

8. Lai CH, Huang HJ. The role of hormones for the treatment of endometrial hyperplasia and endometrial cancer. Curr Opin Obstet Gynecol 2006; 18:29-34.

9. Obel JC, Friberg G, Fleming GF. Chemotherapy in endometrial cancer. Clin Adv Hematol Oncol 2006; 4:458-68.

10. Kanjeekal S, Chambers A, Fung MF et al. Systemic therapy for advanced uterine sarcoma: a systematic review of the literature. Gynecol Oncol 2005; 97:624-37.

11. Sutton G, Kauderer J, Carson LF et al. Adjuvant ifosfamide and cisplatin in patients with completely resected stage I and II carcinosarcomas of the uterus: a Gynecologic Oncolgy Group study. Gynecol Oncol 2005; 96:630-4.

12. Hensley ML, Maki R, Venkatraman E et al. Gemcitabine and docetaxel in patients with unresectable leiomyosarcoma: results of a Phase II trial. J Clin Oncol 2002; 20:2824-31.

13. Long HJ. Management of metastatic cervical cancer: Review of the literature. J Clin Oncol 2007; 25:2966-74.

14. Fiorica J, Holloway R, Ndubisi B et al. Phase II trial of topotecan and cisplatin in persistent or recurrent squamous and nonquamous carcinomas of the cervix. Gynecol Oncol 2002; 85:89-94.

15. Rose PG, Bundy BN, Watkins EB et al. Concurrent cisplatin-based radiotherapy and chemotherapy for locally advanced cervical cancer. N Engl J Med 1999; 340:1144-53.

16. Carteni G, Giannetta L, Ucci G et al. Correlation between variation in quality of life and change in hemoglobin level after treatment with epoetin alfa 400,000 IU administered once-weekly. Support Care Cancer 2007; 15:1057-66.

17. Bokemeyer C, Aapro MS, Courdi A et al. EORTC guidelines for the use of erythropoietic proteins in anemic patients with cancer: 2006 Update. Eur J Cancer 2007; 43:258-70.

18. Crawford J, Althaus B, Armitage J et al. Myeloid growth factors. Clinical practice guidelines in oncology. J Natl Compr Canc Netw 2007; 5:188-202.

19. Baden LR, Segal BH, Brown EA et al. Prevention and treatment of cancer-related infections, Version 1.2008. National Comprehensive Cancer Network Clinical Practice Guidelines in Oncology, 2008. http://www.nccn.org/professionals/physician_gls/f_guidelines.asp.

20. Crawford J, Althaus B, Armitage J et al. Myeloid Growth Factors, Version 1.2008. National Comprehensive Cancer Network Clinical Practice Guidelines in Oncology, 2008. http://www.nccn.org/professionals/physician_gls/f_guidelines.asp.

21. Sung L, Nathan PC, Alibhai SM et al. Meta-analysis: effect of prophylactic hematopcotic colony-stimulating factors on mortality and outcomes of infection. Ann Intern Med 2007; 18:400-11.

CHAPTER 7

Radiation Therapy

Hiram A. Gay, Ron R. Allison and Marcus E. Randall

Introduction

Ionizing radiation has been successfully employed to treat cancer for over a century. The ability to destroy tumor and spare surrounding normal tissue is the essence of radiation oncology. Radiotherapy plays a critical role in delivering the optimal care for women with gynecologic malignancies. However, the benefits of radiotherapy can never be realized without the active cooperation between the gynecologic oncologist and radiation oncologist. This chapter aims to review the fundamentals of radiation oncology for medical students and residents so patients can achieve the best outcomes possible; it is divided into three sections: radiation physics, radiobiology and clinical radiation oncology.

Physics

What Is Ionizing Radiation?

Unlike other forms of electromagnetic radiation, such as light, which do not ionize matter, ionizing radiation ultimately creates free radicals which are highly destructive to living tissue. Ionizing radiation is emitted from naturally occurring radioactive isotopes and can also be artificially generated with the use of electricity. Gamma rays, X-rays, beta particles and electrons are types of radiation that lead to ionization and are often used in clinical radiation oncology.

Teletherapy versus Brachytherapy

Ionizing radiation used clinically can be separated into two categories: teletherapy and brachytherapy. During teletherapy, the radiation source is at some distance from the target. In contrast, during brachytherapy the radiation source is directly in or near the target.

Brachytherapy relies on radioactive isotopes to generate gamma rays, X-rays and/or beta particles (electrons) to treat the target. With brachytherapy, one may permanently or temporarily place a radioactive source into a tumor bed or region at risk. Radioactive isotopes of various energies, sizes and shapes have been created for a myriad of oncologic indications. Using proper techniques, brachytherapy can minimize the irradiation of normal tissues compared to teletherapy. This is because brachytherapy takes advantage of the inverse square law, which states that the intensity of radiation is inversely proportional to the square of the distance from the source. Clinically, this means tissues near radioactive sources receive high doses of radiation while tissue only a few centimeters away may receive minimal doses. This is of critical importance when offering vaginal vault radiation with brachytherapy. The neighboring rectum and bladder may be spared if the procedure is done well. Brachytherapy is utilized in the

Gynecologic Oncology, edited by Paola Gehrig and Angeles Secord.
©2009 Landes Bioscience.

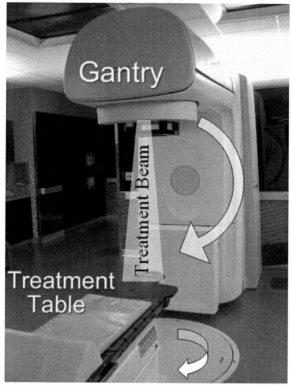

Figure 7.1. LINAC. The upper curved arrow shows the rotation of the gantry which allows a patient to be treated from any angle around a 360° axis. The treatment table is attached to a rotating circular platform on the floor which allows the patient to be treated from additional angles as shown by the lower curved arrow. The treatment beam could be either photons or electrons and is invisible.

gynecologic tract because anatomical cavities like the vagina, endocervical canal and uterus can be easily accessed.

Teletherapy is commonly delivered by linear accelerators or LINACs. LINACs accelerate electrons in a "wave guide" powered by electricity. Unlike radioactive isotopes, radiation generated via electricity can be shut off instantaneously, an important safety feature. These energetic electrons can be used for treatment or can be made to collide with a heavy metal target to generate high-energy X-rays. Figure 7.1 shows a LINAC. Another type of teletherapy unit uses cobalt-60 to produce gamma rays as it decays. Both X-rays and gamma rays are high energy photons. Currently, linear accelerators are the favored source of teletherapy. Teletherapy has the advantage that any location in the body can be targeted and treated with a noninvasive approach. The disadvantage is that normal tissue is treated as the photons or electrons reach their target.

Units of Radiation

Radiation is a form of energy that can be measured by the amount delivered or absorbed. The unit of absorbed dose is the rad and its SI (International System of Units) unit counterpart the Gray (Gy):

1 Gy = 1 Joule/kg = 100 rads = 100 cGy

1 rad = 100 ergs/g = 0.01 Gy = 1 centigray (cGy). Modern radiation oncologists conventionally report doses in cGy rather than rads.

The unit of dose equivalence is the rem and its SI counterpart is the sievert (Sv). These units are explained further in the Radiation Safety section.

The electron volt (eV) is a unit of energy equal to the energy acquired by an electron falling through a potential difference of one volt, approximately 1.602×10^{-19} joules. The therapeutic radiation generated by a LINAC is on the order of MeV or million electron volts.

For historical reasons, mg-Ra eq (milligram equivalents of radium) is used to express the exposure rate of a radioactive source. However, the modern way to calibrate and specify brachytherapy sources is air kerma strength which has the units: $\mu Gy\ m^2\ h^{-1}$.

Photons versus Electrons

The photons generated by a LINAC are in the megavoltage range, unlike diagnostic X-rays which are in the kilovoltage range. At the megavoltage energy level, photons are deeply penetrating. Because of this, photons are frequently used to treat pelvic targets. In addition, deeply penetrating photons only deposit a fraction of their energy on the skin; this is referred to as the "skin sparing" effect. The benefit of the skin sparing effect is that patients do not get severe radiation skin reactions. One drawback is that superficial lesions may be underdosed unless a bolus of tissue equivalent material (usually 0.5-1 cm thick) is placed over them.

On the other hand, electrons penetrate tissue to a limited degree and are mainly used for superficial targets such as the skin or in some cases the inguinal lymph nodes. A rule of thumb for electrons is that they are clinically useful to a depth that equals a fourth of their energy. At this depth, approximately 90% of the maximum dose is delivered. For example, 12 MeV electrons are useful to a 3 cm depth, 9 MeV to 2.25 cm and so on. The inguinal lymph nodes are 6 cm deep on average and an inappropriate selection of electron energy can result in underdosing these nodes. Pelvic CT scanning is useful to accurately determine lymph node depth.

The application of new treatment technologies such as intensity-modulated radiation therapy (IMRT), image-guided radiation therapy (IGRT), image-based brachytherapy, proton beam therapy, treatment planning based on functional imaging such as FDG-PET, adaptive radiotherapy and biologically optimized treatments are just a few of the areas of current research.

Radiobiology

How Does Radiation Work?

Electrons, either produced by a LINAC or generated from the interaction of photons with atoms, are the ionizing particles mostly responsible for the effect of radiation on tumors and tissue. These electrons damage cellular DNA both directly and indirectly. Interestingly, only one-third of the DNA damage is due to the direct interaction of electrons with DNA. Indirect effects are responsible for the majority, approximately two-thirds, of the DNA damage. This indirect action occurs when a nonDNA material is ionized to produce free radicals that diffuse to and damage the DNA. Since the body is mostly water, H_2O is the predominant source of free radical production. The types of DNA lesions that can be induced by radiation are, in order of increasing frequency: DNA-DNA crosslinks, double strand breaks, DNA-protein crosslinks, alkali-labile sites, sugar damages, single strand breaks and base damages. However, it is generally believed that the double strand breaks are the most critical lesions for radiation cell killing in most cell types because they are the least likely to be successfully repaired.

Cell Survival Curves and Dose Fractionation

A course of radiation therapy to the pelvis is often 4500 cGy in 25 fractions of 180 cGy. Why? Part of the answer has to do with cell survival curves. Cell survival curves depict the relationship between radiation dose and the proportion of cells that survive. These are usually obtained from in vitro cell culture experiments. Figure 7.2 shows hypothetical cell survival curves for one normal tissue and one tumor cell line. If one looks at what happens at a dose of 12 Gy (1 Gy = 100 cGy), one sees that a single fraction of 12 Gy yields less survival for both normal tissue and tumor than when one gives 6 fractions of 2 Gy. Sublethal damage repair is the term used for this increase in cell survival when a given radiation dose is split into multiple fractions separated by a time interval. In fact, the multiple 2 Gy daily fraction curves were generated by

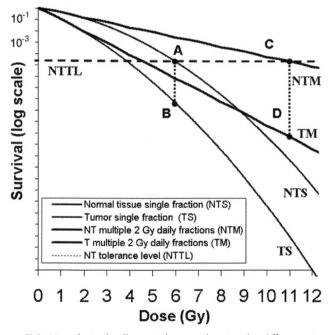

Figure 7.2. Hypothetical cell survival curve showing the difference in cell survival between one normal tissue and one tumor cell line when a single radiation fraction or multiple 2 Gy fractions are delivered. Note that the y-axis, "Survival", has a logarithmic scale while the x-axis, "Dose", has a linear scale (T = tumor; NT = normal tissue).

repeating the 0 to 2 Gy segment of their respective single fraction curve six times to reach a total dose of 12 Gy.

If splitting the radiation dose into daily fractions kills fewer tumor cells than if delivered in one big fraction, what is the advantage of fractionation? The answer lies in what happens to the normal tissue. The dashed black line in Figure 7.2 represents the normal tissue tolerance level; that is, any cell survival below this line would result in unacceptable normal tissue toxicity. A single fraction of 6 Gy delivers this maximum normal tissue dose and is illustrated by point A. The corresponding tumor survival is shown by point B. On the other hand, if one gives 2 Gy daily fractions, 10 Gy can safely be delivered without exceeding the normal tissue tolerance as shown by point C and the corresponding tumor survival is given by point D. As can be seen by comparing points B and D, for a given normal tissue maximum tolerance level, more tumor kill can be achieved by fractionating the treatment. As more fractions are delivered, the gap between the fractionated normal tissue and tumor curves become wider at a faster rate than their corresponding single fraction curves. In other words, after multiple treatment fractions, a larger proportion of tumor cells than normal cells are killed than if we delivered single fractions. This represents a therapeutic gain.

The Four R's

The four R's of radiobiology are: repair of sublethal damage, reassortment of cells within the cell cycle, repopulation and reoxygenation. The four R's gives us a simplified model of what happens between treatment fractions. Repair of sublethal damage was discussed in the previous section, specifically the resulting increase in cell survival when a given radiation dose is split into multiple fractions separated by sufficient time.

Reassortment is related to the cell cycle. The phases of the cell cycle are G_1 (growth 1), S (synthesis), G_2 (growth 2) and M (mitosis). Cycling cells are more radiosensitive than noncycling cells. In general, cycling cells are the most radiosensitive during the M and G_2 phases and most radioresistant during the late S phase. Consequently, the radiosensitivity of a tumor is not constant, but is dynamically changing as the proportion of cells in different phases of the cell cycle changes with time.

Repopulation results from cell division. Rapidly cycling tumor cells often reach mitosis between treatment fractions and reproduce. This rapid reproduction counteracts the killing by radiation and could compromise the chance of eradicating the tumor. Interestingly, the rate of repopulation increases in some normal tissues and tumors approximately 2 weeks after the initiation of radiotherapy. This phenomenon is named accelerated repopulation. For example, in cervical cancer, repopulation and accelerated repopulation are thought to be responsible for the reduction in local control as treatment time is prolonged.

Well-oxygenated cells are more radiosensitive than hypoxic cells by a factor of 2 to 3. Reoxygenation occurs when the radioresistant hypoxic cells in the central portion of tumor become oxygenated between treatments. One explanation is that the well-oxygenated radiosensitive cells at the periphery of the tumor are preferentially killed by radiation. As these peripheral cells die, the tumor diameter decreases and former central hypoxic cells become part of the tumor's well-oxygenated periphery.

Treatment Window

If an appropriate total dose is chosen, at the end of treatment there will be a significant gap between tumor and normal cell kill, clinically translating into a high probability of tumor control with a low probability of complications. Figure 7.3 illustrates this. In general, the dose-response curve for both tumors and normal tissues has

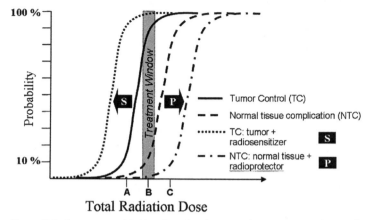

Total Radiation Dose

Figure 7.3. Tumor control (TC) with and without a radiosensitizer and normal tissue complication probability (NTC) with and without a radioprotector, versus total radiation dose. Note that a partial treatment that falls short of the treatment window will result in a poor chance of tumor control unless a radiosensitizer is used. Therefore, it is imperative for patients to complete the intended treatment.

a sigmoidal or "S-shaped" appearance. If one gives a total radiation dose of A, normal tissue complications are less than 5%, but the tumor control is very low, less than 20%. Dose C, on the other hand, has a tumor control approaching 100%, but the chance of complications is high, around 80%. However, dose B has a 90% chance of tumor control, with less than a 15% chance of complications. Treatment B falls in the treatment window, i.e., that range of radiotherapy doses that can achieve acceptable tumor control with low complications.

Radiation Modifiers

There are strategies to improve the therapeutic window. A good radiosensitizer will shift the tumor curve to the left while having a minimal impact on the normal tissue curve. If we look at the "tumor + radiosensitizer" curve and the "normal tissue" curve in Figure 7.3, dose A, which was initially unacceptable, now yields more than 90% tumor control, with less than 5% complications. Cisplatin is a chemotherapy agent that is frequently used as radiosensitizer during pelvic radiotherapy for cervical cancer. Other chemotherapeutic agents that have been used as radiosensitizers in cervical cancer are gemcitabine, carboplatin and paclitaxel.

In contrast, an ideal radioprotector will shift the normal tissue curve to the left without affecting the tumor curve. Therefore, if we look at the "tumor" and "normal tissue + radioprotector curves", dose C, at first unacceptable, now yields close to 100% control with less that 10% complications. The commercially available radioprotector amifostine is used to protect the parotid glands during head and neck radiotherapy, but has no proven clinical use in gynecologic oncology. Retrospective analyses in cervical cancer have identified several poor prognosis biologic markers, including vascular endothelial growth factor (VEGF) and epidermal growth factor receptor (EGFR) that could be targeted and potentially serve as radiation modifiers.

The previous examples are somewhat simplified. In clinical practice, the inherent radiosensitivity of an individual or the radioresistance of tumors is not known. Doses are prescribed based on the clinical data accumulated on thousands of patients over many years.

Dose Rate Effect

Not only is the total dose important in determining the chance of tumor control, but the rate at which it is delivered is also important. This is particularly relevant in brachytherapy where a wide range of dose rates can be achieved depending on the isotope and its activity. The dose rate effect is mostly dependent on the half-times of the four R's of radiobiology. In general, repair has the shortest half-time, followed by reassortment, reoxygenation and repopulation. For example, the dose rate effect caused by the repair of sublethal damage is most dramatic between 0.01 and 100 cGy/min. Cell survival curves for some cell lines have shown that high dose rates result in the least survival, mostly because there is not enough time to repair the sublethal damage. As the dose rate is reduced repair is allowed, resulting in an increase in survival. In some cell lines, as the dose rate is further decreased, reassortment occurs and cells accumulate in G_2, a radiosensitive phase of the cell cycle, resulting in a decrease in survival. This is called the inverse dose rate effect. Finally, as the dose rate is reduced even more, there is a point when repopulation predominates and there is an increase in survival.

Effects of Radiation on the Embryo and Fetus

Radiation to the embryo in weeks 0 to 2 after conception results in a 0.1 to 1% per cGy risk of prenatal death. Malformations are unlikely during this period; it is mostly an "all or nothing" effect.

Radiation during organogenesis, in weeks 2 to 8 after conception, results in malformations, especially in the organs developing at the time of exposure. The risk of radiation-induced malformations is 0.5% per cGy above a threshold dose of 10 to 20 cGy. Note that a single fraction of pelvic photon teletherapy usually delivers 180 to 200 cGy.

Radiation to the fetus in weeks 8 to 25 after conception may result in mental retardation. On weeks 8 to 15, the risk of mental retardation is 0.4% per cGy above a threshold dose of 25 cGy; and there is a reduction of 21 IQ points per 100 cGy above a threshold dose of 5 cGy. In addition, on weeks 16 to 25, the risk of mental retardation is lower at 0.1% per cGy above a threshold dose of 25 cGy; and there is a reduction of 13 IQ points per 100 cGy above a threshold dose of 5 cGy. Finally, the risk of radiation-induced childhood leukemia and solid tumors as a result of radiation exposure after conception is 0.3 to 0.4% per cGy.

If a pregnant individual is exposed to radiation in a medical setting, it is often possible to determine the dose delivered to the fetus. Termination of pregnancy for fetal doses less than 10 cGy is not justified and some believe there is no justification for doses less than 20 cGy. Termination of pregnancy is an individual decision but, to be an informed one, a qualified biomedical or health physicist should assess the situation.

Clinical

Overview

The overriding goal of curative radiation is to deliver sterilizing doses of radiation to the tumor bed and regions at risk for regional spread without causing permanent damage to the surrounding normal tissue. As radiation may be employed in a pre- or postoperative mode, with or without chemotherapy, or as a stand alone procedure, it is critical that an optimal treatment plan be discussed and agreed upon with the gynecologic oncologist and consented to by the patient. If the patient cannot lie still during teletherapy or is unable to undergo brachytherapy, then the best radiation options may not be available. Before a patient starts teletherapy, a sequence of steps must be completed which include: treatment planning, treatment simulation, treatment verification and treatment delivery.

Treatment Planning Overview

Because radiation only treats the volume at which it is aimed, optimal treatment planning is critical to successful radiation therapy (XRT). The radiation oncologist working with trained physicists and dosimetrists can create a custom therapy plan for each patient. Creating a treatment plan requires a working knowledge of which anatomical regions are at risk for disease spread and the individual patient's tumor characteristics. Discussion with the gynecologist can be critical to ensure appropriate coverage of the tumor. Further diagnostic studies such as a contrast-enhanced CT scan, MRI scan and/or PET scan can be used to generate a plan that may include a combination of teletherapy and brachytherapy. Sophisticated and accurate computer planning programs have been developed to optimize this process. Graphical displays are created to assess tumor and normal tissue coverage and radiation dosing. Underdosage of the tumor and overdosage of normal tissues are of critical concern.

Treatment Planning—Target Delineation

Since radiotherapy is a focused treatment, ensuring that the tumor receives the intended dose is one of the most important concerns of the radiation oncologist. Defining the treatment target or volume has to follow a logical and organized approach for it to be successful. The International Commission on Radiation Units (ICRU) has played a critical role in giving radiation oncologists standardized definitions of the various treatment volumes. The ICRU report 62, which has been adopted worldwide, defines these volumes as follows (see Fig. 7.4):

Gross Tumor Volume (GTV): gross demonstrable extent and location of the malignant growth.

Clinical Tumor Volume (CTV): tissue volume that contains a demonstrable GTV and/or subclinical malignant disease that must be eliminated. This volume must be treated adequately in order to achieve the aim of radical therapy.

Planning Target Volume (PTV): geometrical concept used for treatment planning and that takes into account all movements and variations in size and shape of the tumor and daily treatment setup uncertainties.

For example, a radiation oncologist might define the GTV as the visible cervical tumor on a CT scan of the pelvis, the CTV as the GTV plus a margin around it to account for microscopic parametrial extension and the draining lymphatic chains, and the PTV as the CTV plus a margin accounting for tumor motion and daily setup errors.

Treatment Planning—Treatment Fields

Teletherapy treatment fields for gynecologic malignancy generally encompass the pelvis and less commonly the para-aortic region of the abdomen. The pelvis is classically approached by what is termed a four-field box technique. The goal of the box field technique is to encompass the primary tumor bed and draining lymphatics at risk for regional spread. To create the "box", radiation is delivered by matched anteroposterior, posteroanterior, right lateral and left lateral fields. Specifically, the traditional pelvic four field technique is as follows:

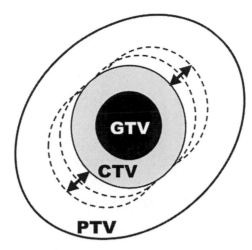

Figure 7.4. Simplified drawing showing a gross tumor volume (GTV), a clinical tumor volume (CTV) and a planning tumor volume (PTV). The GTV is the inner black circle. The CTV encompasses the GTV plus the microscopic tumor infiltration around it shown in gray. The arrows and dashed lines show the CTV's diagonal motion for simplicity. The PTV, an ellipse in this case, encompasses the CTV and takes into account the daily treatment setup errors and CTV's motion.

Anteroposterior/posteroanterior (AP/PA) pelvic fields:
- Superior border: L4-5 interspace to include the external iliac and hypogastric lymph nodes
- Inferior border: mid-portion to inferior border of the obturator foramen
- Lateral borders: 2 cm lateral to the bony pelvis. Figures 7.5 and 7.6 show the pelvic lymphatic drainage and help illustrate the rationale behind the 2 cm lateral margin to the bony pelvis.

Lateral pelvic fields:
- Anterior border: anterior to the pubic symphysis
- Posterior border: cover at least 50% of the rectum. Cover sacral hollow in patients with advanced tumors.
- Superior and inferior borders: same as the anteroposterior/posteroanterior fields to complete the box

For patients at risk of para-aortic spread, the para-aortic area can be treated with separate para-aortic fields or with a four-field technique encompassing both the pelvis and para-aortic nodes. Figures 7.6 and 7.7 illustrate a four-field technique encompassing both the pelvis and para-aortic regions. In this case, the superior border is at the T11-T12 interspace to cover the tumor involved para-aortic lymph nodes. Since the patient had three-dimensional CT based planning, the CT data was used to generate the digitally reconstructed radiographs (DRR) shown in Figure 7.6. Three-dimensional CT based planning is very important in modern pelvic radiation therapy planning because the margins of the traditional pelvic field often provide inadequate coverage of the tumor and lymphatic chains as shown in Figure 7.8. Patients with low-lying tumors, particularly vaginal lesions, often have the inguinal nodes covered by separate electron or photon fields.

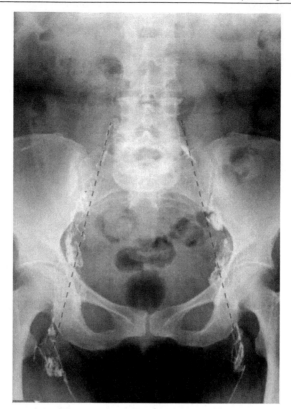

Figure 7.5. Lymphangiogram. Anteroposterior projection of lymph nodes in relation to the bony pelvis (the dashed lines roughly follow the contrast enhanced lymph nodes). Modified from: Zunino S et al. Anatomic study of the pelvis in carcinoma of the uterine cervix as related to the box technique. Int J Radiat Oncol Biol Phys 1999; 44(1):53-9, with permission from Elsevier.

Radiation Dose

Radiation dose can be delivered by external beam alone, brachytherapy alone or the two in combination. Microscopic tumor in the pelvis is usually controlled with 4500 cGy of teletherapy given in 25 daily fractions of 180 cGy each. Clinically evident disease requires higher doses sometimes approaching 9000 cGy. This is often done by combining teletherapy with brachytherapy. Again, this requires good treatment planning since the tolerance of the small bowel is approximately 4500 cGy, and higher doses can result in significant morbidity if not carefully planned.

The Dose Volume Histogram—A Means for Evaluating Treatment Plans

Ensuring the tumor and regions of disease spread receive adequate radiation while not overdosing normal tissue can be a complicated undertaking. Mathematical and physical computer models are available to assist the radiation oncologist reach these goals.

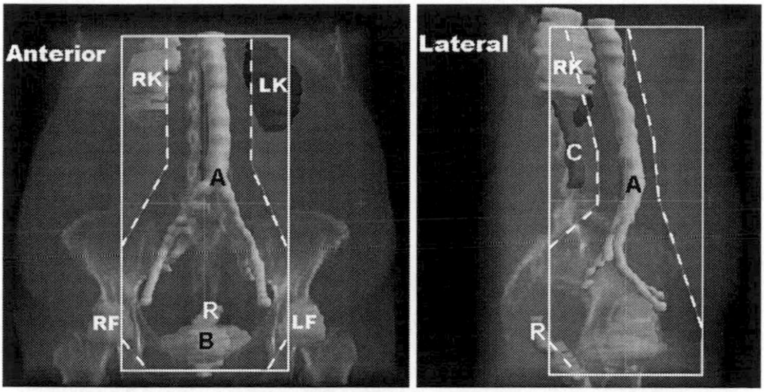

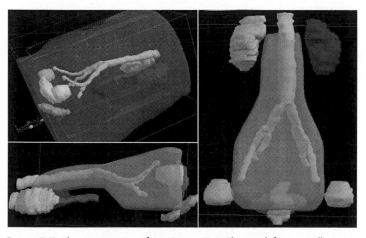

Figure 7.6. Digitally reconstructed radiographs (DRR) of the anteroposterior and right lateral treatment fields encompassing the pelvis and para-aortic area in a patient with endometrial cancer with para-aortic lymph node involvement. The white rectangles show the margins of the treatment fields. The following structures are illustrated: right kidney (RK), left kidney (LK), spinal cord and cauda equina (C), rectum (R), bladder (B), right femoral head (RF), left femoral head (LF) and the descending aorta, common iliac artery, external iliac arteries and internal iliac arteries (A) which also are representative of the lymphatic drainage. The treatment field blocks are translucent gray and bounded by the white dashed lines. Note that the kidneys are partially blocked in the anterior and lateral fields and the spinal cord is completely blocked in the lateral field.

Figure 7.7. The same patient from Figure 7.6. The top left image illustrates the patient's body contour and the projection of the anterior, right lateral and posterior fields. The left lateral field borders were removed for clarity. The bottom left and right images show the volume covered by 100% of the prescription dose. In this case, the calculated dose ratio between the AP/PA fields and lateral fields is 70:30.

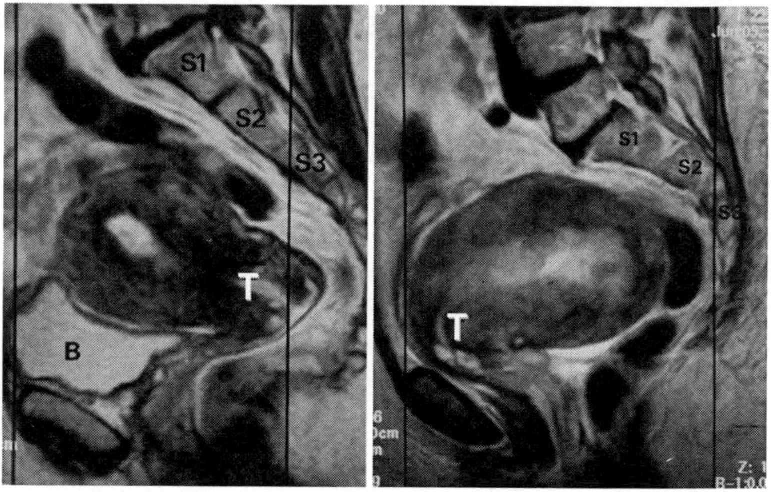

Figure 7.8. The conventional posterior border at the S2-S3 interspace of the lateral fields fails to cover the cervical tumor (left). The traditional anterior border of the lateral fields at the anterior edge of the symphysis pubis does not provide any margin for the cervical tumor in this other patient (right). T = cervical tumor; B = bladder. Reprinted from: Zunino S et al. Anatomic study of the pelvis in carcinoma of the uterine cervix as related to the box technique. Int J Radiat Oncol Biol Phys 1999; 44(1):53-9, with permission from Elsevier.

The cumulative dose volume histogram (DVH) is a plot of the volume of a given structure receiving a specific dose or higher as a function of dose. The DVH provides a quick overview of the treatment plan and is particularly useful when comparing competing plans. For example, there is approximately a 5% chance of clinical nephritis in 5 years if 100% of the kidney receives 2300 cGy, 66% receives 3000 cGy, or 33% receives 5000 cGy. Point a in Figure 7.9 shows that 100% of the right kidney receives 119 cGy or more and that 119 cGy is also the minimum dose received by the entire right kidney. Points b and c show that 66% and 33% of right kidney receive 315 and 749 cGy or more, respectively. These three points are well below the corresponding points associated with a 5% chance of clinical nephritis in 5 years. Consequently, this plan does an excellent job in sparing the right kidney. A quick glance at the DVH shows that the plan also spares well the left kidney. The tolerance of the kidney is dependent on the volume irradiated because it is composed of millions of nephrons in a parallel unit arrangement. In contrast, the spinal cord has a serial unit arrangement and its tolerance is dependent on the maximum dose it receives. That is, breaking one "link" in the spinal cord "chain" is enough to cause complications. Most radiation oncologists like to keep the dose to the spinal cord below 4500 cGy. Point d shows that in this plan the maximum dose to the spinal cord is 3792 cGy which is acceptable. If a tumor target is outlined, a perfect plan will show that 100% of its volume receives the prescribed dose. The plan is approved once the tumor target coverage and normal tissue sparing is deemed satisfactory. These dose-volume histograms also serve as a medical record of treatment.

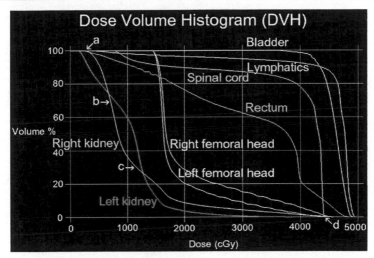

Figure 7.9. Cumulative Dose Volume Histogram for the treatment plan shown in Figures 7.6 and 7.7. Refer to the "Dose Volume Histogram" section for a detailed explanation.

Treatment Simulation

The best treatment plan is useless if it cannot be delivered. The purpose of the treatment simulation is to ensure a highly reproducible therapy that the patient can undergo. A treatment simulator is an apparatus with a diagnostic X-ray tube that duplicates the LINAC in terms of its mechanical, geometrical and optical properties (see Fig. 7.10). Most simulators have fluoroscopic ability and enable real-time visualization of the anatomy. Since the simulator can clearly show the bony anatomy in most cases, it can be very useful when determining the treatment field for bone palliation without the need of sophisticated computer planning. In many cases, the simulation consists of obtaining a CT scan in the treatment position that is imported into the treatment planning computer.

To ensure accurate daily setup, small marks or tattoos are placed on the patient's skin which correspond to the central axis of the treatment fields and setup points. One can instruct the patient to fill her bladder prior to each daily external beam treatment. The expanded bladder may move the radiosensitive small bowel from the field to minimize toxicity (See Fig. 7.11). Bowel exclusion devices, i.e., belly boards, can also be used to reduce the amount of small bowel irradiated.

Treatment Verification and Delivery

Patients undergoing teletherapy may need 25 or more daily treatment visits. Lasers are used in the treatment room to ensure proper alignment of the patient's tattoos with the planned LINAC radiation fields. X-ray images in the treatment position are also employed to check the accuracy of setup guided mostly by bony landmarks. Shielding of normal tissues is accomplished with either a multileaf collimator or Cerrobend (a lead alloy) blocks. Modern LINACS have built-in multileaf collimators, which automatically change the shape the field based on sophisticated computer controls. Cerrobend blocks, on the other hand, are usually manufactured in the radiation department.

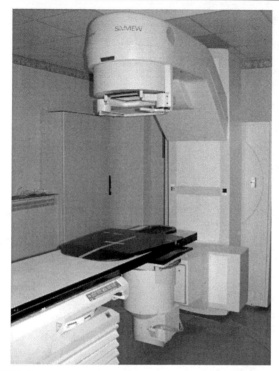

Figure 7.10. Treatment simulator. Compare it to the LINAC in Figure 7.1.

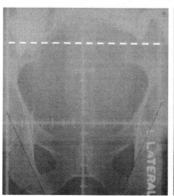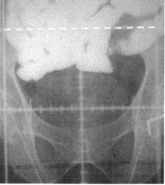

Figure 7.11. Treatment simulation X-ray films. The small bowel is outside the superior border (white dashed line) of the anterior treatment field when the patient has a full bladder (left). After the patient empties her bladder, a significant portion of small bowel enters the treatment field (right). Reprinted from: Hoskins WJ et al. *Principles and Practice of Gynecologic Oncology*. 4th ed. Philadelphia: 2005, with permission from Lippincott Williams and Wilkins.

Treatment Delivery

Radiotherapy for gynecologic malignancies is generally a combination of external beam radiotherapy and brachytherapy. External beam therapy is generally delivered by the four-field box technique as previously outlined, although some centers offer IMRT. Brachytherapy may take the form of high dose rate therapy (HDR) or low dose rate therapy (LDR) depending on the services available.

Treatment Delivery—Intensity Modulated Radiation Therapy

In traditional photon teletherapy, such as the four-field technique, most treatments are delivered with photon beams that are uniform in intensity across the field. On the other hand, in intensity-modulated radiation therapy (IMRT), the intensity profile of the photon beam is varied or modulated to meet the goals of the treatment plan. The modulation of the photon beam is often achieved with the use of a multileaf collimator. IMRT requires careful outlining of the treatment target and careful outlining of normal structures within any part of the irradiated volume. Typically five to nine beams are arranged around the target using strategically chosen gantry angles. The planning computer then optimizes the intensity profile of each beam so that the coverage of the target is maximized while the treatment of normal structures is minimized. This process is computer intensive and is termed "inverse planning". Compared to traditional techniques, IMRT is sometimes advantageous because a higher dose gradient between the target and the normal tissues can be accomplished. This sharper dose gradient could translate into better sparing of normal structures and safer delivery of higher doses to the target than conventional treatments.

Despite the potential benefits of IMRT, there are many concerns regarding its routine use for pelvic malignancies. The location of the target and normal structures can be affected by respiratory motion, weight loss during the course of treatment, daily variations in patient treatment setup and daily changes in bladder distention, bowel and rectal gas (see Fig. 12). Although the larger number of fields used in IMRT can reduce

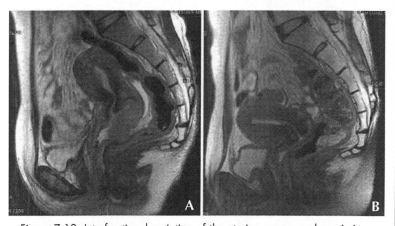

Figure 7.12. Interfractional variation of the uterine corpus and cervix in a 34 year-old patient with FIGO Stage IIb cervical cancer before beginning radiotherapy (left) and after 3060 cGy of whole pelvis irradiation. Reprinted from: Huh SJ et al. Interfractional variation in position of the uterus during radical radiotherapy for cervical cancer. Radiother Oncol 2004; 71(1):73-9 with permission from Elsevier.

the maximum dose normal structures receive, the price paid is that a larger volume of normal tissue is irradiated. This larger volume of irradiated tissue could translate into a higher risk of radiation-induced malignancies which can be an even greater concern in younger patients. IMRT plans rely on CT images to outline the targets and normal structures, which are notoriously weak in showing limited parametrial extension, superficial vaginal extension, etc. To circumvent this limitation, MRI or PET images can be fused with the CT images, but the experience with image fusion is limited. Since treatment with IMRT takes more time to deliver than with conventional techniques, there are radiobiologic concerns that the tumor might repair more of the radiation damage during these protracted treatment times. This effect might be enough to offset the higher total tumor doses possible with IMRT. Because of these and other concerns, IMRT is at present not routinely used for the treatment of gynecologic malignancies.

Brachytherapy

Brachytherapy may be used alone or in combination with teletherapy. This decision is usually made based on the location, size, shape and stage of disease. Brachytherapy can be delivered in a body cavity or in the tissue, which is referred to as intracavitary and interstitial brachytherapy, respectively.

The ICRU defines brachytherapy according to the dose rate as follows:
- Low Dose-Rate (LDR): 40 to 200 cGy/hr
- Medium Dose-Rate (MDR): 200 to 1200 cGy/hr
- High Dose-Rate (HDR): >1200 cGy/hr, although the usual dose rate employed in current HDR brachytherapy units is about 10,000-30,000 cGy/hr

Gynecologic brachytherapy is for the most part either LDR or HDR. LDR brachytherapy has the potential advantage over HDR brachytherapy of allowing more sublethal damage repair of the normal tissues resulting in fewer complications, especially in the late responding tissues. The theoretical disadvantage of LDR brachytherapy is more sublethal damage repair of the tumor and, due to the longer treatment time, more repopulation of the tumor. Both HDR and LDR brachytherapy when used appropriately have shown clinical success.

HDR delivers the treatment in minutes. As such, it is often done in an outpatient setting. As HDR sources are very small, the HDR applicators are also relatively small and can therefore be inserted without general anesthesia. Still, conscious sedation may be required. HDR brachytherapy has been successfully used in the treatment of the vaginal vault in endometrial cancer and in cervical cancer. HDR catheters can also be implanted in the tumor to deliver an interstitial brachytherapy boost. Interstitial brachytherapy in gynecologic oncology is usually facilitated by a Syed-Neblett template (see Fig. 7.13) which helps arrange and immobilize the interstitial needles.

LDR brachytherapy has a long clinical track record since it was developed decades ago. It can also be used to successfully treat cervical cancer and the vaginal vault in endometrial cancer. The Fletcher-Suit LDR technique used to treat cervical cancer is widely used and will be discussed in the next section. Its size usually requires placement under general anesthesia. As treatment time can take a few days, patients require hospitalization and strict bedrest. Since the patient is radioactive during this time, shielded rooms are required in the hospital. Nursing and visiting hours are carefully controlled. Radioactive LDR seeds may be implanted directly into tumors as well.

As the vaginal canal is cylindrical, a device of this shape is commercially available for this application in various diameters. Aptly termed a vaginal cylinder, it is produced for both HDR and LDR brachytherapy. Once loaded with radiation sources, treatment is

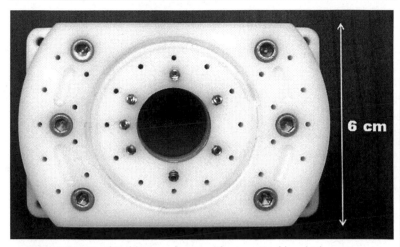

Figure 7.13. Syed-Neblett template used for interstitial brachytherapy.

delivered to the vaginal wall and vault. The nearby bladder and rectum can be relatively spared by this technique.

Fletcher-Suit LDR Brachytherapy Technique

The Fletcher-Suit LDR (low dose rate) brachytherapy technique is commonly used in patients diagnosed with cervical cancer to deliver a high dose boost to the uterine cervix and parametria. It may be used alone in very early disease. When used as a boost, a four-field photon teletherapy technique first treats the pelvis to a dose of 4000 to 5000 cGy. This is typically followed by two Fletcher-Suit brachytherapy treatments. The first Fletcher-Suit treatment is delivered after the completion of pelvic teletherapy and the second is delivered 1 or 2 weeks later, so that the entire treatment course is completed within 8 weeks.

The Fletcher-Suit applicator consists of a central tube, called the tandem, and two lateral ovoids (see Figs. 7.14 and 7.15). A small part called the flange is subsequently attached to the tandem during the implant partly to prevent the tandem from going too far and perforating the uterus. The ovoids are separated from each other by spacers or caps. All the components are held together with special fittings. The radionuclide most commonly used for low dose rate treatments is ^{137}Cs. In the average implant, the tandem is typically loaded with three sources: one at the tip with a strength of 15 mg-Ra eq, followed by two sources of 10 mg-Ra eq. Each ovoid is usually loaded with one source of 15 mg-Ra eq. The dose distribution will be "pear shaped", with the dose ballooning out at the cervix due to the ovoid dose contribution.

The insertion of the Fletcher-Suit applicator is often considered a "minor" surgical procedure. The patient is usually taken to the operating room, put under general anesthesia, placed in the dorsal lithotomy position, and prepped and draped in a sterile fashion. A Foley catheter is placed into the bladder and the balloon is filled with radiopaque contrast material (which can be diluted to avoid obscuring the radiopaque cervical markers) and snugged into the bladder trigone.

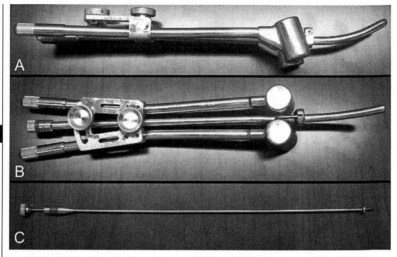

Figure 7.14. A) Side view of the assembled Fletcher-Suit applicator. B) Top view. C) Stylet used to insert radiopaque markers in the cervix using a plunger mechanism. Note the metal disk near the tip to prevent the device from being inserted too deep into the cervix.

Radiopaque markers are placed in the cervix for its radiologic identification. The uterus is sounded to measure the depth of the uterus from the cervical os. The flange is then attached at this distance from the tip of the tandem up to 8 cm. After the endocervical canal is dilated, the tandem is inserted in the cervical os until the flange is in contact with the exocervix. Special care should be taken while sounding the uterus and inserting the tandem to avoid perforation of the uterus. Ultrasound guidance can be very helpful in cases where the uterine canal is difficult to find or perforation is suspected. The ovoids are inserted in the vaginal fornices using the largest caps that can be reasonably accommodated to increase the separation from the sources to the vaginal mucosa. Radiopaque gauze is then packed posteriorly and anteriorly to the tandem and ovoids to move away the bladder and rectum, and to keep the applicator from shifting. Proper packing is critical to minimize the bladder and rectal dose. Anteroposterior and lateral fluoroscopic or X-ray views are obtained to verify the applicator position and for treatment planning purposes. Figures 7.16 and 7.17 show an anteroposterior and a lateral view, respectively, of an actual Fletcher-Suit implant. In an ideal implant:
1. The tandem bisects the ovoids on the lateral view
2. The tandem falls midway between the ovoids and parallel to the patient's sagittal axis on the anteroposterior view
3. The flange is at the level of the exocervix as defined by the cervical markers

After the position of the applicator is deemed acceptable, the treatment planning process starts based on the AP and lateral X-ray images obtained for treatment planning purposes. Treatment planning depends on the dose to point A, the bladder point and the rectal point; these are the recommended definitions by the American Brachytherapy Society (ABS):

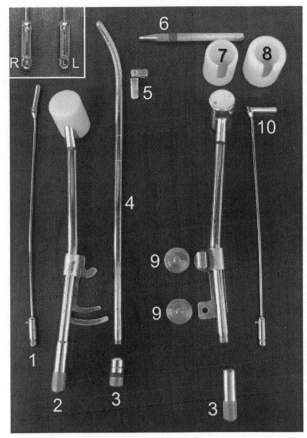

Figure 7.15. Fletcher-Suit applicator. 1. Inserter for colpostat. The top left picture inset shows that the inserters are labeled right and left and usually will only fit their corresponding colpostats to avoid transposing the sources. 2. Assembled colpostat with plastic jacket to increase the separation from the sources to the vaginal mucosa. 3. Metal caps to hold the colpostat's inserter or the tandem's radioactive sources in place. 4. Tandem. The tandem is hollow to typically house three radioactive sources starting at the tip. The tip is curved to conform to the curvature of the uterus. Note the horizontal groves every 1 cm which serve as a ruler when attaching the flange to correspond to the length of the uterus. 5. Flange with "keel" to help stabilize the tandem using radiopaque packing gauze. 6. Small Allen wrench used to fix in place the flange on the tandem. 7. Plastic jacket or spacer to increase the colpostat size. 8. Plastic jacket or spacer to increase the colpostat size to large. 9. Thumb screws used to hold the colpostats and tandem together. 10. Inserter's "bucket" for holding the radioactive source. Although the inserter is initially introduced straight into the colpostats as in 1, the resting configuration of the "bucket" inside the colpostat is as shown in 10 thanks to a hinge.

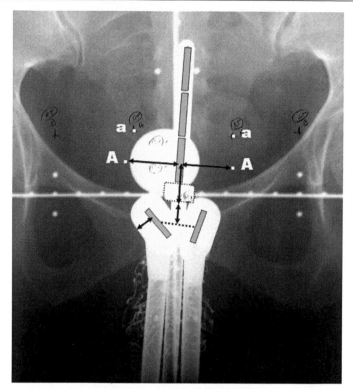

Figure 7.16. Anteroposterior view of a Fletcher-Suit implant showing all five ^{137}Cs sources as gray rectangles. Three sources are in the tandem and one source in each of the colpostats. The small dotted rectangle outlines the flange. The location of point A as suggested by the American Brachytherapy Society is labeled "A". An older, arbitrary, but still commonly used definition for point A is 2 cm above the external cervical os and 2 cm lateral to the midline and is labeled "a".

1. Point A: a line is drawn connecting the middle of each ovoid source on the AP view. From the intersection of this line with the tandem, move superiorly along the tandem 2 cm plus the radius of the ovoids and then 2 cm perpendicular to the tandem laterally. Historically, Point A was originally defined as being 2 cm above the mucus membrane of the lateral vaginal fornix and 2 cm lateral to the center of the uterine canal and was thought to lie in the paracervical area where the uterine artery crosses the ureter. A subsequent, arbitrary, but still commonly used definition for point A is 2 cm above the external cervical os and 2 cm lateral to the midline.

2. Bladder point: defined at the maximum dose on the surface of the Foley's balloon.

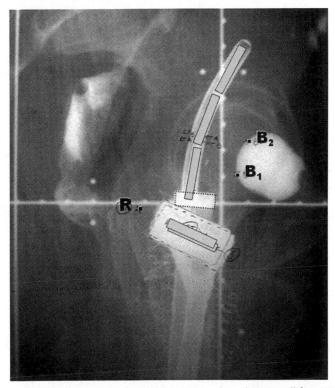

Figure 7.17. Lateral view of a Fletcher-Suit implant showing all five ^{137}Cs sources as gray rectangles. Three sources are in the tandem and one source in each of the colpostats. The small dotted rectangle outlines the flange. The rectal point is labeled "R". Note the radiopaque gauze around the tandem and colpostats in the inferior half of the radiograph. Two bladder points were obtained and labeled "B_1" and "B_2". Note that both bladder points touch the periphery of radiopaque Foley catheter's balloon.

3. Rectal point: defined at the midpoint of the ovoid sources (or at the lower end of the most proximal tandem source) on the AP view and 5 mm behind the posterior vaginal wall on a line drawn between the middle of the ovoid sources on the lateral view.

The prescribed total dose to point A, which includes the dose delivered by pelvic teletherapy, is usually around 8500 cGy. The total dose to the bladder point should be kept below 8000 cGy and to the rectal point below 6500-7000 cGy. Each LDR insertion typically delivers 2000 to 2500 cGy to point A at 50 to 65 cGy per hour. The ^{137}Cs source arrangement and strength are carefully chosen to achieve the goals of the implant. The time necessary to deliver the treatment is calculated by dividing the prescribed dose by the average dose rate obtained from the right and left Point A. Once the treatment plan is approved, the radiation oncologist loads the applicator with the appropriate sources, radiation precautions are initiated and treatment starts.

Sample LDR Fletcher-Suit Orders

- Preop orders:
    IV D5LR@100 mL/hr
    Tap water enema prior to procedure
    Betadine vaginal douche prior to procedure (if not iodine allergic)
    Morphine 2-3 mg IV q6h PRN
    Reglan (metoclopramide HCl) 10 mg IVP
    Zofran (ondansetron) 4 mg IVP
- Postop orders :
    Admit: Dr. _____ Gyn/Onc or Rad/Onc
    Diagnosis: Stage _____ carcinoma S/P tandem and ovoid insertion
    Condition:
    Vitals: q shift
    Activity: Bedrest
    Nursing: Radiation precautions and bedside shield
    Foley to gravity
    I/O's
    Sequential compression device (SCD)/T.E.D. stockings
    Incentive spirometry 10x q1h when awake
    Diet: low residual diet or clear liquids
    IVF: D5NS @ 25 mL/hr
Medications:
    Lomotil (diphenoxylate HCL—atropine sulfate) 2.5 mg PO qid
    Tylenol #3 (codeine 30 mg/acetaminophen 300 mg) 1-2 tabs PO q4h PRN pain
    Lovenox (enoxaparin sodium) 30 mg SQ qd
    If indicated, Bactrim (trimethoprim—sulfamethoxazole) DS 1 tab PO qHS for
        indwelling Foley catheter
- After applicator and radioactive sources are removed:
    D/C Foley
    D/C SCD/T.E.D. stockings
    Urine analysis
    Ambulate
    Optional, betadine vaginal douche, or a 1:16 solution of 3% hydrogen peroxide:
        water douche
    Regular diet
    Heplock IV when tolerating PO

Radiation Safety

Because the biologic effects of radiation depend not only on dose and dose rate, but also on the type of radiation, the dosimetric quantity relevant to radiation protection is the dose equivalent. The dose equivalent is the product of the absorbed dose and the quality factor for the radiation. The types of radiation used in gynecologic oncology are X-rays, gamma rays and electrons, all of which have a quality factor of 1. In contrast, neutrons can have a quality factor ranging approximately from 5 to 20 depending on their energy. For example, 1 Gy of neutrons can be 5 to 20 times biologically more effective than 1 Gy of X-rays. The unit of dose equivalent is the rem and its SI (International System of Units) counterpart the sievert (Sv):

(absorbed dose in rads) · (quality factor) = 1 rem = 0.01 Sv

(absorbed dose in Gy) · (quality factor) = 1 Sv = 100 rem = 1 joule/kilogram

Table 7.1. Annual effective dose equivalent limits in NRCP report no. 91

Type	Dose Equivalent (mSv)	Dose Equivalent (rem)
Radiation Workers:	50	5
• Lens of eye	150	15
• Other organs	500	50
General Public:		
• Infrequent exposure	5	0.5
• Frequent exposure	1	0.1
Students under age 18	1	0.1
Embtyo-Fetus:		
• Total	5	0.5
• After pregnancy is declared	0.5 per month	0.05 per month

In the United States, the National Council on Radiation Protection and Measurements (NCRP) has functioned as the primary standard setting body on radiation protection. Radiation exposure limits are based on the principle of ALARA which stands for As Low As Reasonable Achievable, taking into account the current state of technology and economics of improvement in relation to public health safety.

Table 7.1. shows the NRCP Report no. 91 recommendations on limits for exposure to ionizing radiation.

The amount of radiation exposure depends on three basic concepts: shielding, time and distance. A good gynecologic mnemonic is STD. That is, the shorter the time of exposure, the greater the distance from the radiation source and the greater the shielding around it, the smaller the exposure. Distance deserves further explanation because if the distance from a source is doubled, the exposure will be only one fourth as much. This is due to the inverse square law which states that the intensity of radiation is inversely proportional to the square of the distance from the source:

$$\text{Intensity} = \frac{1}{d^2}$$

Practical Example in Radiation Safety

Assuming a 160 lb female with a Fletcher-Suit LDR loading totaling 65 mg Ra-eq of ^{137}Cs, typical readings detected in air with an ion chamber are shown in Table 7.2.

Occupational workers are provided with radiation monitoring badges to keep record of their exposure to radiation. TV monitors and voice monitors are recommended for patient safety and minimize the personnel's exposure to radiation.

Acute and Late Effects of Radiotherapy

The response of tissues and organs to radiation depends on the radiosensitivity and turnover rate of the cell population. The time interval between irradiation and the expression of side effects depends on the life-span of the mature functional cells and the time stem cells need to reach maturity. Because of this, there are both acute and late effects of radiotherapy. Moreover, if the stem cell population is depleted below the level needed for tissue regeneration, an acute reaction might evolve into a chronic injury. This is called a consequential late effect because it is the result of a severe early effect. Both acute and late toxicities can be magnified by chemotherapy and surgery.

Table 7.2. Approximate exposure readings for a 160 lb female with a Fletcher-Suit implant of 65 mg-Ra eq of ^{137}Cs

Control Point*	Typical Exposure Reading		Occupational Time Limit†	Public Time Limit
	rem/hr	mSv/hr		
Bedside, without shield	0.1	1	12 min/dy	Not allowed
1 meter away from bedside	0.01	0.1	2 hr/dy	Not allowed
Behind shield	0.0002	0.002	10 hr/dy	Not allowed
Doorway or visitor's chair‡	Variable		Variable	Variable, allowed if older than 18 years. No pregnant women.

*See Figure 7.18 which illustrates the control points.
†The annual occupational exposure limit from Table 1 is 50 mSv/yr. Assuming that the patient's nurse works with radioactive patients 50 weeks a year and 5 days a week, the daily exposure limit is:

$$\left(\frac{50 \text{ mSv}}{\text{yr}}\right)\left(\frac{1 \text{ yr}}{50 \text{ wk}}\right)\left(\frac{1 \text{ wk}}{5 \text{ dy}}\right) = 0.2 \text{ mSv/dy}$$

At the bedside, for example, the daily maximum time allowed would be:

$$\frac{\text{daily dose limit}}{\text{exposure reading}} = \frac{\left(\dfrac{0.2 \text{ mSv}}{\text{dy}}\right)}{\left(\dfrac{1 \text{ mSv}}{\text{hr}}\right)\left(\dfrac{1 \text{ hr}}{60 \text{ min}}\right)} = 12 \text{ min/dy}$$

The time limits shown are for illustration purposes and in practice are determined by the radiation safety officer.
‡There is much more variability in the readings at the doorway or visitor's chair since the room dimensions can vary significantly from place to place. Ideally this point is at least 2 meters away form the patient.

The effects of radiation therapy can be worse in patients with genetic radiosensitive disorders like ataxia-telangiectasia and Nijmegen breakage syndrome, or autoimmune disorders like inflammatory bowel disease and collagen vascular disease. The use of radiotherapy in these patients is controversial and treatment should be assessed individually. In patients with cervical cancer who receive pelvic radiotherapy there is a strong correlation between smoking and major late complications, especially of the small bowel.

Radiation side effects are generally predictable and limited to the treatment fields. To predict a radiation side effect, consider the path of the radiation beam.

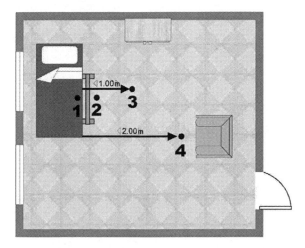

Figure 7.18. Patient room showing four radiation control points: 1: Bedside refers to assisting the patient when circumstances requires personnel to work over the patient while unable to utilize the bed shield. This should only be done in emergency situations. 2: Behind shield care includes routine monitoring, IV maintenance and basic hygiene. Bed sheets should not be changed unless absolutely necessary to avoid inadvertently throwing away a seed or part of the applicator that might have fallen off. In addition, bed linen changing increases unnecessarily the personnel radiation exposure. A pad placed under the patient until the end of treatment is sometimes acceptable. 3: One meter away from bedside is a reasonable minimum distance for healthcare personnel to have discussions with the patient. 4: Doorway or visitor's chair is ideally at least 2 meters away from the patient's bedside. This is usually far enough to significantly reduce the public's exposure to radiation. The door to the public hallway can be left open to help the patient from feeling totally isolated, but a roped sign should be placed across the doorway to prevent inadvertent entering by the public. In general, keeping the door closed will not reduce the exposure levels in the public hallway outside the room.

Acute effects during pelvic radiation therapy include:
1. Skin: desquamation of the skin can manifest as a flaky peeling of the epidermis termed dry desquamation or moist desquamation (see Fig. 7.19). Moisturizers can provide rapid relief of dry desquamation and should not contain chemical irritants. Dry desquamation can progress to moist desquamation when there is total loss of the epidermis. Moist desquamation manifests as serous weeping of the skin and can be painful and associated with an increased risk of infection. Moist desquamation occurs more frequently in areas like the inguinal area because the skin folds act as a bolus and eliminate the photons' skin sparing effect. Silvadene cream or Gentian Violet can be used for moist desquamation. Skin creams, ointments, or moisturizers should not be used before irradiation because the film created can also act as a bolus and reduce the photon skin sparing effect.

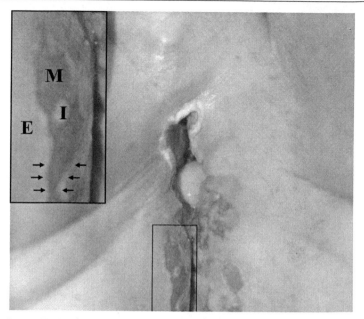

Figure 7.19. This patient with vulvar cancer illustrates some of the acute skin reactions from radiotherapy. The area in the solid rectangle has been magnified in the left picture inset. The "M" highlights one of the many areas of moist desquamation. The "I" is inside an island of new epidermis. The "E" overlies a region of skin erythema. The arrows illustrate how epidermal cells reproduce and move to repair areas of moist desquamation.

2. Gastrointestinal: diarrhea may start by the second week of treatment and is generally increased when chemotherapy is given concurrently. Diarrhea is usually controlled with loperamide or Lomotil (diphenoxylate HCL—atropine sulfate) and can be reduced with dietary modifications and measures to reduce the volume of bowel irradiated.

3. Genitourinary: radiation of the bladder can cause cystitis: urgency, frequency, dysuria and nocturia. Radiation cystitis may start during the third to fifth week of pelvic radiotherapy. Bladder analgesics (for example, phenazopyridine) and/ or antispasmodics (for example, tolterodine) can provide symptomatic relief of radiation cystitis. Vaginitis due to radiation can be treated daily with a very dilute hydrogen peroxide and water mixture or moisturizers. Radiation vaginitis can be worsened by other infectious causes of vaginitis like candidiasis and should be treated.

4. Nausea frequently occurs in patients who receive concurrent pelvic and para-aortic radiotherapy, and if necessary a 5HT3 receptor antagonist or dopamine receptor antagonist could be given prophylactically before each treatment. Suggested doses are ondansetron 8 mg bid or prochlorperazine 10 mg every 6 hours. A patient with uncontrolled radiation-induced diarrhea can become dehydrated and subsequently nauseous.

5. Fatigue due to radiation is poorly understood but rest can provide symptomatic relief. Other common causes of fatigue experienced by cancer patients should be ruled out, including: anemia, dehydration and/or an electrolyte imbalance, depression, pain, sleep disturbances including sleep apnea and insomnia, poor nutrition, physical deconditioning and the presence of other medical comorbidities.

Late effects are much more sensitive to the size of the daily dose of radiotherapy than acute effects. Late effects from pelvic radiation often present months to years after radiotherapy and include proctitis, enteritis, hemorrhagic cystitis, lymphedema, fistula formation, vaginal stenosis and femoral head fracture. Vaginal stenosis may be prevented with the use of a vaginal dilator for a few minutes per day for various months after the completion of treatment. Fistula formation requires immediate surgical consultation. Femoral head fracture should be evaluated by orthopedic surgery.

Clinical Situations Where Radiotherapy Is Used Emergently

Patients with vaginal bleeding can be effectively palliated with teletherapy or brachytherapy. Short or split-courses of photon teletherapy using high-doses per fraction can be effective in palliating pelvic pain from metastatic or recurrent malignancies.

Acknowledgements

The authors would like to thank Karen Fish, C.M.D., Dept. of Radiation Oncology at the Brody School of Medicine in Greenville, NC, for her invaluable contributions to the brachytherapy and radiation safety sections and assistance with the computer generated images; Gerald Connock, Radiation Safety Officer at SUNY Upstate Medical University, for revising the radiation safety section; Dr. Silvia Zunino, Instituto de Radioterapia-Fundación Marie Curie, Córdoba, Argentina; and Dr. Won Park, Department of Radiation Oncology, Samsung Medical Center, Seoul, South Korea for sharing their invaluable clinical images.

Suggested Reading

1. Steel GG. Basic Clinical Radiobiology. 3rd edition. London: Arnold, 2002.
2. Kal HB, Struikmans H. Radiotherapy during pregnancy: fact and fiction. Lancet Oncology 2005; 6:328-33.
3. Hoskins WJ, Perez CA, Young RC et al. Principles and Practice of Gynecologic Oncology. 4th edition. Philadelphia: Lippincott Williams and Wilkins, 2005.
4. International Commission on Radiation Units and Measurements. ICRU Report 62, Prescribing, Recording and Reporting Photon Beam Therapy (Supplement to ICRU Report 50), 1999.
5. Randall ME, Ibbott GE. Intensity modulated radiation therapy for gynecologic cancers: Pitfalls, hazards and cautions to be considered. Seminars in Radiation Oncology, 2006.
6. Nag S, Chao C, Erickson B et al. American Brachytherapy Society. The American Brachytherapy Society recommendations for low-dose-rate brachytherapy for carcinoma of the cervix. Int J Radiation Oncology Biol Phys 2002; 52(1):33-48. Erratum in: Int J Radiation Oncology Biol Phys 2002; 52(4):1157.
7. Nag S, Erickson B, Thomadsen B et al. The American Brachytherapy Society recommendations for high-dose-rate brachytherapy for carcinoma of the cervix. Int J Radiation Oncology Biol Phys 2000; 48(1):201-11.
8. Nag S, Erickson B, Parikh S et al. The American Brachytherapy Society recommendations for high-dose-rate brachytherapy for carcinoma of the endometrium. Int J Radiation Oncology Biol Phys 2000; 48(3):779-90.

CHAPTER 8

Cervical Intra-Epithelial Neoplasia

T. Michael Numnum and Warner Huh

Introduction

Pre-invasive disease of the female genital tract encompasses the cervix, vagina and vulva. The primary goal of screening and treatment of these disease processes is to prevent the progression of pre-invasive disease to invasive carcinoma. This chapter focuses on the etiology, risk factors, screening strategies and treatment modalities for pre-invasive disease of the genital tract, with primary emphasis on pre-invasive cervical disease.

Cervical Intra-Epithelial Neoplasia (CIN)

Cervical cancer accounts for approximately 4,000 deaths per year in the United States. In developing countries, it is the second most common cancer among women.[1] Screening and treating pre-invasive disease of the cervix has significantly reduced the incidence of cervical cancer, and this is supported by the high incidence of cervical cancer in regions of the world void of screening programs. The primary goal of cervical screening is to prevent cervical cancer through the detection and treatment of cervical intra-epithelial neoplasia (CIN). CIN refers to a pre-invasive precursor of cervical cancer. It is believed to be a slowly evolving process, progressing from normal cervical epithelium to mild dysplasia (CIN 1) to moderate dysplasia (CIN 2) to carcinoma in situ (CIN 3). These designations refer to the degree of intra-epithelial neoplasia in relation to the basement membrane of the cervical epithelium. For example, CIN 1 refers to intra-epithelial neoplasia present in the basal third of the epithelium, whereas CIN 3 encompasses greater than two-thirds of the total thickness of the cervical epithelium. (Fig. 8.1) The likelihood of CIN progressing to invasive carcinoma is considered low, especially in CIN 1. In fact, many cases of CIN 1 and even CIN 2 will resolve spontaneously.[2,3]

There are many proposed risk factors for the development of CIN. Human papillomavirus infection is believed to be the strongest risk factor for the subsequent development of CIN, especially certain "high risk" types (i.e., HPV Types 16 and 18).[4-6] In addition, cigarette smoking is believed to be strongly linked to the development of CIN. Other risk factors for CIN include immunosuppression, states of immunodeficiency, sexual activity at an early age, a history of sexually transmitted diseases, sexual promiscuity, or engaging in sexual activity with promiscuous men.[7-10]

Screening for Cervical Cancer

The steady decline in the incidence and mortality rates due to cervical cancer in developed countries is believed to be due to the introduction of the Papanicolaou smear after World War II. Although by no means the perfect test, it is widely recognized as one of the most cost efficient cancer screening tests available. In fact, most cases of invasive cervical cancer occur in women who have either never had a Pap smear or have not received 1 in

---

Gynecologic Oncology, edited by Paola Gehrig and Angeles Secord.
©2009 Landes Bioscience.

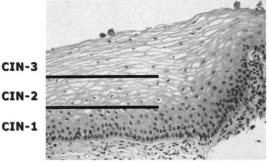

CIN-1 Dysplastic cells occupy lower one-third of mucosa.
CIN-2 Dysplastic cells occupy lower two-thirds of mucosa.
CIN-3 Dysplastic cells extend into upper third of mucosa.

Figure. 8.1. Grading scheme for diysplasia of the mucosa.

5 years antecedent to their diagnosis.[11] Currently, all Pap smears are classified using the Bethesda System, which was introduced in 1988 to standardize terminology for reporting Pap smears. The most recent update to the Bethesda System was developed in 2001 (Table 8.1). The use of the Bethesda System allows clear communication of cytologic findings to clinicians so that management can be based on unambiguous wording.[12]

Although the Pap smear is widely accepted as the standard of care in cervical screening, there is considerable controversy as to when and how often screening should be used. Also, the introduction of HPV DNA testing has led to controversy as to the role it should play in routine cervical screening. Guidelines for cervical screening were published by the American Cancer Society in 2002; there have also been recent publications by the American College of Obstetricians and Gynecologists (ACOG) and the United States Preventive Services Task Force (USPSTF). Screening should begin 3 years after the initiation of sexual intercourse or at the age of 21. Cervical screening should be performed annually if a conventional Pap smear is used or every 2 years if using liquid based cytology. In women who have had three consecutive normal Pap smears and are ≥30 years old, the screening interval can be increased to every 2 or 3 years. Women with a history of CIN 2 or CIN 3, those who are immunodeficient and those with a history of DES exposure should undergo annual screening regardless of past Pap smear results. The use of HPV DNA testing in primary screening has not gained widespread acceptance yet. However, it may be used in women ≥ age 30 as an adjunct to cytology and, if negative, the screening interval may be extended to every 3 years. Currently, HPV DNA testing has its greatest role as a triage test in women with an ASC-US (atypical squamous cells of undetermined significance) Pap smear. There is no consensus among experts as to the upper age limit of screening in women without a history of CIN. In general, women aged 65-70 with three consecutive normal Pap smears and no abnormal Pap smear within 10 years may elect to terminate routine screening. Women with a history of a total hysterectomy may elect to terminate routine screening provided they have no history of CIN 2 or 3 or cervical cancer.[13]

Management of the Abnormal Pap Smear

In the United States, approximately 5-7% of cervical cytology will be reported as abnormal. Since the Pap smear is only a screening test, it cannot be solely relied

---

Table 8.1. The Bethesda system, 2001

Specimen Type: Indicate conventional smear versus liquid based versus other

Specimen adequacy
- Satisfactory for evaluation (describe presence or absence of endocervical/ transformation zone component and any other quality indicators, e.g., partially obscuring blood, inflammation, etc.)
- Unsatisfactory for evaluation ...
  - Specimen rejected/not processed
  - Specimen processed and examined, but unsatisfactory for evaluation of epithelial abnormality because of

General Categorization (optional)
- Negative for intra-epithelial lesion or malignancy
- Epithelial cell abnormality: See Interpretation/Result (specify 'squamous' or 'glandular' as appropriate)
- Other

Interpretation/Result
Negative for Intra-Epithelial Lesion or Malignancy
  Organisms:
  - *Trichomonas vaginalis*
  - Fungal organisms
  - Shift in flora suggestive of bacterial vaginosis
  - Bacteria morphologically consistent with Actinomyces spp.
  - Cellular changes consistent with Herpes simplex virus

  Other non neoplastic findings
  - Reactive cellular changes associated with
    - inflammation
    - radiation
    - intrauterine contraceptive device (IUD)
  - Glandular cells status post hysterectomy
  - Atrophy

  Other
  - Endometrial cells

continued on next page

---

upon in the management of CIN. Guidelines published by the American Society for Colposcopy and Cervical Pathology (ASCCP) assist clinicians in determining the appropriate strategy in managing an abnormal Pap smear.[14] (see Fig. 8.1) An abnormal Pap smear always requires further evaluation, which may include repeat cytology, HPV DNA testing, colposcopy with or without cervical biopsy, endocervical sampling, endometrial biopsy, or an excisional procedure.

In the United States, more than 3 million women each year will be diagnosed with a low grade abnormality on Pap smear (ASC-US or LSIL) The management of the ASC-US and LSIL is best defined by the results of the ALTS trial (ASCUS/LSIL Triage Study). In this trial, women with a diagnosis of ASCUS or LSIL were randomized into three management schemes: 1. repeat cervical cytology, 2. reflex HPV DNA testing and 3. immediate colposcopy. The purpose of the study was to determine the sensitivity and specificity of each of the three management strategies in detecting CIN 3. The results

Table 8.1. Continued

Epithelial Cell Abnormalities
Squamous Cell
- Atypical squamous cells
  - of undetermined significance (ASC-US)
  - cannot exclude HSIL (ASC-H)
- Low grade squamous intra-epithelial lesion (LSIL) encompassing: HPV/mild dysplasia/CIN 1
- High grade squamous intra-epithelial lesion (HSIL) encompassing: moderate and severe dysplasia, (CIN 2/3/CIS)
- Squamous cell carcinoma

Glandular Cell
- Atypical
  - endocervical cells
  - endometrial cells
  - glandular cells
- Atypical
  - endocervical cells, favor neoplastic
  - glandular cells, favor neoplastic
- Endocervical adenocarcinoma in situ
- Adenocarcinoma
  - endocervical
  - endometrial
  - extrauterine
  - not otherwise specified (NOS)

8

indicated that the use of reflex HPV testing in patients with an ASCUS Pap smear is a reasonable option and may reduce the number of colposcopies performed. Also, because the prevalence of a positive HPV test is at least 80% in patients with an LSIL Pap smear, the use of HPV testing in this population would be ineffective as a triage tool. Therefore, it is recommended that patients with LSIL cytology undergo immediate colposcopy. Although colposcopy was considered the reference arm in the ALTS trial, the ultimate sensitivity for the detection of CIN 3 over 2 years was only 56%; therefore, the idea that colposcopy is a gold standard test is now being called into question.[14-16]

Because women with a diagnosis of HSIL (high-grade squamous intra-epithelial lesion) on Pap smear have a >70% risk of CIN 2/3 and a 1-2% risk of invasive carcinoma, all should undergo colposcopy and cervical biopsy. The use of endocervical curettage should be considered also. If there is no pathology present on colposcopy/biopsy, strong consideration should be given to a diagnostic excisional procedure. Also, women with an ASC-H (atypical squamous cells favor high-grade dysplasia) Pap smear may be at higher risk for CIN 2 or 3 than ASC-US or LSIL and subsequently these patients should also undergo colposcopy. A finding of atypical glandular cells (AGC) could represent underlying endocervical or endometrial pathology. These patients should undergo colposcopy, endocervical curettage and endometrial biopsy if they are >35 years old or if they are at risk for endometrial hyperplasia (obesity, history of abnormal uterine bleeding). Any Pap smear reported as invasive carcinoma or adenocarcinoma should undergo colposcopy (if there is not an obvious gross lesion) and biopsy to obtain a definitive diagnosis. Pregnant women with abnormal cytology should be managed in the same manner as nonpregnant patients. However, ECC should be avoided and diagnostic excisional procedures should be reserved for patients with suspected invasive cancer.[14]

Colposcopy

The use of colposcopy to evaluate abnormal cervical cytology has been considered the "gold standard" for many years. However, colposcopy is still a subjective measure and should always be taken in context within the given clinical situation. Colposcopy can be easily performed in the office setting without the use of local anesthesia. The colposcope (Fig. 8.2) should, in general, be set at a magnification of 7.5-15X. 3% acetic acid should be applied to the cervix once the transformation zone has been fully visualized by the colposcopist. Since the transformation zone is the most likely area to contain CIN or microinvasive carcinoma, this area should be closely inspected. Abnormal areas of the transformation zone have many different appearances, but most commonly appear dull-white after application of acetic acid. The presence of small vessels in a mosaic or punctate pattern may also indicate the presence of CIN (Fig. 8.3). Any area in the transformation zone that has this appearance should be biopsied. Lugol's iodine solution may also be used at the time of colposcopy. Neoplastic epithelium does not contain glycogen and, therefore will be nonstaining when iodine is applied. If the transformation zone is not fully visible or if a lesion cannot be seen in its entirety, then the colposcopy is considered unsatisfactory. The use of endocervical curettage should be considered in the nonpregnant patient who has a lesion extending into the endocervical canal or in the patient with abnormal glandular cytology. Also, if an endometrial biopsy is indicated (for the evaluation of an AGC diagnosis), it can be performed at the time of colposcopy.[14]

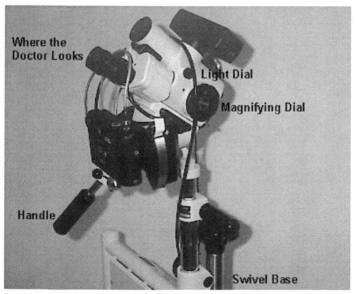

Figure. 8.2. The colposcope.

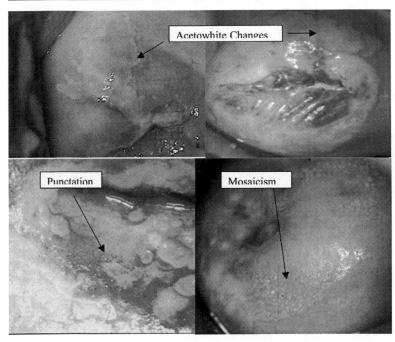

Figure. 8.3. Examples of colposcopic findings.

Treatment of Biopsy Proven CIN

Once the diagnosis of CIN is established by colposcopy, treatment decisions can be made based on the severity of CIN present. In 2001, the ASCCP established evidence based guidelines for biopsy proven CIN management and follow-up[17] (see Fig. 8.2). In general, CIN 1 lesions are managed conservatively, given that up to one-half will spontaneously regress. Close follow-up is indicated and consists of either cytology at 6 and 12 months or HPV DNA testing at 12 months. Colposcopy and cytology at 12 months is also considered an acceptable form of follow-up. If the result of follow-up testing is negative, then the patient may return to the previous screening method. If the severity of CIN worsens in the follow-up period, then that patient should undergo treatment. The management of persistent CIN 1 is controversial, given that the likelihood of progression to cancer is low. Follow-up and treatment are both viable options and the ultimate decision to treat persistent CIN 1 may ultimately come down to physician and patient preference. If the patient desires treatment, either excisional or ablative therapy is considered appropriate.[15-18]

In general, a diagnosis of CIN 2 or 3 requires some form of treatment.[17] The treatment of pre-invasive cervical disease generally falls into two categories: ablative therapy and excisional therapy. A recent meta-analysis of treatment modalities for CIN suggests that all therapies are largely equivalent in treating pre-invasive disease.[19] Ablative therapy, however, is contraindicated in the presence of endocervical disease or if prior colposcopy was unsatisfactory. Regardless of the modality employed, the entire transformation zone should be treated.[17]

Ablative therapies include cryosurgery, CO_2 laser ablation and electrofulgaration of the transformation zone. To perform an ablative procedure, there must be no discrepancy between cytologic and histologic diagnoses, no evidence of glandular/endocervical disease and colposcopy must be satisfactory, including the ability to visualize the entire extent of disease. Also, if there is any evidence of microinvasion on colposcopy, an excisional procedure must be performed.[17]

Cryosurgery is a cost-effective and inexpensive ablative procedure that can be performed as an outpatient. It involves the use of compressed nitrous oxide and a metal probe placed tissue to be destroyed. Cryosurgery is typically most effective for small lesions that are well-visualized. It should not be performed in the face of inadequate colposcopy or if the entire lesion cannot be visualized. The most common side effect is a watery discharge that may persist for several weeks. Cure rates can exceed 90%, although women with larger lesions and moderate to severe CIN tend to recur at higher rates. Possible long-term complications of cryosurgery of the cervix include infertility and cervical stenosis, although this has not been confirmed by the literature.[19,20]

CO_2 laser ablation is another safe, cost-effective ablative modality. Tissue should be vaporized to a depth of 7-9 mm to ensure adequate destruction of the basal epithelium. Laser ablation can be performed on an outpatient basis. Patients may complain of cramping and watery discharge lasting up to 2 weeks. An obvious disadvantage of laser therapy, as with any ablative procedure, is the lack of tissue for a definitive diagnosis.[21,22]

Excisional therapies include cold knife conization, loop electrosurgical excisional procedure (LOOP/LEEP/LEETZ), laser conization and hysterectomy. Excisional procedures can be used as primary therapy for CIN, suspected microinvasion, endocervical disease, or for recurrence after an ablative procedure. They can also be used as a diagnostic procedure if there is a pathologic discrepancy between cytologic and histologic diagnosis. The most obvious advantage of an excisional therapy over ablative therapy is that it produces a specimen for pathologic analysis.

A cold knife conization is a technique by which a cone shaped portion of the cervix, including the transformation zone, is removed with a scalpel. This is the oldest technique used to evaluate CIN. Its main advantage is that it produces a clean surgical specimen without any electrocautery artifact. Therefore, it is very advantageous in cases of suspected microinvasion, adenocarcinoma-in-situ, or in patients with extensive endocervical disease. However, it cannot be done in an office setting and carries higher postsurgical complications than other excisional therapies.

The loop electrosurgical excision procedure (LEEP) is the most commonly used therapy for the management of high grade CIN. The LEEP procedure uses a wire loop with a blended electrical current passing through it to excise the transformation zone of the cervix, the area at highest risk for dysplasia. This can be done in an office setting under local anesthesia. Complications are rare, but may include bleeding, infection and cervical stenosis. Because of the cautery artifact created by a LEEP, the incidence of positive margins is higher than when using cold knife conization.[22-24]

The use of hysterectomy is not considered a cost-effective option for the primary treatment of CIN. However, it may be considered a reasonable option in patients with recurrent CIN 2 or 3. In addition, hysterectomy may be a reasonable option for therapy in patients with other uterine pathology, such as menorrhagia unresponsive to nonsurgical therapy.[17]

The use of LEEP has been studied in the past as part of a "see and treat" protocol. Patients with high grade cervical cytology may be treated without a histologic diagnosis. This strategy has been employed in "high-risk" patient populations who are unreliable and noncompliant with follow-up. Colposcopy must be performed at the time of "see and treat" to exclude an obvious invasive carcinoma. Although patients must be counseled on the risk of overtreatment with this strategy, this may be a viable option in the patient at risk for noncompliance.[25]

Pregnant women with biopsy confirmed CIN should be managed conservatively until after delivery. CIN tends to regress at a high rate with delivery, sparing the patient the risks associated with therapy during pregnancy. However, pregnant women with a diagnosis of microinvasion should undergo a diagnostic conization even during pregnancy and further treatment planning should be based on the results of conization.[17]

The optimal mode of management of adolescents with biopsy proven CIN is unknown. However, it is generally agreed that adolescents with biopsy proven CIN 2 or less can be observed conservatively with colposcopy and cytology at 4 to 6 month intervals, provided that colposcopy is satisfactory, there is no endocervical disease and there is a reasonable expectation of compliance. CIN 3 should be managed with an ablative or excisional procedure.[17]

A vaccine directed against HPV is currently commercially available. Randomized clinical trials have demonstrated that the use of a vaccine directed against prevalent HPV types will not only reduce the incidence of HPV infection but also reduce the incidence of HPV related CIN.[26,27] In June 2006, the Food and Drug Administration (FDA) announced the approval of Gardasil® for the prevention cervical cancer, precancerous genital lesions and genital warts due to human papillomavirus (HPV) Types 6, 11, 16 and 18. The vaccine is approved for use in females 9-26 years of age and is administered in three separate injections given over a 6-month period.

Cure rates in patients treated for CIN are variable and tend to be higher in patients receiving excisional therapy. Patients with positive margins after excisional therapy tend to have higher recurrence rates. Observed cure rates for ablative and excisional therapies are similar and can be >80%. Recurrence rates are higher in patients with prior therapy, those with large lesions, the immunocompromised and patients over the age of 30.[19,22,23] Follow-up after treatment should consist of HPV DNA testing at 12 months or cytology at 6 and 12 months. A combination of cytology and colposcopy is also considered acceptable.[18]

Vulvar Intra-Epithelial Neoplasia (VIN)

Pre-invasive disease of the vulva is, like CIN, considered a precursor lesion of vulvar carcinoma. Like CIN, VIN is classified as mild (VIN 1), moderate (VIN 2), or severe (VIN 3). Its age distribution is considered bimodal, with peak incidences occurring in the third and seventh decades of life. Like CIN, there is a strong association between high risk HPV DNA infection and VIN.[28] Lichen sclerosis and other nonneoplastic epithelial disorders of the vulva are also associated with the development of VIN, as are cigarette smoking, HIV infection and immunosuppression. The potential for VIN to progress to invasive carcinoma is considered low; however, progression to carcinoma is higher in the older and immunosuppressed populations.[29-31] Up to one-half of VIN lesions are symptomatic, with pruritis being the most common symptom. However, because of the asymptomatic nature of VIN lesions, there is often a delay in diagnosis. VIN lesions may appear as red, white, or pigmented areas on the vulva. Any abnormal appearing lesion on the vulva suspicious for VIN should be biopsied without hesitation. A Keyes biopsy instrument under local anesthesia is the preferred method of biopsy because it allows determination of the depth of disease present (Fig. 8.4). Colposcopy may not only aid in establishing the

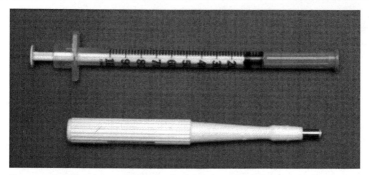

Figure 8.4. Tuberculin syringe (top) and Keyes punch biopsy (bottom).

diagnosis, but also help identify nonvisible pathology. Colposcopy can be performed using acetic acid or with the use of toludine blue, which will stain abnormal areas in the presence of acetic acid.

Once diagnosis is established, there are several modalities of therapy for the treatment of VIN. Therapy should be aimed at the prevention of cancer and any possibility of vulvar carcinoma should be ruled out before proceeding with definitive therapy. Because of the potential sexual and psychological morbidity associated with vulvar surgery, therapy should be as conservative as possible, especially in the younger patient. Localized lesions are best managed with the use of superficial excision with primary closure of the defect.[29] Care should be taken to restore the anatomy of the vulva to its original form if possible; the clitoral anatomy should also remain intact, especially in younger and sexual active patients. CO_2 laser surgery can be used for women with multifocal disease, thereby avoiding a large excisional procedure. Appropriate use of the CO_2 laser can eradicate VIN in a single treatment and allow appropriate healing without scarring. However, cure rates with the laser tend to be somewhat lower than excisional surgery; as many as 33% of patients will require more than one treatment to control disease.[32,33] Topical 5-fluorouracil (5-FU) has been used in the past, particularly in younger women with multifocal disease. It is, however, associated with significant painful inflammation, edema and superficial ulcerations.[34] Imiquimod is an immune response modifier that has potential benefit in the patient with VIN, especially if vulvar condylomas are present.[35] 5% Imiquimod can be applied topically three times a week for up to 16 weeks and can achieve up to 90% total response rate. Other options for therapy include cryotherapy, Trichloroacetic acid (TCA) and podophyllin. These are general reserved for low grade lesions (VIN-1) and vulvar condyloma. In general, VIN is associated with a high recurrence rate. Therefore, carcinoma should always be ruled out at the time of recurrence and patients should be counseled that new vulvar symptoms should be reported to their physicians.

Vaginal Intra-Epithelial Neoplasia (VAIN)

Vaginal intra-epithelial neoplasia (VAIN) occurs less frequently than VIN and CIN. However, its incidence has been steadily rising over the past several decades, in large part due to increased awareness and screening. The severity of VAIN is designated the same is CIN and VIN, with VAIN 1 considered mild dysplasia, VAIN 2 moderate dysplasia and VAIN 3 designated severe dysplasia or carcinoma-in-situ. The majority of women diagnosed with VAIN either have concurrent CIN or have a history of CIN.

This is the result of women with VAIN sharing common risk factors also associated with VIN and CIN such as HPV infection. Also, women that had a hysterectomy for CIN or early cervical carcinoma may have also had dysplastic extension into the vagina that wasn't recognized at the time of surgery or subsequently developed afterwards. Given that vaginal cancer is a relatively rare malignancy, the malignant potential of VAIN is considered relatively low.[36]

VAIN is usually asymptomatic and detected at the time of cytologic screening. Sometimes VAIN will present as an abnormal vaginal discharge or vaginal spotting. Colposcopy of the vagina followed by biopsy of any abnormality is required for diagnosis. Colposcopy should be performed in the same manner as that performed for the cervix. Lugol's solution may also be used at the time colposcopy with biopsies performed on a nonstaining area of the vagina.

Treating VAIN requires complete eradication of the abnormal vaginal epithelium. Surgical excision (partial vaginectomy) is the most common treatment modality employed. This can be difficult, however, in cases where large, multifocal disease is present. Also, the thin walls of the vagina and the close proximity of the bladder and rectum may preclude safe surgical excision. Therefore, the CO_2 laser is also a commonly used safe technique to treat VAIN.[36-39] Intravaginal 5-FU has also been used successfully to treat widespread VAIN. As with treatment of VIN, side effects of topical 5-FU are troublesome and may prompt discontinuation of therapy.[40]

Conclusion

Since the advent of the Pap smear, there have many advances in the diagnosis and management of pre-invasive disease of the female genital tract. The recent development of HPV directed vaccine therapy will possibly have a dramatic impact on the incidence of pre-invasive disease over the next several decades. Future research in this area will be directed towards targeting additional HPV types in a single vaccine. Access to vaccine therapy and other therapeutics is a major problem worldwide, and future resources will need to be utilized to provide therapy to the world's underserved areas, such as southeast Asia and sub-Saharan Africa. New detection strategies, such as optical detection systems, are currently being studied worldwide in an effort to improve the sensitivity and specificity of diagnosis of CIN that colposcopy cannot attain. Also, advancement is underway at the genomic level in an effort to determine who is at risk for progression of CIN to carcinoma, thereby targeting therapy for patients at greatest risk for the development of cervical cancer.

Suggested Reading

1. Saslow D et al. American cancer society guideline for the early detection of cervical neoplasia and cancer. CA Cancer J Clin 2002; 52(6):342-62.
2. Solomon D et al. The 2001 bethesda system: terminology for reporting results of cervical cytology. JAMA 2002; 287(16):2114-9.
3. Wright TC Jr et al. 2001 consensus guidelines for the management of women with cervical cytological abnormalities. JAMA 2002; 287(16):2120-9.
4. Results of a randomized trial on the management of cytology interpretations of atypical squamous cells of undetermined significance. Am J Obstet Gynecol 2003; 188(6):1383-92.
5. Wright TC Jr et al. 2001 consensus guidelines for the management of women with cervical intraepithelial neoplasia. Am J Obstet Gynecol 2003; 189(1):295-304.
6. Martin Hirsch PL, PE, Kitchener H. The cochrane database of systematic reviews. Surgery for cervical intraepithelial neoplasia. The Cochrane Library 2003; 3:1-40.

References

1. Parkin DM, Bray F, Ferlay J et al. Global cancer statistics, 2002. CA Cancer J CLin 2005; 55(2):74-108.

2. Holowaty P, Miller AB, Rohan T et al. Natural history of dysplasia of the uterine cervix. J Natl Cancer Inst 1999; 91(3):252-8.

3. Arends MJ, Buckley CH, Wells M. Aetiology, pathogenesis and pathology of cervical neoplasia. J Clin Pathol 1998; 51(2):96-103.

4. Khan MJ, Partridge EE, Wang SS et al. Socioeconomic status and the risk of cervical intraepithelial neoplasia grade 3 among oncogenic human papillomavirus DNA-positive women with equivocal or mildly abnormal cytology. Cancer 2005; 104(1):61-70.

5. Schiffman MH, Bauer HM, Hoover RN et al. Epidemiologic evidence showing that human papillomavirus infection causes most cervical intraepithelial neoplasia. J Natl Cancer Inst 1993; 85(12):958-64.

6. Ho GY, Kadish AS, Burk RD et al. HPV 16 and cigarette smoking as risk factors for high-grade cervical intra-epithelial neoplasia. Int J Cancer 1998; 78(3):281-5.

7. Clarke EA, Hatcher J, McKeown Eyssen GE et al. Cervical dysplasia: association with sexual behavior, smoking and oral contraceptive use? Am J Obstet Gynecol 1985; 151(5):612-6.

8. Herrero R, Brinton LA, Reeves WC et al. Sexual behavior, venereal diseases, hygiene practices and invasive cervical cancer in a high-risk population. Cancer 1990; 65(2):380-6.

9. Azocar J, Abad SM, Acosta H et al. Prevalence of cervical dysplasia and HPV infection according to sexual behavior. Int J Cancer 1990; 45(4):622-5.

10. La Vecchia C, Franceschi S, Decarli A et al. Cigarette smoking and the risk of cervical neoplasia. Am J Epidemiol 1986; 123(1):22-9.

11. Janerich DT, Hadjimichael O, Schwartz PE et al. The screening histories of women with invasive cervical cancer, connecticut. Am J Public Health 1995; 85(6):791-4.

12. Solomon D, Davey D, Kurman R et al. The 2001 bethesda system: terminology for reporting results of cervical cytology. JAMA 2002; 287(16):2114-9.

13. Saslow D, Runowicz CD, Solomon D et al. American cancer society guideline for the early detection of cervical neoplasia and cancer. CA Cancer J Clin 2002; 52(6):342-62.

14. Wright TC Jr, Cox JT, Massad LS et al. 2001 consensus guidelines for the management of women with cervical cytological abnormalities. JAMA 2002; 287(16):2120-9.

15. Results of a randomized trial on the management of cytology interpretations of atypical squamous cells of undetermined significance. Am J ObstEt Gynecol 2003; 188(6):1383-92.

16. A randomized trial on the management of low-grade squamous intraepithelial lesion cytology interpretations. Am J Obstet Gynecol 2003; 188(6):1393-400.

17. Wright TC Jr, Cox JT, Massad LS et al. 2001 consensus guidelines for the management of women with cervical intraepithelial neoplasia. Am J Obstet Gynecol 2003; 189(1):295-304.

18. Guido R, Schiffman M, Solomon D et al. Postcolposcopy management strategies for women referred with low-grade squamous intraepithelial lesions or human papillomavirus DNA-positive atypical squamous cells of undetermined significance: a two-year prospective study. Am J Obstet Gynecol 2003; 188(6):1401-5.

19. Martin Hirsch PL, Paraskevaidis E, Kitchener H. Surgery for cervical intraepithelial neoplasia. Cochrane Database of Systematic Reviews (Online) 2000(2):CD001318.

8

20. Creasman WT, Hinshaw WM, Clarke Pearson DL. Cryosurgery in the management of cervical intraepithelial neoplasia. Obstet Gynecol 1984; 63(2):145-9.
21. Burke L. The use of the carbon dioxide laser in the therapy of cervical intraepithelial neoplasia. Am J Obstet Gynecol 1982; 144(3):337-40.
22. Mitchell MF, Tortolero Luna G, Cook E et al. A randomized clinical trial of cryotherapy, laser vaporization and loop electrosurgical excision for treatment of squamous intraepithelial lesions of the cervix. Obstet Gynecol 1998; 92(5):737-44.
23. Montz FJ, Holschneider CH, Thompson LD. Large-loop excision of the transformation zone: effect on the pathologic interpretation of resection margins. Obstet Gynecol 1993; 81(6):976-82.
24. Prendiville W, Cullimore J, Norman S. Large loop excision of the transformation zone (LLETZ). A new method of management for women with cervical intraepithelial neoplasia. Br J Obstet Gynaecol 1989; 96(9):1054-60.
25. Numnum TM, Kirby TO, Leath CA et al. A prospective evaluation of "see and treat" in women with HSIL Pap smear results: is this an appropriate strategy? J Low Genit Tract Dis 2005; 9(1):2-6.
26. Koutsky LA, Ault KA, Wheeler CM et al. A controlled trial of a human papillomavirus type 16 vaccine. N Engl J Med 2002; 347(21):1645-51.
27. Harper DM, Franco EL, Wheeler C et al. Efficacy of a bivalent L1 virus-like particle vaccine in prevention of infection with human papillomavirus types 16 and 18 in young women: a randomised controlled trial. Lancet 2004; 364(9447):1757-65.
28. van Beurden M, ten Kate FW, Tjong AHSP et al. Human papillomavirus DNA in multicentric vulvar intraepithelial neoplasia. Int J Gynecol Pathol 1998; 17(1):12-6.
29. Thuis YN, Campion M, Fox H et al. Contemporary experience with the management of vulvar intraepithelial neoplasia. Int J Gynecol Cancer 2000; 10(3):223-7.
30. Jones RW, Rowan DM. Spontaneous regression of vulvar intraepithelial neoplasia 2-3. Obstet Gynecol 2000; 96(3):470-2.
31. Kaufman RH. Intraepithelial neoplasia of the vulva. Gynecol Oncol 1995; 56(1):8-21.
32. Reid R. Superficial laser vulvectomy. I. The efficacy of extended superficial ablation for refractory and very extensive condylomas. Am J Obstet Gynecol 1985; 151(8):1047-52.
33. Wright VC, Davies E. Laser surgery for vulvar intraepithelial neoplasia: principles and results. Am J Obstet Gynecol 1987; 156(2):374-8.
34. Krupp PJ, Bohm JW. 5-fluorouracil topical treatment of in situ vulvar cancer. A preliminary report. Obstet Gynecol 1978; 51(6):702-6.
35. van Seters M, Fons G, van Beurden M. Imiquimod in the treatment of multifocal vulvar intraepithelial neoplasia 2/3. Results of a pilot study. J Reprod Med 2002; 47(9):701-5.
36. Sillman FH, Fruchter RG, Chen YS et al. Vaginal intraepithelial neoplasia: risk factors for persistence, recurrence and invasion and its management. Am J Obstet Gynecol 1997; 176(1 Pt 1):93-9.
37. Yalcin OT, Rutherford TJ, Chambers SK et al. Vaginal intraepithelial neoplasia: treatment by carbon dioxide laser and risk factors for failure. Eur J Obstet Gynecol Reprod Biol 2003; 106(1):64-8.
38. Diakomanolis E, Rodolakis A, Boulgaris Z et al. Treatment of vaginal intraepithelial neoplasia with laser ablation and upper vaginectomy. Gyn Obstet Invest 2002; 54(1):17-20.
39. Diakomanolis E, Stefanidis K, Rodolakis A et al. Vaginal intraepithelial neoplasia: report of 102 cases. Eur J Gynaecol Oncol 2002; 23(5):457-9.
40. Sillman FH, Sedlis A, Boyce JG. A review of lower genital intraepithelial neoplasia and the use of topical 5-fluorouracil. Obstet Gynecol Surv 1985; 40(4):190-220.

8

CHAPTER 9

Cervical Cancer

Cecelia H. Boardman

Introduction

Cervical cancer is a highly preventable cancer that is steadily declining in incidence in the United States. However, it remains a significant problem worldwide. Cervical cancer is clinically staged and is treated either surgically or with a combination of radiation and chemotherapy. Recurrent cervical cancer is rarely curable.

Epidemiology

Cervical cancer is the third most common gynecologic cancer in the United States, with 9,700 new cases annually.[1] In the US, approximately one-third of women with cervical cancer will succumb to their disease. Since the advent of Pap smear screening, the incidence of cervical cancer has steadily declined. Mortality due to cervical cancer has declined by 70% in the United States. Unfortunately, in developing nations, where Pap smear screening is not as prevalent, cervical cancer is one of the leading causes of cancer mortality among women. Even in the United States, 50% of women with newly diagnosed cervical cancer have never had a Pap smear and an additional 10% have not had a Pap smear within the preceding 5 years. Interestingly, however, these women are not always uninsured or without access to health care. In a study of cervical cancer cases diagnosed in the membership of a large HMO in California, 70% of women with newly diagnosed cervical cancer had been seen by a health care provider at least once in the 3 years prior to diagnosis and 42% had been seen three or more times. Missed opportunities for screening allow for premalignant changes to progress undetected and untreated to invasive malignant disease. The latency period from dysplasia to invasive cancer is variable. It has been reported to be as long as 5 to 15 years, although it is thought to be shorter in women who are immunocompromised. It is critical that primary care providers ensure that their patients are up to date with all recommended screening tests. While new recommendations provide for the option to discontinue Pap smear screening after age 70 in selected women, it should be remembered that 25% of all cases and 40% of deaths from cervical cancer occur in women ages 65 and over.[2] Pap smear screening has proven to be the most cost effective screening test available to date. The false negative rate of a single Pap smear is 10%, although this has improved recently with the liquid based collection systems.

Risk Factors and Etiology

Risk factors for cervical cancer include early age at first sexual intercourse, increasing number of sexual partners, a history of HPV infection, a history of other sexually transmitted diseases, a high-risk male partner, low socio-economic status and smoking. Smoking is the single most modifiable risk factor. Women with pre-invasive disease that undergo treatment are more likely to develop recurrence of pre-invasive disease if

Gynecologic Oncology, edited by Paola Gehrig and Angeles Secord.
©2009 Landes Bioscience.

they continue to smoke, while those who quit smoking are much less likely to develop recurrence. Carcinogens found in cigarette smoke are found to be concentrated in cervical mucous 40 to 100 times higher than in the alveoli.[3] Cervical cancer is truly a smoking-related cancer and efforts should be directed at smoking cessation with the first abnormal Pap smear.

Human papilloma virus (HPV) has been proven to be an etiologic agent in the development of cervical cancer.[4] More than 90% of cervical cancers have detectable HPV DNA. However, not all women infected with HPV will develop cervical cancer.

By age 30, 80% of the US population has been exposed to HPV. Most individuals clear HPV viral infection over 6 to 18 months following exposure. Those patients who have persistence of viral infection are more likely to develop premalignant and malignant changes. No intervention exists to eradicate the virus; it is cleared through the action of the host immune system. Type specific HPV infection can be prevented through vaccination. Currently, there is one FDA approved HPV vaccine, Gardasil® (Quadrivalent Human Papillomavirus [HPV Types 6, 11, 16, 18] Recombinant Vaccine) It has been shown to be highly effective at preventing VIN 2 and 3, VAIN 2 and 3, CIN 1/2/3, AIS and genital warts. The efficacy for preventing cervical cancer is inferred through the prevention of the necessary precursor lesions. HPV infection precedes premalignant change which precedes invasive disease. HPV is the key cofactor in malignant transformation. Key proteins from the HPV particle have been shown to be transforming and immortalizing, critical to mediate malignant change. Early protein 6 (E6) binds p53, modulates apoptosis and activates telomerase. E7 binds Rb and induces transcriptional activation of several cellular promoters.

Presentation and Diagnosis

Classically, cervical cancer presents with postcoital spotting. Any woman with a complaint of bleeding with or after intercourse should be thoroughly evaluated for cervical cancer. Other common presenting symptoms are noted in Table 9.1. Symptoms of pain and weight loss are concerning for advanced stage disease, as is hematuria or hematochezia. Patients may also be completely asymptomatic, with either an abnormal pap or an obvious lesion noted on routine exam (Fig. 9.1). It should be remembered that a Pap smear is a screening test that should be applied to an asymptomatic patient. For the symptomatic patient or the patient with an obvious lesion, diagnostic testing should be employed. Diagnostic testing includes colposcopy and directed biopsy. Colposcopy involves the use of a microscope to inspect the cervix after the application of solutions (3% acetic acid and Lugol's iodine), to allow for the detection of the abnormal cells. Diagnosis is made on confirmatory biopsy. Squamous cell carcinoma accounts for 80% and adenocarcinoma accounts for almost 20% of all cervical cancers. Less than 5% of cases of cervical cancer are neuroendocrine or small cell carcinomas. Rare primary tumors of the cervix include melanoma, lymphoma and sarcoma. It is rare for tumors to metastasize to the cervix, although the most common primary cause of metastatic disease to the cervix is the uterus.

Table 9.1. Common presenting symptoms of cervical cancer

Abnormal uterine bleeding	Hematochezia
Postmenopausal bleeding	Weight loss
Abnormal vaginal discharge	Fatigue
Pelvic pain	Leg or back pain
Hematuria	

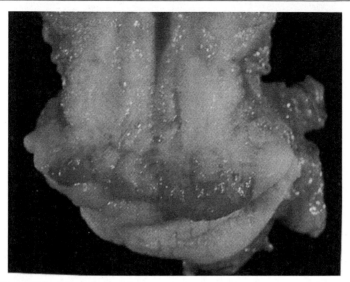

Figure 9.1. Cervical cancer.

Staging and Prognosis

Cervical cancer is staged clinically because a significant proportion of women in the US with cervical cancer will not undergo primary surgical management and the majority of women worldwide with cervical cancer do not have access to advanced radiographic imaging or surgical techniques. Modalities permitted to assess clinical stage include physical exam, chest radiograph and laboratory evaluation including a complete blood count, comprehensive metabolic panel and urinalysis. Physical examination includes assessment for supraclavicular and inguinal adenopathy, ascites and omental cake (uncommonly found) and lower extremity edema. Pelvic examination is critical for assessment for local extension of disease. Palpation of the cervix is performed to assess tumor size and extension onto the vaginal mucosa. The cardinal and uterosacral ligaments are often best palpated on rectovaginal exam to assess for tumor extension into these structures. Assessment of the urinary and lower gastrointestinal system can be accomplished via intravenous pyelogram (IVP) and barium enema (BE) as well as cystoscopy and proctoscopy. While IVP and BE are the only allowable imaging studies to assess stage, in clinical practice in the US, more definitive imaging is often employed to make treatment decisions. CT scanning of the abdomen and pelvis is 97% specific but only 25% sensitive.[5] MRI has the advantage of improved delineation of the pelvic soft tissues and has improved sensitivity in the detection of nodal spread. Recently there has been a growing body of evidence and interest in PET scanning in the initial staging of cervical cancer. PET has improved sensitivity and specificity relative to MRI or CT, although its true clinical value has yet to be assessed in a prospective fashion in newly diagnosed cervical cancer. The value of abdominal imaging (CT or MRI) is that it provides all of the information from an IVP and BE in addition to being able to assess for nodal disease to allow for treatment planning. Abnormalities noted on initial evaluation should be followed by further diagnostic testing, e.g., a bone scan if bony pain or elevated serum calcium, liver MRI if elevated LFTs, or cystoscopy if microscopic hematuria on urine analysis. Cystoscopy and proctoscopy is performed at the discretion of the attending physician but should definitely be performed on the patient with microscopic or gross blood in the urine or stool on presentation.

Clinical staging is summarized in Table 9.2. Conceptually, Stage I is disease confined to the cervix. Stage II represents disease that has begun to spread locally,

Table 9.2. Staging and treatment

Stage		Site of Involvement	Treatment	Five-Year Survivorship (%)
I		**Disease confined to the cervix**		81-96
	Ia	Confined to the cervix, microscopic		
	Ia1	Invasion <3 mm, width <7 mm	Cone or simple hysterectomy	
	Ia2	Invasion ≤5 mm, width <7 mm	RH	
	Ib	Confined to the cervix, macroscopic		
	Ib1	Tumor ≤4 cm	RH	
	Ib2	Tumor >4 cm	RH or RT-Chemo	
II		**Tumor spread beyond cervix but not to distal vagina or pelvic sidewall**		65-87
	IIa	Tumor involving proximal vagina	RH if small lesion or RT-Chemo	
	IIb	Tumor involving medial parametria	RT-Chemo	
III		**Tumor spread to distal vagina or pelvic sidewall**		35-50
	IIIa	Tumor involving distal vagina	RT-Chemo	
	IIIb	Tumor involving pelvic sidewall or hydronephrosis	RT-Chemo	
IV		**Metastatic disease**		15-20
	IVa	Involvement of bladder or rectal mucosa	RT-Chemo	
	IVb	Spread to distant organs (e.g., liver, lung)	Chemo	

Abbreviations: RH: modified radical or radical hysterectomy; RT-Chemo: radiation therapy with chemosensitization; Chemo: chemotherapy.

Table 9.3. *Five-year survivorship for cervical cancer*

Stage	Survivorship	Survivorship with Chemosensitization
Ib	85	NA
High risk Ib	74	83
IIa	73-80	NA
IIb	68	77
IIIa	45	63
IIIb	36	63
IVa	15	63
IVb	2	NA

NA = not available due to lack of patients in this substage.

either vaginally or into the parametria. Stage III disease is comprised of tumors that spread to the distal vagina or to the pelvic sidewall. Ureteral obstruction, even in the absence of bulky parametrial disease, is Stage III disease by definition. Stage IV disease is either locally advanced with direct invasion into the bladder or rectal mucosa or metastatic to distant sites such as the lungs. Common sites of metastatic disease are lymph nodes (pelvis, para-aortic, mediastinal and supraclavicular) in up to 65%, liver parenchyma in 20%, pulmonary parenchyma in 35% and bones in 25% of patients with recurrent disease. It should be reiterated that identification of metastatic disease to the lymph nodes by radiographic imaging cannot be used to change the patient's clinical stage.

Prognosis for women with cervical cancer is clearly dependent upon stage of disease at diagnosis. Additional negative epidemiologic prognostic factors include younger age, African-American race, lower socioeconomic status, anemia (Hgb <10) at diagnosis or during treatment and HIV positivity. Stage for stage, African-American women have a poorer survivorship when treated for cervical cancer. The reasons for this are as yet unclear but represent an area of intense study. Pathologically, tumor size, depth of stromal invasion, lymphvascular space involvement, extension into the endometrium, parametrial involvement, vaginal margin involvement and spread to lymph nodes are prognostic factors.

Survivorship with cervical cancer is excellent, even for patients with advanced disease. Survivorship statistics are summarized in Table 9.3. The advent of chemoradiation has dramatically improved the survivorship data for women with cervical cancer and represents the significant progress that can be achieved in cancer care through clinical trials.

Treatment

Overview

Treatment options for squamous cancers and adenocarcinomas of the cervix usually involve a choice of surgery or combination chemotherapy and radiation. The choice between the two is determined by the stage of disease, the patient's comorbidities and desire for future childbearing, if appropriate and willingness to accept the risk of certain side effects. Surgical options include conization, simple hysterectomy, modified radical (Type II) and radical (Type III) hysterectomy, radical trachelectomy and exenteration (please refer to chapter on surgical procedures). Risks and side effects of surgery vary according to the procedure selected. Conization and simple hysterectomy generally are associated with a low risk of complications and minimal blood loss. Radical hysterectomy is associated with an average blood loss of 750 cc, a 1-2% risk of ureterovaginal or vesicovaginal fistulas and an overall slightly higher risk of postoperative complications than seen with simple hysterectomy. Bladder dysfunction and inability to void is common after radical hysterectomy, requiring bladder retraining and intermittent self-catheterization. Radiation is associated with loss of ovarian function in all patients and a 5-10% rate of bowel and bladder complications, which include hemorrhagic cystitis, radiation proctitis and enteritis, malabsorption and bowel obstruction. Radiation complication rates are higher in patients with multiple prior abdominal surgeries, thin patients, smokers and patients with peripheral vascular disease and certain collagen vascular diseases. Chemotherapy may add mild to moderate nausea and bone marrow suppression, particularly anemia, but these are generally easily managed. The added benefit in terms of response and survival rates that chemotherapy provides (discussed later) far outweighs the increased risk of toxicity.

Surgery

Stage I cervical cancer is usually managed surgically. Squamous cervical cancer that presents with only microscopic disease (Stage Ia1) in a motivated patient who desires to preserve fertility and is willing to be compliant with follow-up, can be managed with a conization alone. Conservative management of early adenocarcinoma of the cervix is more controversial because of the risk of skip lesions involving the endocervix. However, there is literature support for this approach, provided the conization margins are negative and endocervical curettage above the level of the conization is also negative. For the patient who has completed childbearing, a simple hysterectomy is appropriate. If lymphvascular space invasion is noted histologically, the risk of spread to lymph nodes becomes clinically apparent and therefore the nodes must be assessed pathologically. For larger lesions (Stage Ia2 and Ib), a modified radical or radical hysterectomy is performed. A newer procedure which preserves the uterus for childbearing in the future, a radical trachelectomy, can be performed for lesions under 2 cm in the motivated patient. Radical implies removal of the uterus and a rim of surrounding normal tissue, not removal of the ovaries. A Type II differs from a Type III hysterectomy in where the margin of normal tissue is transected. In a Type II, or modified radical, hysterectomy the cardinal ligament is transected at the level of the ureter and the uterosacral ligament is transected one-third of the way back from the cervix. In a Type III, or radical, hysterectomy the cardinal ligament is transected at the pelvic sidewall and the ureter carefully dissected out from this tissue and the uterosacral ligament is transected two-thirds of the way back. In both types, the ovaries can be spared or removed. If the ovaries are left in situ and the patient requires postoperative radiation, the ovaries can be transposed or fixed, usually high in the paracolic gutters, outside of the radiation treatment field so as to prevent ovarian failure or sterilization from radiation, which occurs at around 20 cGy. Additionally, a rim, usually 2 cm, of normal vaginal tissue is taken and a pelvic lymphadenectomy is performed. Centers nationally are exploring laparoscopic, robotic and vaginal approaches to these traditional surgical options.

Unfortunately, even patients with early stage tumors can sometimes be found to have metastatic disease at the time of radical surgery. Patients who have positive parametrial margins, positive vaginal margins, or positive lymph nodes are at high risk for pelvic relapse and should receive adjuvant chemoradiation once they have recovered from

Table 9.4. Factors influencing risk of recurrence of cervical cancer after radical hysterectomy

High Risk Factors	Intermediate Risk Factors
Positive lymph nodes	>1/3 stromal invasion
Positive parametria	Lymphvascular space invasion
Positive vaginal margin	Tumor size >2 cm

surgery.[6] Additionally, patients found to have two or more intermediate pathologic risk risk factors, including lymphvascular space invasion, tumor size >2 cm and deep stromal invasion, are also at risk for pelvic failure and should also be treated with adjuvant radiation after surgery.[7] Refer to Table 9.4 for intermediate and high risk factors.

Chemoradiation

For patients with advanced stage disease, or those with medical comorbidities that preclude surgical management, treatment combines radiation and chemotherapy. Radiation therapy consists of daily weekday external radiation treatments for 5-6 weeks. External radiation is designed to deliver a total dose of 40 and 45 cGy in 1.8 to 2.0 cGy fractions. However, this amount of radiation will only treat microscopic tumor; doses of up to 85 cGy are required to sterilize macroscopic tumor. Therefore, additional radiation is delivered to the tumor itself in the form of a radiation implant or brachytherapy. Brachytherapy can be delivered through two major implant systems, interstitial or a Syed template and tandem and ovoids (please refer to radiation oncology chapter). The selection of a particular brachytherapy system is determined by the geometry of the patient's tumor as well as by institutional expertise. A known quantity of either cesium or iridium is used to load the device and left in place for a period of time to achieve the desired dose in the tumor bed. Between two and five applications are employed. The dose delivered to the surrounding tissues falls off according to the inverse square law, meaning that the dose rapidly declines as the distance away from the source increases. This limits the dose delivered to the surrounding organs such as the bladder, rectum and small intestine so as to minimize toxicity while allowing for maximization of the dose delivered to the tumor itself.

Chemotherapy, generally single agent cisplatin, is given once weekly for four to six treatments along with the radiation. While the goal is radiosensitization—improving the efficacy of the radiation in the treatment field—this approach has been shown to decrease the risk of distant recurrences. Chemotherapy may help to overcome the hypoxic tumor microenvironment. More importantly, this approach has also been shown to decrease the risk of death due to disease by up to 50%. These significantly positive results from five major prospective, randomized clinical trials effectively changed the standard of care in cervical cancer and prompted the National Cancer Institute to release a Clinical Announcement in 1999 strongly encouraging the use of chemotherapy in patients who require radiation therapy for the management of cervical cancer.[8] Several chemotherapy regimens have been evaluated, including cisplatin alone, cisplatin in combination with 5-flourouracil (5-FU) and cisplatin with 5-FU and hydroxyurea. Toxicities are greater with the combination regimens and the response rates are not that drastically different. Therefore, most institutions tend to use single agent cisplatin on a weekly basis for a total of four to six treatments, although some institutions still favor the combination of 5FU and cisplatin.

Several important treatment related prognostic factors should be kept in mind during radiation for cervical cancer. Duration of treatment is typically between 6 and 8 weeks. Patients whose duration of therapy extends beyond 8 weeks have higher rates of disease persistence and recurrence. Therefore, treatment time should be kept to a minimum and treatment breaks should be avoided. Anemia during treatment is also an important prognostic indicator. Patients whose hemoglobin level falls to 10.0 at the time of diagnosis or lower at any time during radiation therapy have lower response rates, higher recurrence rates and lower overall survival. Anemia should be aggressively treated during radiation, although the type of treatment is at the treating physician's discretion. It should be remembered that erythropoietin stimulating factors take approximately 1 month to raise hemoglobin levels by 1 gram. In addition, there is some preclinical data which suggests that growth factors may negatively impact radiation therapy. Therefore, there is often a role acutely for transfusion in these patients to achieve a hemoglobin level above 10.0.

Radiation complications can be seen in 5-10% of patients and can present at any time following radiation, even up to 20 years later. Most commonly, however, they present within 6 to 18 months following treatment. Radiation toxicity affecting the bladder can include urinary urgency, frequency and hematuria. Bowel toxicity includes proctitis, hematochezia, malabsorption and even bowel obstruction and perforation. Late radiation complications can include radiation-induced sarcomas.

Chemotherapy is generally well tolerated with only mild to moderate nausea and bone marrow suppression. As noted previously, maintainence of hemoglobin at or above 10.0 gm/dl is of critical importance from a response standpoint, so this must be carefully monitored. Mild to moderate neutropenia and thrombocytopenia is also observed. Cisplatin is renally cleared and can be directly nephrotoxic, so renal function must be carefully monitored and adequate hydration and renal clearance assured.

Contemporaneous administration of nephrotoxic drugs such as gentamicin should be avoided. Additionally, cisplatin can lead to a distal renal tubular acidosis with wasting of potassium and magnesium. Cisplatin can also cause ototoxicity and neurotoxicity, although not commonly at the dose used in radiosensitization in cervical cancer. There is generally no alopecia associated with this chemotherapy. Mild to moderate nausea is effectively managed with the 5-HT3 drugs such as ondansetron. For patients with ureteral obstruction consideration should be given to the placement of nephrostomy tubes or ureteral stents (often not feasible) to improve renal function. For patients who opt not to have the kidneys decompressed or who have medical renal disease with inadequate renal function, 5FU can be used as a radiation sensitizer as it does not rely on the kidneys for clearance.

Small Cell Carcinomas

Small cell carcinomas of the cervix should be considered systemic at diagnosis and are therefore not treated with locally directed therapy like squamous or adenocarcinomas. These are best approached like small cell carcinoma of the lung with multi-agent chemotherapy combining etoposide and a platinum drug as well as prophylactic whole brain radiation, with additional treatment directed specifically at the pelvis if appropriate.

Special Considerations

Stage IVa cervical cancer was traditionally managed with surgical removal of the bladder, uterus, cervix, upper vagina, parametria and pelvic lymph nodes, a procedure known as an anterior exenteration. However, patients with Stage IVa disease have an

80% risk of developing recurrent disease within the first 2 years of diagnosis if managed initially with surgery or radiation alone. Therefore, it is more common to manage these patients with chemoradiation initially and reserve surgery for management of treatment-related fistulas or recurrence. Vesicovaginal fistulas can develop in up to 10% of patients treated for Stage IVa cervical cancer. While this is a recognized complication of the treatment itself, it is usually indicative of persistent or recurrent disease and the actual incidence is probably much lower. Primary anterior exenteration is therefore only uncommonly performed and usually only for a fistula at presentation. Management of Stage IVb disease at presentation is more problematic. Some feel that aggressive treatement is inappropriate, as it is most likely palliative in nature and therefore favor a short course of radiation alone to the pelvis only to palliate bleeding and pain. However, some patients with Stage IVb disease will achieve a durable response and, rarely, some are even cured. Therefore, some physicians favor aggressive initial managment, either with combination chemotherapy followed by pelvic radiation if good response is obtained or with chemoradiation followed by more systemic combination chemotherapy if pelvic control is obtained. Patients with persistent pelvic disease despite chemoradiation are not curable and will ultimately succumb to their disease. Another approach is to enroll such patients on Phase III clinical trials as they are not technically treatable with curative intent and therefore are the most appropriate patients to enroll on clinical trials in an effort to improve the standard of care.

In summary, the choice of the type of initial management for cervical cancer is dependent upon the stage, the patient's desire for future childbearing, her medical comorbidities and the toxicities she is willing to accept. See Table 9.1 for an overview. As a general rule, if one approach is used for initial management, the other can be employed for management of pelvic recurrence.

Recurrent Disease

Management of recurrent disease depends primarily upon the site of recurrence and the treatment modality initially employed. Recurrent disease involving the pelvis centrally has the potential to be cured through the treatment modality that was not employed at primary presentation. For patients who underwent surgery primarily, chemoradiation can be used with curative intent. For the patient who underwent primary chemoradiation, a central pelvic recurrence can be managed with a total pelvic exenteration. In an exenterative procedure, the uterus, cervix, parametria, bladder and rectum are removed and the fecal and urinary streams diverted. An exenteration is only considered when the tumor can be completely removed surgically. Ureteral obstruction or para-aortic adenopathy imply that the tumor is unresectable.

The classic presentation of a pelvic sidewall recurrence is ureteral obstruction, unilateral lower extremity edema and pain. For those unfortunate women who develop recurrence of cervical cancer involving the pelvic sidewall, upper abdomem, or lung, the prognosis is poor. Median survivorship in these patients is 6 months. Traditionally patients were treated with single agent cisplatin chemotherapy. However, if a patient has previously received cisplatin, i.e., with prior chemoradiation, the likelihood of responding to cisplatin alone is low (<10%). Increasingly, the use of combination chemotherapy regimens such as topotecan plus cisplatin or paclitaxel plus cisplatin are being used to improve response rates. While an advantage in overall survivorship has not been seen with the paclitaxel combination as compared to single agent cisplatin, the topotecan cisplatin combination has been shown to extend median survival to greater than 9 months. This remains an area of intense research.

For the patient with a pelvic recurrence and bilateral ureteral obstruction, consideration should be given to a palliative approach. Death due to uremia from renal failure is relatively painless and one of the more comfortable ways to succumb to this disease. Interventions to relieve renal obstruction, most commonly nephrostomy tubes, may prolong the patient's life but if the disease cannot be effectively treated, may result in increased suffering due to pain from pelvic disease. Pelvic recurrence of cervical cancer can result in significant pain and may require aggressive treatment with high doses of narcotics and other adjuvants for relief.

Future Directions

Clinical research has made tremendous improvements in cervical cancer care in the past decade. Current areas of study include newer radiation techniques, the use of biologics either along with chemoradiation or in the setting of recurrence and chemotherapy combinations in recurrent disease. Intensity modulated radiation therapy (IMRT) uses multiple radiation beams to try to more specifically target the cancerous tissues and minimize the toxicity to the normal surrounding tissues. The efficacy and safety of this approach in whole pelvic radiation therapy for cervical cancer is being actively studied. In an effort to make primary radiation therapy more effective, the Gynecologic Oncology Group is studying the use of tirapazamine along with cisplatin for chemosensitization during primary radiation therapy. Cetuximab in combination with cisplatin is also being evaluated for primary chemosensitization as well as in advanced and recurrent cervical cancer. Treatment of recurrent cervical cancer continues to be a significant challenge. Various chemotherapy combinations, such as gemcitabine-cisplatin, navelbene-cisplatin, paclitaxel-cisplatin and topotecan-cisplatin are being studied. In addition the biologic agents such as bevacizumab are also being evaluated in advanced and recurrent disease by the GOG. We hope that continued progress in the fight against cervical cancer is made through clinical research efforts.

Suggested Reading

1. Moore DH. Cervical cancer. Obstet Gynecol 2006; 107:1152-61.
2. National Cancer Institute. NCI issues clinical announcement on cervical cancer: chemotherapy plus radiation improves survival. U.S. Department of Health and Human Services, Public Health Service, National Institutes of Health. February, 1999. Available at: http://www.cancer.gov/newscenter/cervicalcancer.

References

1. American Cancer Society. Cancer facts and figures. Atlanta Ga 2005:1-60.
2. Jemal A, Siegel R, Ward E et al. Cancer statistics. CA Cancer J Clin 2006; 56(2):106-30.
3. Hellberg D, Nilsson S, Haley NJ et al. Smoking and cervical intraepithelial neoplasia: nicotine and cotinine in serum and cervical mucus in smokers and nonsmokers. Am J Obstet Gynecol 1988; 158(4):910-3.
4. Walboomers JMM, Jacobs MV, Manos MM et al. Human papillomavirus is a necessary cause of invasive cervical cancer worldwide. J Pathol 1999; 18:912-9.
5. Camilien L, Gordon D, Fruchter RG et al. Predictive value of computerized tomography in the presurgical evaluation of primary carcinoma of the cervix. Gynecol Oncol 1988; 30(2):209-15.

6. Peters WA 3rd, Liu PY, Barrett RJ 2nd et al. Concurrent chemotherapy and pelvic radiation therapy compared with pelvic radiation therapy alone as adjuvant therapy after radical surgery in high-risk early-stage cancer of the cervix. J Clin Oncol 2000; 18(8):1606-13.
7. Sedlis A, Bundy BN, Rotman MZ et al. A randomized trial of pelvic radiation therapy versus no further therapy in selected patients with stage IB carcinoma of the cervix after radical hysterectomy and pelvic lymphadenectomy: A gynecologic oncology group study. Gynecol Oncol 1999; 73(2):177-83.
8. National Cancer Institute. NCI issues clinical announcement on cervical cancer: chemotherapy plus radiation improves survival. U.S. Department of Health and Human Services, Public Health Service, National Institutes of Health. Available at: http://www.cancer.gov/newscenter/cervicalcancer 1999.

9

CHAPTER 10

Uterine Epithelial Cancer

Israel Zighelboim and Matthew A. Powell

Introduction

Epidemiology

Carcinoma of the endometrium represents the most common gynecologic malignancy in the United States and accounts for more than 90% of all malignant neoplasms affecting the uterine corpus. This disease has a particular geographic distribution, with highest incidences found in North America and Europe and much lower rates in eastern Asia and Africa. In 2008, approximately 40,100 new cases will be diagnosed and 7,400 women will die of this disease in the United States.

Risk Factors

Risk factors for endometrial cancer include unopposed stimulation of estrogenic receptors in the uterus by estrogens or certain selective estrogen receptor modulators (SERMS) such as tamoxifen, chronic anovulation, truncal obesity, diabetes mellitus, hypertension, nulliparity and late menopause (Table 10.1).

Unopposed estrogenic stimulation of the endometrium is thought to cause endometrial hyperplasia. The World Health Organization classifies endometrial hyperplasias in the following groups: simple without atypia, complex without atypia, simple with atypia and complex with atypia. If untreated 1% of simple hyperplasias and 3% of complex hyperplasias without atypia would progress to endometrial cancer. In the presence of atypia this figure approaches 25-30%. Furthermore, up to 42% of patients with atypical hyperplasia may have concurrent endometrial cancer upon further evaluation.

More recently, the association of certain endometrial cancers with the hereditary nonpolyposis colonic cancer (HNPCC) and other familial cancer syndromes has been recognized. In fact, the most common manifestation of the HNPCC syndrome in women is endometrial cancer. These women have a 40-60% lifetime risk of developing endometrial cancer.

Historically, patients with endometrial carcinomas have been categorized in two groups. The largest group includes Type I or estrogen-dependent. These patients tend to be younger at diagnosis with clear evidence of unopposed hyperestrogenic stimulation as their main risk factor. Tumors of this type usually arise in association with endometrial hyperplasia and carry an overall better prognosis. Type II patients are on average older at diagnosis and lack evidence of sustained unopposed estrogenic endometrial exposure as their main risk factor. Tumors in this group tend to be poorly differentiated and include more uncommon and aggressive histologic subtypes including papillary serous and clear cell.

Gynecologic Oncology, edited by Paola Gehrig and Angeles Secord.
©2009 Landes Bioscience.

Table 10.1. Endogenous and exogenous risk factors

Characteristic	RR [X]
Obesity	
>30 lbs	3
>50 lbs	10
Hypertension	1.5
Nulliparous	2
Diabetes mellitus	2.8
Late menopause	4
Unopposed estrogen	9.5
Atypical complex hyperplasia	29

Histology

Types

Approximately 80-85% of uterine cancers are endometrioid adenocarcinomas and 20% of these may have areas with squamous differentiation. Other varieties include papillary serous and clear cell carcinomas. These less common and more aggressive types have higher rates of extrauterine spread at the time of diagnosis and frequently carry with them histologic features associated with poor prognosis such as deep myometrial and lymphovascular space invasion and cervical involvement, though neither of these are required for them to present in advanced stages. Table 10.2 shows the histologic classification and frequencies of epithelial tumors of the uterus.

Grades

Endometrioid adenocarcinomas are categorized based on their histologic grade. The International Federation of Gynecology and Obstetrics (FIGO) classifies these tumors in three grades (Fig. 10.1):

G1 Well-differentiated adenocarcinoma with ≤5% nonsquamous or nonmorular solid growth pattern.

G2 Adenocarcinomas with some differentiation and 6-50% nonsquamous or nonmorular solid growth pattern.

G3 Undifferentiated carcinoma with >50% nonsquamous or nonmorular solid growth pattern.

Papillary serous and clear cell carcinomas are automatically classified as Grade 3.

Table 10.2. Histologic classification of epithelial uterine tumors

Endometrioid adenocarcinoma	75-80%
Mixed adenocarcinoma	10%
Uterine papillary serous carcinoma	<10%
Clear cell carcinoma	4%
Mucinous adenocarcinoma	1%
Squamous cell carcinoma	<1%
Undifferentiated	1-5%

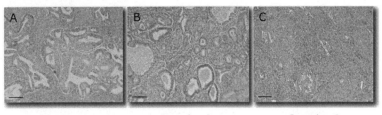

Grade 1   Grade 2   Grade 3

Figure 10.1. Grading of endometrial adenocarcinoma depends on the architectural pattern. A) Well-differentiated tumors show cells and glands that closely resemble normal endometrium, with a well-preserved glandular pattern, scanty stroma and some lack of uniformity of the epithelial cells. More than 95% of the tumor is composed of glands and less than 5% of the tumor, shows a solid growth pattern. B) Grade 2 tumors show a less well-defined glandular pattern, although still discernible (moderately differentiated). The tumor cells are more irregular with multilayering and large, pleomorphic nuclei. Six to 50% of the tumor is solid. C) Adenocarcinoma that show more than 50% solid growth are graded as poorly differentiated. The glandular pattern may be hardly recognizable and the cells are arranged in solid sheets and cords. Nuclear pleomorphism is marked with irregular nucleoli and widespread mitotic activity and many abnormal mitotic figures. In the vast majority of cases, the architectural and nuclear grades correspond. In the rare cases where they are at variance and the nuclei show bizarre nuclear atypia ('notable nuclear atypia' WHO), the nuclear grade is the more reliable indicator of prognosis and raises the grade by one. Scale bars A-C: 200μm. (Courtesy of Jochen KM Lennerz, MD; Lauren V. Ackerman Laboratory of Surgical Pathology, Washington University School of Medicine, St. Louis, MO.)

Spread Patterns

Endometrial cancer may spread by direct extension, transtubal exfoliation of malignant cells with subsequent seeding into the peritoneal cavity, lymphatic or hematogenous dissemination.

Clinical Features

Screening

Only about 50% of cases of endometrial cancer will have abnormalities on a Pap smear. Furthermore, noninvasive evaluation using imaging modalities is costly and has an unacceptably low sensitivity and specificity. In short, evaluation of the endometrial cavity requires tissue sampling. Several devices have been developed for this purpose. The Pipelle biopsy device allows sampling in the office setting with sensitivity greater than 90%. However, tissue sampling is costly and not devoid of patient discomfort. For all the above, screening for endometrial cancer at the general population level is currently not recommended. Biopsy of the endometrium for screening purposes should be reserved for women with risk factors such as postmenopausal women who have been on unopposed estrogen therapy, family history of HNPCC, premenopausal women with prolonged untreated chronic anovulation.

Clinical Features

Most cases of endometrial cancer occur in peri- and postmenopausal women and more than 90% of cases will present with abnormal uterine bleeding. Therefore, the presence of postmenopausal bleeding should prompt immediate and thorough evaluation to rule out the presence of a gynecologic malignancy. Similarly, pre- or peri-menopausal women with history of chronic anovulation or other risk factors deserve careful evaluation. Pap smear evidence of atypical glandular cells (AGC) of any subcategory in an anovulatory woman or in women over age 35 should include evaluation of the endometrium. One should also consider evaluation of women with endometrial cells on the Pap smear in postmenopausal women. The presence of atypical endometrial cells on a Pap smear should be followed by endometrial sampling. More infrequently patients may present with a pelvic mass, pyometra or signs and symptoms related to the presence of metastatic disease.

Evaluation and Surgical Staging

A detailed history should include assessment of risk factors, family history and functional evaluation that could suggest advanced disease. Physical examination commonly provides supportive evidence of chronic anovulation. The abdominal exam can discover masses consistent with peritoneal or omental extension or ascites. Careful pelvic and rectal exams will allow complete evaluation of the vulva, vagina, cervix and other pelvic structures. This will assist in ruling out differential diagnoses and assessing the presence of extrauterine extension. Office endometrial biopsy is very accurate in detecting endometrial carcinoma. Patients with a nondiagnostic office biopsy or negative biopsies in the context of high clinical suspicion should be evaluated further with a dilatation and curettage (D & C), ideally with direct visual evaluation of the entire cavity with hysteroscopy. Consider endocervical curettage in patients with possible cervical involvement; however be aware this test has a very high false positive rate.

Once the diagnosis is confirmed, the patient should be evaluated for other malignancies as appropriate based on age and family history (breast, colon, urinary tract). Additional preoperative evaluation usually includes investigation of liver and renal function, baseline hematologic parameters and radiologic imaging as clinically indicated to evaluate for advanced disease and should include, at minimum, a chest radiograph. Determination of perioperative CA125 levels may be useful in certain cases. If elevated, this marker suggests extrauterine disease and may assist in evaluating response to treatment.

The current surgical staging for cancer of the uterine corpus was adopted by FIGO in 1988 and is summarized in Table 10.3. All patients should undergo surgical exploration with complete staging unless precluded by other medical conditions. The surgical staging procedure should include collection of fluid or washings for cytologic evaluation, careful visual inspection and/or palpation of peritoneal surfaces with directed biopsies as indicated, extrafascial hysterectomy with bilateral salpingo-oophorectomy, pelvic and para-aortic lymph node sampling. Lymph node dissection can be omitted in cases of well-differentiated adenocarcinoma without myometrial invasion. Intraoperative assessment with visual inspection and frozen sectioning are invariably inaccurate and appropriate clinical judgment should dictate when full staging is indicated. Minimally invasive procedures (laparoscopic and robotic)

Table 10.3. FIGO surgical staging of cancer of the uterine corpus, 1988

Stage/Grade	Description
IA G123	Tumor limited to endometrium
IB G123	Invasion to less than one-half of the myometrium
IC G123	Invasion to more than one-half of the myometrium
IIA G123	Endocervical glandular involvement only
IIB G123	Cervical stromal invasion
IIIA G123	Tumor invades serosa and/or adnexal and/or positive peritoneal cytology
IIIB G123	Vaginal metastasis
IIIC G123	Metastasis of pelvic and/or para-aortic lymph nodes
IVA G123	Tumor invasion of bladder and/or rectal mucosa
IVB G123	Distant metastases including intra-abdominal extension and/or inguinal lymph nodes

10

are becoming increasingly common for the initial surgical staging and treatment of endometrial cancer.

Therapy

General Therapy Overview

Because most cases of endometrial cancer present with early symptoms, the disease is diagnosed in Stage I in 72% of the cases. Long-term survivorship exceeds 78% and 85% in registries in Europe and the United States, respectively.

Patients with localized and well-differentiated disease are usually cured by hysterectomy and bilateral salpingoophorectomy alone. However, other patients may benefit from additional therapy in the form of radiation, cytotoxic chemotherapy or hormonal manipulation following their initial staging procedures.

It is important to keep in mind that adjuvant radiotherapy in patients with early-stage endometrial cancer has not been proven to increase overall survival. Its main role is that of preventing local recurrences which indirectly can have an important impact in the quality of life of these patients. Therapeutic modalities include the use of vault brachytherapy (vaginal cylinders), standard external pelvic and/or extended field irradiation and intensity modulated radiotherapy. These techniques have specific indications, and their use should be tailored to the particular clinical characteristics by a radiation oncologist with experience in the treatment of gynecologic malignancies. The incidence of major complications after radiotherapy approaches 4-5% and can be even higher following transperitoneal lymphadenectomies.

Many cytotoxic chemotherapeutic agents have been evaluated in patients with endometrial cancer. The objective response rates to several cytotoxic agents have varied widely: cisplatin (20-35%), carboplatin (30%), adriamycin (20-35%), epirubicin (25%), paclitaxel (35%). The combination of doxorubicin with cisplatin has yielded the highest response rates to date (40-66%). The addition of paclitaxel to doxorubicin and cisplatin (TAP regimen) further improves objective response rates, progression-free survival and overall survival and represents the current standard of care in patients with advanced endometrial cancer. In general, response to cytotoxic chemotherapy in endometrial cancer is disappointing, and these agents are only considered to be palliative.

Hormonal manipulation with high dose progestins in the form of either medroxyprogesterone acetate 200 mg daily or megestrol acetate 160 mg daily approaches response rates of 20% in the presence of estrogen and progesterone receptors (ER/PR). An alternative regimen alternating megestrol acetate 160 mg daily and tamoxifen 80 mg daily every 3 weeks has demonstrated an overall response rate of 27% (90% in patients with histologically confirmed Grade 1 tumors, 24% in those with Grade 2 and 22% among patients with Grade 3 disease). These modalities are often used for patients with advanced or recurrent disease whose tumors tested positive for these receptors. Undifferentiated tumors rarely express ER/PR (<25%).

Treatment for Stage I

Hysterectomy and bilateral salpingoophorectomy is curative for patients with low grade endometrioid histology. Some advocate treating the vaginal cuff with radiotherapy using a vaginal cylinder to further reduce the risk of local recurrence. The following risk factors certainly increase the risk of recurrence in patients with Stage I disease: increasing age, moderate to poorly differentiated tumors (Grades 2 and 3), presence of lymphovascular invasion and outer third myometrial invasion. The Gynecologic Oncology Group has defined a group of patients called "high-intermediate risk." These patients meet one of the following criteria: are at least 70 years old with one risk factor, are at least 50 years old with two additional risk factors or have three risk factors regardless of their age. The use of adjuvant whole pelvic radiation in these patients reduces the risk of recurrence by 58%.

High dose progestins administered either orally, intramuscularly or topically via medicated intrauterine devices have been used occasionally in young patients who desire to maintain their fertility or those who are very poor surgical candidates carrying Grade 1 endometrioid cancers. Response rates range from 58-100%. Most of these lesions will likely recur after therapy is withdrawn in these hyperestrogenic patients. Repeated biopsies in these cases are obligatory to rule out recurrence or progression. This modality is by no means standard of care and should be reserved for selected cases until data of long-term outcomes become available.

Treatment for Stage II

Most patients should receive postoperative radiotherapy. If the involvement of the cervix is known before the procedure, some advocate a modified radical hysterectomy with complete pelvic and para-aortic lymphadenectomy at the time of staging laparotomy. If the lymph nodes prove negative, no further treatment is needed. Others recommend preoperative radiotherapy followed by staging and completion hysterectomy.

Treatment for Stage III

Most patients receive radiotherapy after surgery. Usually, these patients are treated with combination of external beam and brachytherapy with boosts to vaginal vault or nodal areas as appropriate. Patients who are not good candidates for radiotherapy may benefit from progestational agents as previously discussed. Increasingly patients are being treated with cytotoxic chemotherapy for Stage III tumors. These are treated with chemotherapy alone or in combination or sequentially with radiotherapy.

Treatment for Stage IV

The treatment in these cases should be highly individualized according to disease characteristics, comorbidities and physician/patient preference. Treatment modalities may include whole abdominal radiation (which is especially useful after completely resected upper abdominal disease), palliative radiation and/or chemotherapy or hormonal manipulation as appropriate. All patients with advanced disease should be considered for participation in clinical trials. Current evidence would support cytotoxic chemotherapy as front-line therapy following surgical resection for Stage IV cancers rather than whole abdominal radiotherapy.

Recurrent Disease

Localized nodal or distant recurrences may be treated with palliative radiotherapy. In some instances when no prior radiation has been administered, the use of pelvic irradiation and/or surgical resection may prove curative for isolated central recurrences at the vaginal vault. Patients who meet criteria should be offered hormonal manipulation. As previously mentioned, a 20% response rate should be anticipated and those patients who respond have shown improved overall survivorship after this treatment. There is some evidence that supports the use of Tamoxifen in patients who do not respond to standard progestin therapy. Response rates of 20% have been documented under these circumstances. Cytotoxic chemotherapy is appropriate for many of these patients. Low levels of ER/PR may predict a higher chance of response to cytotoxic chemotherapy. Patients with recurrent endometrial cancer should be offered participation in clinical trials.

Table 10.4 summarizes current treatment recommendations for patients with endometrial cancer.

10

Table 10.4. Treatment of endometrial cancer

Condition	Possible Therapies to Consider
Stage IA or IB and Grade 1 or 2	No further therapy or vaginal brachytherapy
Any Stage I Grade 3 or IC of any grade*	No further therapy versus vaginal brachytherapy versus whole pelvic radiotherapy
Stage II*	Vaginal brachytherapy versus pelvic RT versus modified radical hysterectomy + pelvic lymphadenectomy
Stage IIIA (positive peritoneal cytology)*	Treat based on uterine disease risk factors as above versus progestins versus intraperitoneal P-32
Stage IIIB/IIIC*^	Whole pelvic or extended field or abdominal radiotherapy +/– vaginal or nodal boost; consider hormonal manipulation
Stage IV and recurrent disease (extrapelvic)^	Systemic chemotherapy versus whole abdominal radiation versus palliative radiation versus hormonal therapy versus combinations
Recurrent disease (pelvic)	Radiotherapy in the patient without prior RT versus possible surgical resection (exenteration) versus chemotherapy

*Consider chemotherapy if the tumor is of papillary serous or clear cell histology
^Would not recommend whole abdominal radiotherapy if the patient is suboptimally debulked

Complications

Complications of surgical staging are rare. The inclusion of lymph node dissection in the procedure adds very little perioperative morbidity when performed by an experienced gynecologic oncologist. Lymphedema of the lower extremities and lymphocysts are uncommon complications of lymphadenectomy. Acute effects of radiation may include gastrointestinal side effects, hematologic and/or metabolic disturbances. Long-term effects after abdomino-pelvic radiation are usually related to bowel and bladder dysfunction.

Follow-Up

Typically patients are evaluated clinically with pelvic exam and Pap smear (controversial) every 3 months the first 2 years, then twice a year for 2 years and then annually. A yearly chest X-ray is commonly obtained although controversial. Other imaging studies such as CT and MRI may be useful in isolated cases. If a patient has an elevated CA125 at the time of diagnosis, this can also be used to follow her for a disease recurrence (similar to women with ovarian cancer).

Prognosis

As previously discussed, there are several factors associated with prognosis in patients with endometrial cancer. These include: histologic type, grade, stage, depth of myometrial invasion, lymphovascular invasion and cervical involvement. Overall, the survival by FIGO stage in endometrial cancer approaches: 85% for Stage I, 75% for Stage II, 45% for Stage III and 25% for Stage IV disease. However, these figures can vary considerably depending on grade and other prognostic features. For example, 5-year survival in patients with Stage IA FIGO Grade 1 tumors reaches 95% while Stage IC FIGO Grade 3 lesions have a 5-year survival of only 42%. Mortality from endometrial cancer is higher in African-American women. It is unclear whether this difference is due to a delayed diagnosis or a higher incidence of poor prognostic features in the tumors of these women.

Suggested Reading

1. Amant F, Moerman P, Neven P et al. Endometrial cancer. Lancet 2005; 366:491-505.
2. Markman M. Hormonal therapy of endometrial cancer. Eur J Cancer 2005; 41:673-5.
3. Benshushan A. Endometrial adenocarcinoma in young patients: evaluation and fertility-preserving treatment. Eur J Obstet Gynecol Reprod Biol 2004; 117:132-7.
4. Ramirez PT, Frumovitz M, Bodurka DC et al. Hormonal therapy for the management of Grade 1 endometrial adenocarcinoma: a literature review. Gynecol Oncol 2004; 95:133-8.
5. Clark TJ, Mann CH, Shah N et al. Accuracy of outpatient endometrial biopsy in the diagnosis of endometrial cancer: a systematic quantitative review. BJOG 2002; 109:313-21.

References

1. Jolly S, Vargas C, Kumar T et al. Vaginal brachytherapy alone: an alternative to adjuvant whole pelvis radiation for early stage endometrial cancer. Gynecol Oncol 2005; 97:887-92.
2. Sutton G, Axelrod JH, Bundy BN et al. Whole abdominal radiotherapy in the adjuvant treatment of patients with stage III and IV endometrial cancer: a gynecologic oncology group study. Oncol 2005; 97:755-63.

3. Thigpen JT, Brady MF, Homesley HD et al. Phase III trial of doxorubicin with or without cisplatin in advanced endometrial carcinoma: a gynecologic oncology group study. J Clin Oncol 2004; 22:3902-8.

4. Fleming GF, Filiaci VL, Bentley RC et al. Phase III randomized trial of doxorubicin + cisplatin versus doxorubicin + 24-h paclitaxel + filgrastim in endometrial carcinoma: a gynecologic oncology group study. Ann Oncol 2004; 15:1173-8.

5. Fleming GF, Brunetto VL, Cella D et al. Phase III trial of doxorubicin plus cisplatin with or without paclitaxel plus filgrastim in advanced endometrial carcinoma: a gynecologic oncology group study. J Clin Oncol 2004; 22:2159-66.

6. Keys HM, Roberts JA, Brunetto VL et al. Gynecologic oncology group. A phase III trial of surgery with or without adjunctive external pelvic radiation therapy in intermediate risk endometrial adenocarcinoma: a gynecologic oncology group study. Gynecol Oncol 2004; 92:744-51.

7. Montz FJ, Bristow RE, Bovicelli A et al. Intrauterine progesterone treatment of early endometrial cancer. Am J Obstet Gynecol 2002; 186:651-7.

8. Creutzberg CL, van Putten WL, Koper PC et al. Surgery and postoperative radiotherapy versus surgery alone for patients with stage-1 endometrial carcinoma: multicentre randomised trial. PORTEC study group. Post operative radiation therapy in endometrial carcinoma. Lancet 2000; 355:1404-11.

9. Randall ME, Filiaci VL, Muss H et al. Randomized phase III trial of whole-abdominal irradiation versus doxorubicin and cisplatin chemotherapy in advanced endometrial carcinoma: a gynecologic oncology group study. J Clin Oncol 2006; 24:36-44.

10

CHAPTER 11

Uterine Sarcomas

Amy L. Jonson and Patricia L. Judson

Introduction

Over 96% of cancers from the uterine corpus are epithelial tumors and arise from the endometrium. The remaining 4% of uterine cancers are a group of mesenchymal tumors that arise from either the myometrium or endometrium (Fig. 11.1). These rare tumors are collectively referred to as uterine sarcomas.

There are many different ways to classify uterine sarcomas. Most classification systems are based on histopathologic characteristics of the tumors and can be complicated and tedious to use. Clinically, most uterine sarcomas fall under one of four major types (listed in order of decreasing incidence): carcinosarcomas, leiomyosarcomas, endometrial stromal sarcomas and adenosarcomas.

Epidemiology

The largest original epidemiologic study looking at uterine sarcomas was published in 1986 by Harlow et al.[1] This study utilized data collected by the Surveillance, Epidemiology and End Result (SEER) registry between 1973 and 1981. The incidence of carcinosarcomas was found to be 8.2 per 1 million women per year. The incidence for leiomyosarcomas and endometrial stromal sarcomas was 6.4 and 1.8, respectively. This analysis was also one of the first studies that identified a much higher incidence of uterine sarcomas in black women. This is in stark contrast to what is seen with the more common epithelial tumors of the endometrium, where the risk is higher in white women.

A more recent review of 2,677 cases of uterine sarcoma reported in the SEER registry, between 1989-1999, confirmed previously published studies that found black women to be at higher risk for uterine sarcoma than white women.[2] The age adjusted incidence in white women was 3.6 per 100,000 as compared to 7.0 in black women. The racial difference in the incidence of uterine sarcomas was only statistically significant for carcinosarcomas and leiomyosarcomas, not for endometrial stromal sarcomas or adenosarcomas. The study also found that black women were significantly younger than white women at the time of diagnosis with a mean age of 62.7 years, compared to 64.2 years.

Both SEER database studies found that the incidence of carcinosarcoma increases dramatically with age. Most women are in their seventh decade of life when they are diagnosed. In the most recent SEER analysis, for example, the incidence of carcinosarcomas in black women between the ages of 35-64 was 2.1 per 100,000 compared to 11.1 in black women ≥65 years of age. Leiomyosarcomas, however, usually occur at an earlier age. Average age at diagnosis is 53, and the incidence remains relatively stable throughout middle age.

Gynecologic Oncology, edited by Paola Gehrig and Angeles Secord.
©2009 Landes Bioscience.

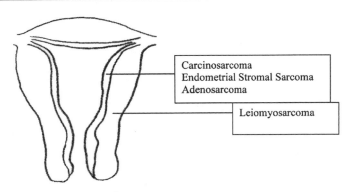

Figure 11.1. Anatomy of uterine sarcomas.

11

Epidemiologic factors that are felt to be independently associated with an increased risk for carcinosarcoma include exposure to radiation and Tamoxifen use. Evidence looking at prior radiation exposure has varied greatly. The relative risk (RR) has been estimated at 5.4.[3] One study found that 12-30% of patients diagnosed with carcinosarcoma had received prior radiation.[4] When identified, these sarcomas usually occur between 5-25 years after the radiation exposure. A recent analysis of 39,451 breast cancer patients found that women who had received Tamoxifen therapy had a RR of 4.62 for developing a subsequent uterine carcinosarcoma compared to women not receiving Tamoxifen.[5]

Clinical Presentation

As with the more common endometrial cancers, the most likely symptom for women with all histologic subtypes of uterine sarcoma is abnormal vaginal bleeding. Between 75-95% of patients will present with complaints of postmenopausal bleeding or menometrorrhagia. Approximately 10% of women will complain of discharge without bleeding and 10% of women will complain of pain and/or a mass.

The triad of vaginal bleeding, pain and tumor fungating from the external cervical os has been found to be strongly associated with the diagnosis of carcinosarcoma. As for leiomyosarcomas, a commonly stated, but yet unproven, association is the finding of a rapidly growing fibroid. One study found that only 0.27% of women undergoing hysterectomy for a rapidly enlarging fibroid were found to have a leiomyosarcoma.[6]

Diagnosis

Most uterine carcinosarcomas are diagnosed preoperatively by either an endometrial biopsy or dilatation and curettage. They may also be commonly diagnosed by a simple biopsy if there is tumor prolapsing from the cervix. Occasionally carcinosarcomas are not diagnosed until after hysterectomy because the preoperative endometrial sampling only identifies the carcinomatous component of the mixed tumor. Leiomyosarcomas, conversely, are diagnosed preoperatively in less than half of the cases. Most cases are diagnosed incidentally on postoperative pathology. Endometrial sampling most likely misses the tumor because it arises within the muscle wall of the uterus and not the lining of the uterus which is sampled by the endometrial biopsy.

Studies which can be informative prior to surgery when a uterine sarcoma is diagnosed include: chest radiograph (CXR) to rule out pulmonary metastasis and

Table 11.1. FIGO staging for uterine cancer (please refer to Chapter 10)

Stage	Description
IA	Tumor limited to endometrium
IB	Tumor invasion <1/2 of total myometrial thickness
IC	Tumor invasion ≥1/2 of total myometrial thickness
IIA	Endocervical gland involvement
IIB	Cervical stromal invasion
IIIA	Tumor invades serosa and/or adnexa and/or positive peritoneal cytology
IIIB	Vaginal metastasis
IIIC	Metastasis to pelvic and/or para-aortic lymph nodes
IVA	Tumor invades bladder and/or bowel mucosa
IVB	Distant metastasis including intra-abdominal and/or inguinal lymph nodes

11

computed tomography (CT) or magnetic resonance imaging (MRI) to assess extra-uterine spread of disease. The tumor marker CA125 may also be tested. If the level is elevated at the time of diagnosis, it can be a helpful tool to measure response to treatment or to identify disease recurrence.

Treatment

The mainstay of treatment for all uterine sarcomas is surgery. Surgery includes a total hysterectomy and staging procedure. In most cases, it is also appropriate to perform a bilateral salpingo-oophorectomy. This is important, not only for identifying metastatic disease, but also because many sarcomas are hormone dependent or responsive. A sample of free fluid from the pelvis or a collection of washings should also be performed. Additional surgical procedures and subsequent adjuvant therapies depend on the sarcoma being treated and are discussed in further detail in the subsections below.

Staging

Uterine sarcomas are surgically staged tumors. The surgical staging procedure performed is dependent on the type of sarcoma identified. The final stage is not assigned until the surgical pathology is complete. There is no independent staging system for uterine sarcomas; the most common staging system used is the International Federation of Gynecology and Obstetrics (FIGO) staging criteria for endometrial cancer (Table 11.1).

Prognosis

With the exception of low-grade endometrial stromal sarcomas, uterine sarcomas are aggressive tumors with a poor prognosis (Table 11.2). Most patients die from their disease within 2 years of diagnosis. This is in stark comparison to endometrial cancer where 5-year survival is greater than 75%.

Carcinosarcoma

Carcinosarcoma is the most common type of uterine sarcoma, accounting for approximately 50% of all sarcomas. Carcinosarcomas are commonly referred to as a MMMT (Malignant Mixed Mesodermal Tumor or Malignant Mixed Mullerian Tumor). The key pathologic feature and therefore namesake, is the mixed nature of the tumor. Carcinosarcomas are characterized by the presence of both malignant epithelial cells (carcinoma) and malignant stromal cells (sarcoma).

Table 11.2. Uterine sarcoma: survival by histology and clinical stage

Histology	Total Cases	Stage I n (% survival)	Stage II n (% survival)	Stage III n (% survival)	Stage IV n (% survival)
LGESS	65	55 (89)	4 (75)	6 (67)	0 (-)
HGESS	100	90 (78)	1 (0)	7 (14)	2 (0)
LMS	108	91 (48)	3 (67)	5 (0)	9 (0)
MMMT	399	245 (36)	55 (22)	69 (10)	30 (6)

Data from Berchuck et al (1988, 1990), Covens et al (1987), De Fusco et al (1989), Doss et al (1984), Kahanpää et al (1986), Larson et al (1990, 1990), Macasaet et al (1985), Mantravadi et al (1981), Norris and Taylor (1966, 1966), Norris et al (1966), Piver et al (1984), Spanos et al (1984), Wheelock et al (1985) and Yoonessi and Hart (1977).

LGESS = low grade endometrial stromal sarcoma

HGESS—high grade endometrial stromal sarcoma

LMS = leiomyosarcoma

MMMT = malignant mixed mesodermal tumor = carcinosarcoma

Adapted from: Morrow CP, Curtin JP. Synopsis of Gynecologic Oncology. 5th Ed. 1998:188, with permission from Elsevier.

Carcinosarcomas can be further sub-categorized into homologous or heterologous tumors. Homologous tumors are composed of malignant cells arising from tissues normally found within the uterus. In heterologous carcinosarcomas, tumor cells from tissues not usually found in the uterus are identified. Some of the most common heterologous components are osteosarcoma (bone), rhabdomyosarcoma (skeletal muscle) and chondrosarcomas (cartilage).

The etiology of carcinosarcomas has been a topic of debate. There are three classic theories first described by Meyer: collision theory, combination theory and composition theory.[7,8] The collision theory is based on the premise that the carcinoma and sarcoma arise separately from one another, but in close enough proximity that the malignant cells eventually collide and become mixed with one another. In contrast, the combination theory suggests that both the carcinoma and sarcoma arise from the same pluripotential stem cell precursor. Lastly, the composition theory is characterized by the carcinoma and sarcoma arising from different cell lines at the same site. In this setting, a sarcoma arises from the stromal tissue within a pre-existing carcinoma. Recent advancements in immunohistochemical and molecular genetic studies, however, have updated what is known about the histopathogenesis of carcinosarcomas. Evidence has shown that carcinosarcomas should be considered high-grade variants of carcinoma. This most supports the combination theory of a monoclonal origin with subsequent divergent differentiation.[9]

Carcinosarcomas should be staged surgically similar to endometrial cancers. The recommended procedure includes total hysterectomy, bilateral salpingo-oophorectomy and bilateral pelvic and para-aortic lymph node dissection with a sampling of peritoneal washings (Fig. 11.2). A modified radical hysterectomy may need to be performed when there is evidence of tumor extension outside of the uterus. Some authorities would advocate more extensive surgical staging, to include omentectomy or other peritoneal biopsies.

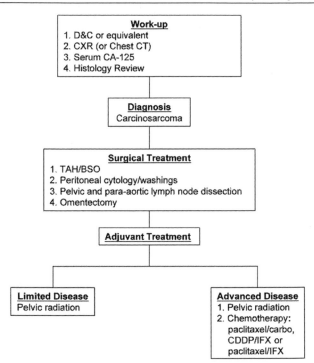

Figure 11.2. Management algorithm for carcinosarcoma.
Abbreviations: CXR: Chest radiograph; CT: Computed tomography; TAH: Total abdominal hysterectomy; BSO: Bilateral salpingo-oophorectomy; carbo: carboplatin; CDDP: cisplatin; IFX: ifosfamide.

Definitive standards for adjuvant therapy have not been established for uterine carcinosarcoma. Historically, radiation therapy has been the primary adjuvant modality. Radiation has been shown to reduce disease recurrence in the pelvis, but it has not been shown to improve overall survival.[10-12] Recent studies, however, have supported the role of adjuvant chemotherapy.[13-15] Common chemotherapy agents used include: cisplatin and ifosfamide,; paclitaxel and carboplatin, paclitaxel and ifosfamide (GOG#161); or gemcitabine and docetaxel.

Leiomyosarcoma

Leiomyosarcomas make up 40% of uterine sarcomas. They arise from the smooth muscle cells of the myometrium. Two of three pathologic criteria must be identified in order to make the diagnosis of a leiomyosarcoma: coagulative tumor cell necrosis, diffuse moderate to severe cytologic atypia and/or mitotic index greater than 10/10 hpf.[16]

The standard surgical treatment for leiomyosarcomas is total hysterectomy, bilateral salpingo-oophorectomy and sampling of peritoneal washings (Fig. 11.3). As opposed to carcinosarcomas, there has been no benefit found to performing routine lymph node dissection in these patients; however there is retrospective data which supports resecting and cytoreducing patients with gross nodal disease. These tumors tend to

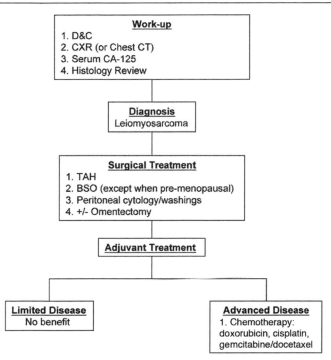

Figure 11.3. Management algorithm for Leiomyosarcoma.
Abbreviations: CXR: Chest radiograph; CT: Computed tomography; TAH: Total abdominal hysterectomy; BSO: Bilateral salpingo-oophorectomy.

spread hematogenously, and past studies have found that almost all patients with lymph node metastasis already have evidence of intraperitoneal disease spread.[17,18] If an enlarged lymph node is identified on preoperative imaging or found at the time of surgery, a selective biopsy may be performed to confirm metastatic disease or as previously mentioned, one could consider surgical cytoreduction which may be associated with improved outcomes.

Adjuvant radiation therapy has not been found to provide a clear benefit in the case of leiomyosarcoma of the uterus.[19,20] Two-thirds of recurrences occur outside of the pelvis and therefore outside of the radiation field. Radiation therapy is generally reserved for those cases where disease recurrence is isolated to the pelvis. Chemotherapy may be used in cases of late stage or recurrent disease. The most commonly used agents include doxorubicin, cisplatin, docetaxel and gemcitabine.

Endometrial Stromal Sarcoma

Endometrial stromal sarcomas account for only 8% of the uterine sarcomas. Endometrial stromal sarcomas arise from the cells within the stroma of the endometrium. There are two subtypes, low-grade endometrial stromal sarcoma (LGESS) and high-grade endometrial stromal sarcoma (HGESS). LGESS are characterized by proliferative phase endometrial stromal differentiation. HGESS have neoplastic cells that are similar to the sarcomatous portion of carcinosarcomas.

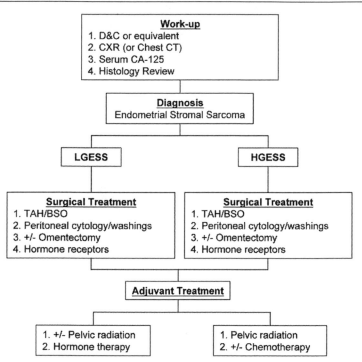

Figure 11.4. Management algorithm for endometrial stromal sarcoma. Abbreviations: CXR: Chest radiograph; CT: Computed tomography; TAH: Total abdominal hysterectomy; BSO: Bilateral salpingo-oophorectomy.

LGESS distinguish themselves from all of the other uterine sarcomas in their clinical behavior. Even though up to 50% of Stage I cases will recur, the recurrence usually does not occur until late, and the tumors grow very slowly. It is not uncommon for LGESS to recur 5 to 15 years after initial diagnosis. Because the prognosis of LGESS is so different from that of the other uterine sarcomas, including HGESS, most studies exclude them from analysis.

Surgical staging of endometrial stromal sarcomas involves collection of peritoneal washings and removal of the uterus, cervix, Fallopian tubes and ovaries (Fig. 11.4). If the diagnosis is made preoperatively, a discussion regarding removal of the ovaries is imperative secondary to the hormonal responsiveness of these tumors.

A modified radical hysterectomy is performed in the setting where there is evidence of tumor extension outside of the uterus. For HGESS, pelvic and para-aortic lymph node dissection should be performed to complete the staging procedure.

For endometrial stromal sarcoma, adjuvant pelvic radiation therapy has been found to decrease the risk of pelvic recurrence. Furthermore, endometrial stromal sarcomas (both LGESS and HGESS) have a high prevalence of progesterone receptors. All tumors should be tested for receptor status and patients who test positive should receive adjuvant progestin therapy. For patients who have receptor-negative tumors, chemotherapy with doxorubicin can be used.

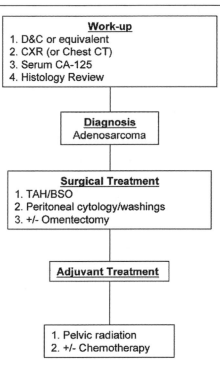

Figure 11.5. Management algorithm for adenosarcoma.
Abbreviations: CXR: Chest radiograph; CT: Computed tomography; TAH: Total abdominal hysterectomy; BSO: Bilateral salpingo-oophorectomy.

Adenosarcoma

Adenosarcomas are the least common type of uterine sarcoma and make up less than 2% of all sarcomas. Adenosarcomas are generally sessile or pedunculated polypoid masses that arise from the endometrium. These tumors often grow enough to fill the uterine cavity and like carcinosarcomas, can be found protruding from the cervical os. Adenosarcomas are characterized by a mixture of sarcomatous stromal cells with benign Müllerian epithelial cells.

The recommended treatment for adenosarcomas is total hysterectomy, bilateral salpingo-oophorectomy and sampling of peritoneal washings (Fig. 11.5). A modified radical hysterectomy may need to be performed when there is tumor extension to the parametrial tissue or if there is suspicion of involvement of the lower uterine segment or cervix.

Due to the extreme rarity of this tumor, there are no formal studies looking at adjuvant treatment for adenosarcomas. It is felt that adjuvant radiation therapy may provide a decrease in the risk of pelvic recurrence. This is based on findings for carcinosarcomas and high-grade endometrial sarcomas which have similar malignant stromal components.

Suggested Readings

1. Hoskins W, Perez C, Young R. Principles and Practice of Gynecologic Oncology. New York: Lippincott Williams and Wilkins, 2000.
2. Morrow CP, Curtin JP. Synopsis of Gynecologic Oncology. New York: Churchill Livingstone, 1998.
3. Gershenson DM, McGuire WP, Gore M et al. Gynecologic Cancer Controversies in Management. Philadelphia: Elsevier Churchill Livingstone, 2004.

References

1. Harlow BL, Weiss MS, Luften S. The epidemiology of sarcomas of the uterus. J Natl Cancer Inst 1986; 76:399.
2. Brooks SE, Zhan M, Cote T, Banquet C. Surveillance, epidemiology and end results of 2677 cases of uterine sarcoma 1989-1999. Gynecol Oncol 2004; 93:204-8.
3. Kempson RL, Bari W. Uterine sarcomas, classification, diagnosis and prognosis. Human Pathol 1970; 1:331-49.
4. Norris HJ, Taylor HB. Postirradiation sarcomas of the uterus. Obstet Gynecol 1965; 26(5):689-94.
5. Curtis RE, Freedman MD, Sehrman ME, Fraumeni Jr JFR. Risk of malignant mixed Müllerian Tumors after tamoxifen therapy for breast cancer. J Natl Cancer Inst 2004; 96(1):70-4.
6. Parker WH, Fu YS, Berek JS. Uterine sarcoma in patients operated on for presumed Leiomyosarcoma and rapidly growing leiomyoma. Obstet Gynecol 1994; 83:414.
7. Fenn M, Abell M. Carcinosarcoma of the ovary. Am J Obstet Gynecol 1971; 110:1066-74.
8. Sreenan JJ, Hart WR. Carcinosarcomas of the female genital tract. Am J Surg Pathol 1995; 19:666-74.
9. Ronnett BM, Zaino RJ, Ellenson LH, Kurman RJ. Blaustein's Pathology of the female genital tract. New York; Springer-Verlag, 2002.
10. Reed NS, Mangioni C, Malmstrom H et al. First results of a randomized trial comparing radiotherapy versus observation postoperatively in patients with uterine sarcomas. An EORTC-GCG Study. Int J Gynecol Cancer 2003; 12:4.
11. Salazar OM, Bonfilio TA, Patten SF et al. Uterine sarcomas: Analysis of failures with special emphasis on the use of adjuvant radiation therapy. Cancer 1978; 42:1161-70.
12. Hornback NB, Omuna G, Mahor FJ. Observations on the use of adjuvant radiation therapy in patients with stage I and II uterine sarcoma. Int J Radiat Oncol Biol Phys 1986; 12:2127-30.
13. Toyoshima M, Akahira J, Matsunaga G etal. Clinical experience with combination paclitaxel and carboplatin therapy for advanced or recurrent carcinosarcoma of the uterus. Gyencol Oncol 2004; 94:774-8.
14. Ramondetta LM, Burke TW, Jhingran A et al. A phase II trial of cisplatin, ifosfamide and mesna in patients with advanced or recurrent uterine malignant mixed müllerian tumors with evaluation of potential molecular targets. Gynecol Oncol 2003; 90:529-36.
15. Manolitsas TP, Wain GV, Williams KE et al. Multimodality therapy for patients with clinical stage I and II malignant mixed müllerian tumors of the uterus. Cancer 2001; 91 1437-43.
16. Bell SW, Kempson RL, Hendrickson MR. Problematic uterine smooth muscle neoplasms: a clinicopathologic study of 213 cases. Am J Surg Pathol 1994; 18:535.
17. Leibsohn S, D' Ablaing G, Mishell DR, Schlaerth JB. LMS in a series of hysterectomies performed for presumed uterine leiomyomas. Am J Obstet Gynecol 1990; 162:968.

18. Goff BA, Rue LW, Fleischhacker D et al. Uterine leiomyosarcoma and endometrial stromal sarcoma: lymph node metastases and sites of recurrence. Gynecol Oncol 1993; 50(1):105-9.

19. Hornback NB, Omura G, Major FJ. Observations on the use of adjuvant radiation therapy in patients with stage I and II uterine sarcoma. Int J Radiat Oncol Biol Phys 1986; 12:2127.

20. Berchuck A, Rubin SC, Hoskins WJ et al. Treatment of uterine leiomyosarcoma. Obstet Gynecol 1988; 71:845.

11

CHAPTER 12

Epithelial Ovarian Cancer

Jason A. Lachance and Laurel W. Rice

Introduction

Epithelial ovarian cancer is the leading cause of morbidity and mortality from gynecologic malignancies in the United States. Disease is most often detected at an advanced stage, which is associated with less favorable outcomes. Currently, there is no reliable screening test to detect early stage disease and primary prevention is unproven and controversial.

Epidemiology

Approximately 21,650 women in the United States will be diagnosed with epithelial ovarian cancer in 2008 and 15,520 women will die of disease. Ovarian cancer is the eighth most common cancer affecting women in the US today and the fourth most common cancer-related cause of death, after lung, breast and colon cancers. The lifetime risk of ovarian cancer for women in the United States is approximately 1 in 70 with a median age at diagnosis of 63 years (Table 12.1). While ovarian cancer incidence in African-American women is slightly lower than in Caucasians (10 cases per 100,000 women versus 13 cases per 100,000 women), it is more common in North American and Northern European women compared to Japanese women. These differences are multifactorial and related to genetics, diet, or environmental factors.

Risk Factors

The molecular and biological mechanisms of epithelial ovarian cancer are poorly understood. One commonly believed theory is that uninterrupted or incessant ovulation increases the probability of genetic errors occurring during repair of the ovarian epithelium leading to subsequent malignant proliferation. Risk factors are divided into patient-related or exposure-related.

A. Patient-Related

Increasing age is the strongest patient-related risk factor, with the incidence of ovarian cancer climbing with each year of life. In general, ovarian cancer is a disease of postmenopausal women with the median age of diagnosis at 63. The next strongest risk factor is family history. Approximately 5-10% of women with ovarian cancer have inherited genetic changes that predisposed them to ovarian cancer. While the risk of developing ovarian cancer in the general population is 1.8%, a woman with one first degree family member affected by this disease has a lifetime risk of 5% and with two first-degree relatives, the lifetime risk climbs to 25-50%. *BRCA1* and *BRCA2* mutations, which are rarely somatic, are the most completely described genetic abnormalities associated with ovarian cancer and as well, predispose to breast cancer (Figs. 12.1 and 12.2). Certain ethnic groups, such as Ashkenazi Jews,

Gynecologic Oncology, edited by Paola Gehrig and Angeles Secord.
©2009 Landes Bioscience.

Table 12.1. Independent prognostic factors for survival identified by cox multivariable proportional hazard model

Prognostic Factors			
Age (y)	**HR**	**95% CI**	**P**
<45	1.00		
45-54	1.22	0.70, 2.14	0.49
55-64	1.24	0.73, 2.12	0.42
65-74	1.97	1.17, 3.31	0.01
≥75	2.80	1.61, 4.86	<0.001
FIGO Stage			
I	1.00		
II	4.25	2.51, 7.20	<0.001
III	8.03	5.04, 12.77	<0.001
IV	11.75	6.99, 19.76	<0.001
Residual Disease			
<1cm	1.00		
≥1cm	1.72	1.33, 2.22	<0.001

Abbreviations: HR: hazard ratio; CI: confidence interval; FIGO: International Federation of Gynecology and Obstetrics. From: Tingulstad. Prognostic factors in ovarian cancer. Obstet Gynecol 2003, with permission.

more frequently carry *BRCA1* and *BRCA2* mutations. Up to 90% of all hereditary cases of ovarian cancer are associated with mutation of the *BRCA1* gene located on chromosome 17q21. Mutations of the *BRCA2* gene, located on chromosome 13q22, occur less commonly. Hereditary nonpolyposis colorectal syndrome (HNPCC), or Lynch Type II Cancer Syndrome, is responsible for the remaining 10% of hereditary ovarian cancers. In this syndrome, mutations of DNA mismatch repair genes leads to an elevated risk of proximal colon cancer, as well as synchronous malignancies of the stomach, uterine corpus, or ovary.

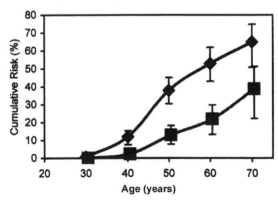

Figure 12.1. Cumulative risk of breast (♦) and ovarian (■) cancer in *BRCA1*-mutation carriers. From: Antonioul. Am J Hum Genet 2003, with permission.

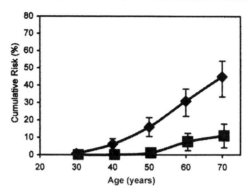

Figure 12.2. Cumulative risk of breast (♦) and ovarian (■) cancer in *BRCA2*-mutation carriers. From: Antonioul. Am J Hum Genet 2003, with permission.

Reproductive factors including nulliparity, infertility, late menopause, early menarche and first childbirth after age 35, all result in prolonged periods of uninterrupted ovulation, increasing the risk of epithelial ovarian cancer.

B. Exposure-Related

Exposure-related risk factors, including diet, tobacco, caffeine and alcohol, have not been implicated in ovarian carcinogenesis. Conflicting reports have implicated talc powder, which historically contained asbestos. Some data suggests that infertility agents, such as clomiphene citrate, increase the number of ovulations and thus may increase the risk of ovarian cancer. This has yet to be substantiated and may be confounded by factors such as infertility as described above.

Screening

Greater then 80% of women with ovarian cancer present with advanced stage disease. Presently, there is no reliable, cost-effective, clinically applicable screening program available, although significant effort continues to be directed towards identifying such a tool (please refer to Chapter 2). Screening tests for ovarian cancer are statistically limited by the overall low prevalence of the disease. Physical examination, CA125 laboratory blood testing and transvaginal ultrasounds are the cornerstones for ongoing efforts in the screening for ovarian cancer. Whether studied independently or in conjunction, none of the three have lowered the morbidity or mortality associated with this malignancy.

Physical examination, although inexpensive, has extremely poor sensitivity, specificity and positive predictive value. Extraovarian spread has frequently occurred by the time a mass can be appreciated by bimanual pelvic examination. Many palpable ovarian masses are benign, resulting in a high false positive rate when physical exam is used alone for screening.

Transvaginal ultrasound has an excellent sensitivity for detecting ovarian masses but an unacceptably high false positive rate secondary to the presence of many benign ovarian lesions. Scoring systems using morphology indices or Doppler flow have failed to establish reliable predictors for malignancy.

CA125 serum testing also has a high false positive rate because it is frequently elevated by numerous other conditions including but not limited to endometriosis, fibroids, diverticulitis, liver disease, or abdominopelvic infections.

While ovarian cancer screening for the general, asymptomatic, population is not recommended at this time, patients at high risk to develop ovarian cancer including women with *BRCA1* or *BRCA2* mutations, HNPCC carriers and women with two first degree relatives are more likely to benefit from screening, secondary to the increased prevalence of disease. The American College of Obstetrics and Gynecology Committee Opinion issued in 2002 points out there is no data to support screening in this population, aside from genetic testing to identify them as high risk. However, many practitioners perform transvaginal ultrasounds and serum CA125 tests at regular intervals in this high risk population, with consideration given to prophylactic oophorectomy (see below).

Histology

Approximately 65-70% of all ovarian malignancies are epithelial, with germ cell tumors (25%), sex cord stromal tumors (5%) and metastatic tumors to the ovaries (<5%) accounting for the remainder. Epithelial ovarian malignancies arise from the glandular epithelium of the ovary and are classified as adenocarcinomas. Histologic subtypes include serous, mucinous, endometrioid, transitional-cell, or clear-cell histology. Serous tumors are the most common subtype, comprising approximately 46% of all epithelial ovarian carcinomas, with mucinous tumors being the second most common. Clinically, mucinous tumors may grow quite large and can be associated with appendiceal tumors; appendectomy is recommended at the time of surgical resection of mucinous ovarian tumors, particularly for right sided-ovarian lesions. Over 10% of endometrioid tumors, resembling endometrial glands, are associated with endometriosis and some have suggested that they are the byproduct of the intense inflammatory reaction associated with endometriosis. Brenner tumors, or transitional cell neoplasms, are almost all benign (98%) and therefore have a favorable prognosis. Clear-cell tumors, which comprise 3% of epithelial ovarian carcinomas, have the poorest prognosis and are the only histological subtype that independently predicts poorer survival. Rarely, ovarian neoplasms demonstrate squamous differentiation or mixed differentiation.

Approximately 15-20% of all epithelial neoplasms are tumors of low malignant potential (LMP). These tumors are characterized by some degree of nuclear atypia, mitotic activity, cellular stacking or stratification and architecturally complex glands. The prognosis and treatment of this histologic subtype is clearly different from the frankly malignant invasive carcinomas. Nearly 75% of these tumors are Stage I at the time of diagnosis. In one series, the 5-, 10-, 15- and 20-year survival rates of patients with low malignant potential tumors (all stages), as demonstrated by clinical life table analysis, were 97%, 95%, 92% and 89%, respectively. Mortality was stage-dependent: 0.7%, 4.2% and 26.8% of patients with Stages I, II and III, respectively, died of disease. Chemotherapy is only occasionally prescribed in the treatment of women with LMP tumors and is dependent on certain histologic features. Disease recurrence can occur many years after the initial diagnosis, reflecting the indolent biologic behavior of this tumor.

Signs and Symptoms

Approximately 70-80% of women with ovarian cancer have advanced stage disease at the time of diagnosis, including large volume intra-abdominal tumor and massive ascites. Prior to diagnosis, women with ovarian cancer frequently have vague, nonspecific symptoms. The most common symptoms include abdominal bloating, early satiety, heartburn, constipation and nausea, as well as genitourinary symptoms including frequency, urgency, or incontinence. Shortness of breath from malignant pleural effusions or diaphragmatic disease is also common. All of these symptoms are frequently reported in numerous benign conditions including reflux, irritable bowel syndrome, cholelithiasis and urinary tract infections, adding to the challenge of diagnosing ovarian cancer in its earliest stages. Research has shown that the severity of these symptoms may be higher in women with ovarian cancer.

Differential Diagnosis

Several other conditions mimic ovarian carcinoma in their presenting signs and symptoms, and these must be ruled out prior to proceeding with the treatment for a presumed ovarian cancer. Gastrointestinal (GI) malignancies, including colon, gastric and appendiceal cancers, often present with similar signs and symptoms. When a GI tumor metastasizes to the ovaries, it is referred to as a Krukenberg tumor. Primary breast cancer and lymphoma may also present with signs and symptoms associated with ovarian metastases. It is imperative, when possible, to preoperatively determine the primary site of disease using history, physical exam, tumor markers, imaging studies and tissue biopsy. However, in a significant number of cases surgical evaluation with frozen section pathologic analysis is necessary to secure the diagnosis and establish appropriate surgical management.

A number of benign conditions including endometriosis, pelvic inflammatory disease (PID) and diverticulitis must also be excluded. All three can present with an elevated CA125 and a pelvic mass. A detailed menstrual history and sexual history may help to differentiate endometriosis and PID from ovarian cancer.

Evaluation

Initial work-up should include a comprehensive medical history, physical exam, laboratory work-up and imaging studies. Important components of the medical history include medical comorbidities, prior surgeries, obstetric and gynecologic history, as well as family history. Physical exam should include a careful evaluation of the lungs, heart and abdomen and a thorough pelvic and rectal exam. The history and physical exam are both used to estimate the likelihood the patient actually has ovarian cancer, to estimate the extent of disease and to assess whether or not the patient is an operative candidate. Laboratory work-up should include a CBC, coagulation panel and complete metabolic panel, including liver function tests. Bloating and early satiety leading to malnutrition can result in anemia, electrolyte abnormalities due to fluid shifts associated with ascites and coagulopathies. Tumor markers, including CA125, CEA and CA-19-9 must be considered. CA125 should be obtained whenever there is a suspicion for ovarian cancer. Although it has limited utility as a screening tool, CA125, if elevated in the presence of ovarian cancer, can be used as a tool to assess the patient's response to treatment, as well as an early marker for disease recurrence.

Standard preoperative work-up, including chest X-ray (rule out malignant effusions) and 12-lead EKG (rule out cardiac disease), must also be considered. Assessment of the gastrointestinal tract, including screening for colorectal carcinoma, as well as assessment of the genitourinary tract, is frequently indicated. The patient's other comorbid conditions dictate further diagnostic evaluation.

Initial radiographic evaluation when ovarian carcinoma is suspected almost always includes CT evaluation, either spiral or PET. Evaluating the extent and location of disease is critical in ascertaining tumor resectability. Unfortunately, preoperative radiographic evaluation is limited in predicting a surgeon's ability to achieve optimal cytoreduction (residual tumor <1 cm) of metastatic ovarian neoplasms.

Paracentesis and/or thoracentesis are many times necessary for symptomatic relief or in preparation for a surgical procedure. In patients where the diagnosis is uncertain or is not a surgical candidate, cytological analysis of either fluid can facilitate differentiating a benign from a malignant process, as well as contributing to staging information if malignant. Biopsy techniques including fine needle aspiration and core biopsy are sometimes necessary for the same purposes as delineated above for ascites or pleural effusions.

Treatment

Surgical Staging and Surgical Debulking

The nonspecific symptoms and the lack of a reliable screening test account for the high frequency with which patients are diagnosed with Stage III-IV. Reports from multiple investigators suggest that the number of women with advanced stage disease at the time of diagnosis exceeds 70-80%, with high stage correlating with poor survival. Of women with Stage I disease, 5-year survival approaches 95%, while for those with Stage III/IV, the survival falls to 20-30% (Fig. 12.3).

For both early and advanced stage ovarian cancer, surgery is a mainstay of treatment and has two specific goals: complete staging and cytoreduction. Establishing the stage of a malignancy, as well as the full extent of disease, is an essential component to treatment, including eligibility for treatment protocols. In 1986 the Federation of International Gynecologic Oncologists (FIGO) established strict criteria for the surgical staging of ovarian cancer.

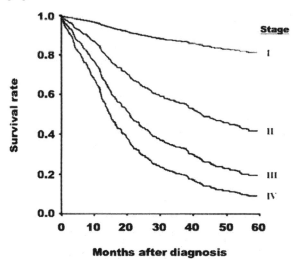

Figure 12.3. Estimated 5-year survival by International Federation of Gynecology and Obstetrics stage after adjusting for age and residual tumor. From: Tingulstad. Prognostic factors in ovarian cancer. Obstet Gynecol 2003, with permission.

In those ovarian cancer cases where extraovarian tumor is not obvious, complete surgical staging includes a specific sequence of tissue procurement, designed to rule out occult or microscopic disease. The rationale is based on basic knowledge and understanding of the mechanisms of ovarian cancer metastases. Ovarian cancer most commonly metastasizes by shedding cells from the ovarian epithelium. Thus, upon opening the abdomen when ovarian carcinoma is suspected, the surgeon should first obtain peritoneal washings or ascitic fluid to be sent for cytology. Following pathologic confirmation of malignancy, most commonly with bilateral salpingo-ophorectomy and total hysterectomy, multiple biopsies of the peritoneum are taken from the bladder serosa, the pelvic sidewalls, the anterior and posterior cul de sac, the paracolic gutters and the diaphragm. Lymph node metastases are also common. Pelvic and para-aortic lymph node sampling are mandatory when there is no evidence of disseminated disease. Hematogenous spread to the liver parenchyma, lung and distant sites is extremely rare and is almost always ruled out at the time of preoperative radiographic evaluation.

For advanced stage disease, the goal of surgery is to perform aggressive surgical cytoreduction, or "debulking" defined as "optimal" if the largest remaining tumor nodule is less than 1 in diameter. Bowel resection, splenectomy, diaphragmatic stripping and ablation of implants are frequently necessary to achieve an optimal cytoreduction. Numerous studies, including those by the Gynecology Oncology Group (GOG), have shown higher chemotherapy response rates, longer progression-free intervals and improved median survival in patients with an "optimal" surgical cytoreduction (Fig. 12.4).

In the 1970s it was common practice to perform a "second look laparotomy" to help identify patients at high risk for recurrence and those who would benefit from consolidation therapy, assuming that occult disease is often not detectable by conventional surveillance methods. A number of investigators have failed to show any survival benefit when this procedure is routinely performed. Patients with negative second-look

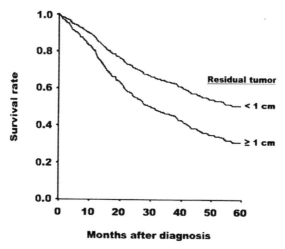

Figure 12.4. Estimated 5-year survival by residual tumor after adjusting for age and International Federation of Gynecology and Obstetrics stage. From: Tingulstad. Prognostic factors in ovarian cancer. Obstet Gynecol 2003, with permission.

laparotomies have a recurrence rate of 50%, supporting limited utility for this procedure. Thus, second look laparotomy or laparoscopy should be reserved for patients on particular clinical trial protocols that mandate this procedure.

Chemotherapy

Patients with Stage IA or IB disease and well or moderately differentiated tumors are considered to have a favorable prognosis and need no further treatment after their primary surgery. These patients have a 5-year survival rate that exceeds 95% without chemotherapy. Following surgery, all of these patients enter surveillance with history, physical exam and CA125 testing. Patients with Stage IA and IB tumors that have Grade 3 histology or any patients with Stage IC disease carry an unfavorable prognosis and it is recommended that they receive at least three cycles of platinum-based chemotherapy with paclitaxel. Some experts would recommend six cycles of the same chemotherapeutic agents for this patient population.

For patients with advanced stage disease, with or without optimal cytoreduction at the time of diagnosis, the current standard induction treatment is 6-8 cycles of platinum and taxane-based chemotherapy. Approximately 80-90% of women with epithelial ovarian carcinoma will initially respond to these chemotherapeutic agents. The 10-20% of patients who are resistant to platinum and taxane-based chemotherapy represent a subset of patients with an extremely poor prognosis. Second-line chemotherapy or enrollment in Phase I/II clinical trials are the only available options. At the completion of chemotherapy, a normal clinical exam, a normal CA125 and a negative CT scan support disease remission. At this juncture, patients can consider investigational protocols or enrollment in periodic surveillance, including physical exam and serum CA125, every 4 months for 2 years. The use of routine annual CT scans can be considered, but it is required in the presence an abnormal physical exam, a rising CA125, or a symptomatic patient. After 2-3 years of negative surveillance, the screening interval can be increased to every 6 months for 2-3 more years and then annually. If residual disease is documented after the completion of induction chemotherapy, either by physical exam, CA125 or CT evaluation, options include the administration of the same chemotherapeutic agents or second-line chemotherapeutic agents; investigational protocols should always be considered. When remission is achieved, there are no well-established guidelines for consolidation chemotherapy, and presently this type of treatment is considered investigational.

Successfully achieving optimal cytoreduction at the time of initial surgery varies between 20-80%, although it is generally higher when performed by subspecialists in gynecologic oncology. In a recent meta-analysis reviewing 53 studies, the average rate of optimal cytoreduction was 42%, leaving a significant percentage of patients in the suboptimal surgical category, with its associated poor prognosis. The rate of optimal cytoreduction is dependent on tumor volume, tumor location, surgeon and institutional philosophies. Patients who have a suboptimal cytoreductive procedure incur the morbidity and mortality of surgery without reaping the compensatory survival benefits.

In patients who are successfully cytoreduced to optimal status, the use of intraperitoneal chemotherapy has emerged as a successful adjuvant for postoperative consolidation. The concept of intraperitoneal chemotherapy has existed for several years but has been limited by difficulty with delivery of the agents, as well as historical bias including the simultaneous emergence of platinum and taxane based intravenous therapy. In this therapeutic method, an indwelling catheter is placed through the abdominal wall in order to deliver chemotherapeutic agents directly into the abdominal cavity where they directly contact cancer cells on peritoneal surfaces. Recent improvements in catheters and delivery techniques have brought this concept to the forefront once more, and recently a large cooperative group conducted a randomized controlled trial which demonstrated a marked survival advantage of 16 months when compared to conventional intravenous chemotherapy. At present, this approach can only be recommended for optimally cytoreduced patients as smaller diameter tumor nodules are more amenable to penetration of chemotherapeutic agents; however it provides hope for continued progress in treatment of this disease.

Several investigators have reported on the use of "neoadjuvant chemotherapy" followed by cytoreductive surgery as an option for patients with a high preoperative probability of a suboptimal initial surgical procedure. Proponents of "neoadjuvant chemotherapy" have suggested that this treatment alternative reduces the morbidity and mortality of an "unsuccessful" surgical procedure while possibly retaining the survival benefits of initial surgery. Several studies have shown less operative morbidity, higher rates of optimal surgical cytoreduction and comparable survival in patients treated with "neoadjuvant chemotherapy". There are often preoperative clinical findings that predict an increased likelihood of a suboptimal debulking, including disease in the bowel mesentery, liver parenchyma, or disseminated peritoneal disease. When unresectability is suspected, clinicians will obtain a tissue diagnosis via fine needle aspiration or core biopsy of easily accessible tumor to confirm the diagnosis. Patients then receive three or four cycles of carboplatin and paclitaxel. If they respond favorably to the chemotherapy, an interval debulking surgery is performed, with an increased likelihood of achieving an optimal result. After surgery, the patient receives an additional 3-6 cycles of chemotherapy. To date, no prospective randomized, controlled trials have been published evaluating the efficacy of "neoadjuvant chemotherapy" compared to initial surgery, but clinical trials are currently ongoing.

Recurrent Disease Survival/Prognosis

The 5-year survival for women with advanced stage ovarian carcinoma is 20-25% overall, and with disease recurrence the probability of cure is negligible. The signs of recurrence include, pain, nausea, reflux, or constipation. Surveillance with history/physical exam, CA125 and CT imaging has failed to improve overall survival although disease recurrence can be documented at an earlier point.

Recurrent disease is classified as either early or late. Patients with early recurrence are those who recur less than 6 months after completion of platinum based chemotherapy and are considered "platinum-resistant." When second line therapy is compared to supportive care in platinum-resistant patients, no studies have shown improvement in progression-free interval or quality of life. Treatment in this patient population includes second line, or "relapse chemotherapy" including etoposide, topotecan, doxorubicin, liposomal doxorubicin, taxotere, gemcitabine, or weekly paclitaxel. Less than 10% of patients who recur after a prior complete remission will reach complete remission once more, and their median survival is less than 1 year.

Patients who recur more than 6 months after completion of their platinum based therapy are classified as "platinum-sensitive." Relapse chemotherapy for these patients includes re-introduction to a platinum-based regimen. Response rates range from 20-60% and the probability of response is more likely with a prolonged progression-free interval between completion of primary therapy and disease recurrence.

Palliation

The poor prognosis associated with recurrent ovarian cancer, including platinum-sensitive and -resistant disease, must be discussed openly in the context of end of life issues. A false sense of hope should not be conveyed in this situation given the small number of women who are successfully cured. It is a patient's right to have a reasonable understanding and expectation of their disease course. The risk/benefit ratio of multiple surgeries, as well as third and fourth line chemotherapy, must be discussed in detail, recognizing the potential morbidity, diminished quality of life and possible lack of survival benefit. The recognized cost can also include long hospitalizations with diminished time with family and friends. Topics such as code status, hospice care and pain management need to be included in the dialogue between physician and patient. In the setting of recurrent disease, quality of life and minimization of toxicity need to be key goals of any therapy.

Prevention

Chemoprevention

The prevention of ovarian cancer is an area of increasing interest. In genetically uncharacterized women in the United States, the life-time risk of developing ovarian cancer is approximately 1.5%. *BRCA1* mutations increase the risk to 30-60%, while women who harbor a mutant *BRCA2* gene have an estimated risk of 10-30%. Prevention efforts can be best assessed in the context of genetic predisposition. Biochemical markers of transformation, including cell cycle progression and apoptosis, serving as surrogate endpoint markers, have potential utility in future ovarian cancer chemoprevention trials.

Oral contraceptive pills (OCPs) are estimated to reduce the overall risk of ovarian cancer by approximately 40%. Little is known about the mechanism of the protective effect of OCPs against ovarian cancer, although it has been postulated that a major mechanism of OCP protection relates to a decrease in ovulatory cycles. There is data to suggest that an increased rate of apoptosis in aberrant epithelial cells secondary to the progestational component may also play an important role. The degree of protection and the length of protection appear to be associated with the duration of OCP usage. One of the largest case-control studies to date is the Centers for Disease Control Cancer and Steroid Hormone Study. In this study, 546 women with ovarian cancer and 4228 control subjects from eight population-based cancer registries were compared. Women with a history of OCP use had a 40% reduced risk of epithelial ovarian cancer when compared to women with no history of OCP use. This protective effect was evident with as little as 3 months of OCP use and continued for up to 15 years following discontinuation of OCP use. It has been estimated that more than half of all ovarian cancers in the United States could be prevented by OCP usage for at least 4-5 years. Identical risk reduction for OCPs has been shown for both high-estrogen/high-progesterone content and low-estrogen/low-progesterone content pills. The lifetime risk of ovarian cancer is approximately 45% in *BRCA1* carriers and 25% among *BRCA2* carriers and in this patient population, the use of OCPs as a chemopreventive agent against ovarian cancer should be considered.

Surgery

The low overall incidence of ovarian cancer in the general population accounts for the limited role of prophylactic oophorectomy for ovarian cancer risk reduction in this patient population. However, in *BRCA1* and *BRCA2* carriers, the risk of developing ovarian carcinoma is significantly greater, as described above. The American College of Obstetrics and Gynecology, as well as numerous cancer societies, advocate prophylactic oophorectomy after the age of 35 in women with these genetic syndromes who have completed child-bearing.

Future Directions

Given the current outlook for women with advanced stage ovarian cancer, extensive research efforts are currently being directed towards targeted therapies. Borrowing from models in other malignancies such as colon cancer, lymphoma and melanoma, investigators are actively pursuing therapies aimed and reducing toxicity and improving survival. These biologic agents fall into several categories. Targeted therapies disrupt known carcinogenic pathways—such as humanized monoclonal antibodies to growth factor receptors, tyrosine kinase inhibitors, small molecule inhibitors and other signal transduction modulators. Anti-angiogenesis is another pathway, and monoclonal antibodies to vascular endothelial growth factor (VEGF) have shown efficacy in treatment of metastatic colon cancer and lung cancer. Passive and active immunotherapies are also under investigation, including the study of vaccines as well as antibodies to the CA125 molecule. Therapeutic genetic strategies remain far from clinical use, but research continues towards the goal of individually tailored cancer therapy, specific to an individual's tumor.

References

1. The reduction in risk of ovarian cancer associated with oral-contraceptive use. The cancer and steroid hormone study of the centers for disease control and the National Institute of Child Health and Human Development. N Engl J Med 1987; 316:650-5.
2. FLGO (International Federation of Gynecology and Obstetrics) annual report on the results of treatment in gynecologic cancer. Int J Gynaecol Obstet 2003; 83(Suppl 1:)ix-xxii, 1-229.
3. Antoniou A, Pharoah PD, Narod S et al. Average risks of breast and ovarian cancer associated with BRCA1 or BRCA2 mutations detected in case Series unselected for family history: a combined analysis of 22 studies. Am J Hum Genet 2003; 72:1117-30.
4. Barnhill DR, Kurman RJ, Brady MF et al. Preliminary analysis of the behavior of stage I ovarian serous tumors of low malignant potential: a gynecologic oncology group study. J Clin Oncol 1995; 13:2752-6.
5. Gadducci A, Landoni F, Sartori E et al. Analysis of treatment failures and survival of patients with Fallopian tube carcinoma: a cooperation task force (CTF) study. Gynecol Oncol 2001; 81:150-9.
6. Goff BA, Mandel LS, Melancon CH et al. Frequency of symptoms of ovarian cancer in women presenting to primary care clinics. JAMA 2004; 291:2705-12.
7. Jemal A, Murray T, Ward E et al. Cancer statistics. CA Cancer J Clin 2005; 55:10-30.
8. McGuire WP, Hoskins WJ, Brady MF et al. Cyclophosphamide and cisplatin compared with paclitaxel and cisplatin in patients with stage III and stage IV ovarian cancer. N Engl J Med 1996; 334:1-6.
9. Ness RB, Grisso JA, Vergona R et al. Oral contraceptives, other methods of contraception and risk reduction for ovarian cancer. Epidemiology 2001; 12:307-312.
10. Randall TC, Rubin SC. Cytoreductive surgery for ovarian cancer. Surg Clin North Am 2001; 81:871-83.
11. Tingulstad S, Skjeldestad FE, Halvorsen TB et al. Survival and prognostic factors in patients with ovarian cancer. Obstet Gynecol 2003; 101:885-91.
12. Whitehouse C, Solomon E. Current status of the molecular characterization of the ovarian cancer antigen CA125 and implications for its use in clinical screening. Gynecol Oncol 2003; 88:S152-7.

CHAPTER 13

Fallopian Tube Carcinoma

Edward Tanner, Russell S. Vang and Robert L. Giuntoli, II

Introduction

Fallopian tube carcinoma is a rare gynecological malignancy encompassing less than 1% of all those arising from the genital tract.[1] Characterization of the disease and its treatment are hindered by the small numbers of patients affected. In a review of the literature, only six articles include more than 100 patients. Similarities to other gynecologic malignancies exist however, most notably with regards to ovarian cancer. These similarities have been used to infer prognosis and guide treatment. Fortunately, unlike epithelial ovarian cancer, Fallopian tube carcinoma is frequently diagnosed at an earlier stage, leading to comparatively improved outcomes. However, one needs to consider the possibility that many "ovarian" cancers may in fact be Fallopian tube carcinomas that could not be accurately diagnosed secondary to the extent of disease.

Epidemiology

In the United States, Fallopian tube carcinoma has an average annual incidence of approximately 3.6 cases per million women.[2] The peak incidence is approximately 56 years of age with cases described in a range from 18 to 88. Some evidence suggests that the incidence of Fallopian tube carcinoma is increasing, at least in part due to the aging population.[3] Proposed risk factors for the development of Fallopian tube carcinoma include chronic inflammation due to pelvic infections, pelvic tuberculosis and tubal endometriosis.[4] As with ovarian cancer, a definite association between Fallopian tube carcinoma and *BRCA* germ line mutations has been noted with a 16% rate of *BRCA1* or *2* mutations in patients with Fallopian tube carcinomas. This rate is far higher that the 0.24% prevalence of *BRCA1* mutations in the general population.[5,6] Prophylactic bilateral salpingo-oophorectomy has been advocated to reduce the risk of ovarian and Fallopian tube cancer in *BRCA* positive women. The surgical quandary involves whether or not to remove the uterus as a small portion of the Fallopian tube may remain imbedded in the uterine cornua.

Presentation

Many cases of Fallopian tube carcinoma are symptomatic when the disease remains in an early stage; however, the majority of these symptoms are nonspecific. Most commonly, complaints include postmenopausal vaginal bleeding (50%) and colicky abdominal pain (20-40%).[7,8] Latzko's classic triad, defined as the presence of a profuse watery vaginal discharge, pelvic pain and pelvic mass, is found in less than 15% of patients. An uncommon finding of Fallopian tube carcinoma is hydrops tubae profluens which is characterized by intermittent profuse watery or serosanguinous discharge occurring at the time of intermittent decompression of an obstructed tube. While pathognomonic, this feature is noted in only 5% of cases.[9]

Gynecologic Oncology, edited by Paola Gehrig and Angeles Secord.
©2009 Landes Bioscience.

Findings on pelvic examination are nonspecific and can include palpation of an adnexal mass or leucorrhea. While an endometrial malignancy is more likely, the presence of adenocarcinoma on Papanicolaou smear leads to the diagnosis of Fallopian tube carcinoma in approximately 14% of cases.[10] This diagnosis should be considered especially if endometrial and endocervical curettage is negative.

A variety of tumor markers have been evaluated regarding the early detection of Fallopian tube malignancies. Cancer antigen 125 (CA125) values greater than 65 U/mL correlate with either an ovarian, primary peritoneal or Fallopian tube malignancy with a sensitivity of 98% and specificity of 75%. However, benign conditions can also lead to elevated CA125 levels including endometriosis, fibroids, pregnancy and pelvic inflammatory disease. In addition, CA125 levels are only consistently elevated in advanced-stage disease when the opportunity for early detection and cure has been lost. CA125 therefore does not function well as an independent preoperative predictor of an early-stage Fallopian tube malignancy or as a screening test for the disease. CA125 levels can be used to guide treatment as will be discussed later.

Several imaging modalities have been evaluated with regards to preoperative predictive value including computerized tomography (CT), magnetic resonance imaging (MRI) and sonography. A sausage-shaped mass with papillary projections or adnexal mass are most commonly described on sonography but do not frequently discriminate between malignancy and other benign pathologies such as tubo-ovarian abscess and hydrosalpinx. Color doppler, which can detect increased vascularity associated with a malignancy, may further aid in discriminating between these etiologies.[11] Although case reports exist in the literature, CT is unlikely to specifically predict a Fallopian tube malignancy; however, the presence of a solid or complex adnexal mass with or without ascites would still likely lead to appropriate surgical exploration.[12] MRI can also be used to evaluate for invasion into surrounding tissues such as the bowel and bladder. Despite these modalities, a correct diagnosis of Fallopian tube carcinoma is made in at most only 6% of cases preoperatively.[13]

Pathology

Ovarian cancer may involve the Fallopian tube and vice versa. In cases of advanced disease, determination of the site of tumor origin may not be readily apparent. Currently, immunohistochemistry studies are unable to consistently differentiate between serous adenocarcinomas arising from the ovary and Fallopian tubes. In cases of carcinoma involving both the ovary and Fallopian tube, it is not known whether an intra-epithelial component in the Fallopian tube indicates a true Fallopian tube primary, secondary Fallopian tube involvement by an ovarian primary (with extension into the Fallopian tube in an intra-epithelial fashion), or that both sites represent independent tumors involving the ovaries, Fallopian tubes and peritoneum in a multifocal fashion.[14]

In 1978, Sedlis proposed specific diagnostic criteria to define primary Fallopian tube carcinoma.[15]

a. The main tumor lies in the tube and arises from the endosalphinx;
b. Histologically, the pattern resembles tubal mucosa and is often papillary;
c. If the wall is involved, transition from benign to malignant epithelium must be demonstrated; and
d. The ovaries and uterus are normal or contain less tumor than the tubes.

The vast majority of Fallopian tube carcinomas arise from coelomic epithelium with 12% occurring bilaterally.[16] As in ovarian malignancies, papillary serous adenocarcinoma encompasses the majority of these tumors (90%). A gross photograph of

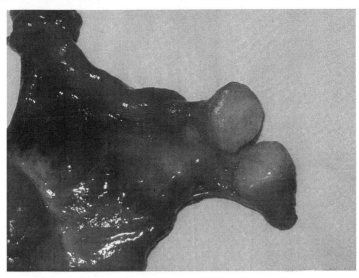

13

Figure 13.1. Photomicrograph of a primary Fallopian tube carcinoma specimen.

a primary Fallopian tube carcinoma specimen is shown in Figure 13.1. Less common histological subtypes include endometrioid, clear cell and mucinous tumors.[15] Serous and clear cell histologies confer a worse prognosis. Germ cell tumors and sarcomas have also been reported. The fimbriated end is the most common site of early serous carcinoma of the Fallopian tube in women with *BRCA* mutations, thus allowing for the recommendation of a bilateral salpingo-oophorectomy without hysterectomy in these high-risk women.[17]

Surgical Staging

Definitive management of Fallopian tube carcinoma is guided by the course of disease spread. Initially, malignant cells grow and invade through the Fallopian tube lumen as well as directly into the remainder of the adnexa and uterus. Exfoliation of malignant cells by either transcoelomic spread or passage through the tubal fimbriae leads to seeding of the abdominal cavity. In contrast, lymphatic drainage leads to involvement of the retroperitoneal lymph nodes. Hematological spread is usually a late and relatively uncommon finding.

The surgical staging of Fallopian tube carcinoma is identical to that for ovarian cancer. As a result, intraoperative pathologic differentiation between ovarian and Fallopian tube cancer is not vital. Staging includes total hysterectomy, bilateral salpingo-oopherectomy, tumor debulking, omentectomy, sampling of ascites or peritoneal washings, peritoneal, diaphragmatic, bowel and bladder biopsies and retroperitoneal lymph node dissection to evaluate all areas at risk for disease spread. Based on ovarian cancer data, improved survival from cytoreductive surgery begins to become apparent with tumor debulking to <2 cm for any individual implant. Optimal debulking is defined as removal of all disease ≥1 cm. Optimal cytoreduction with no gross residual disease affords maximum survival benefit. The International Federation of Gynecology

Table 13.1. Fallopian tube cancer staging[8]

Stage	Criteria
Stage 0	Carcinoma in situ (limited to tubal mucosa)
Stage I	Growth limited to the Fallopian tubes
Ia	Growth limited to one tube with extension into submucosa and/or muscularis
Ib	Growth limited to both tubes with extension into submucosa and/or muscularis
Ic	Tumor either Stage 1A or 1B with tumor extension through or onto the tubal serosa or with ascites containing malignant cells or with positive peritoneal washings
Stage II	Growth involving one or both Fallopian tubes with pelvic extension
IIa	Extension and/or metastasis to the uterus and/or ovaries
IIb	Extension to other pelvic tissues
IIc	Tumor either Stage IIA or IIB and with ascites containing malignant cells or with positive peritoneal washing
Stage III	Tumor involving one or both Fallopian tubes with peritoneal implants outside of the pelvis and/or positive retroperitoneal or inguinal nodes
IIIa	Tumor grossly limited to the true pelvis with negative nodes but with histologically confirmed microscopic seeding of abdominal peritoneal surfaces
IIIb	Tumor involving one or both tubes with histologically confirmed implants of abdominal peritoneal surfaces, none exceeding 2 cm in diameter
IIIc	Abdominal implants greater than 2 cm in diameter and/or positive retroperitoneal or inguinal nodes
Stage IV	Growth invading one or both Fallopian tubes with distant metastases. If pleural effusion is present, there must be positive cytology to be Stage IV. Parenchymal liver metastases equal Stage IV.

13

and Obstetrics (FIGO) adopted a staging classification for Fallopian tube carcinoma in 1992. The poor prognosis of muscular wall invasion has been recognized since that time, leading to the acceptance of a modified FIGO staging criteria (Table 13.1).[18] Additional modifications have been proposed for Stage I tumors involving only the fimbriated end of the Fallopian tube. As these tumors are exposed to the peritoneal cavity, the prognosis is worse when compared to other minimally invasive tumors occurring more proximally in the tube. The designation Stage I(F) has been proposed to account for these cases.[19]

Clinical data suggest an improved prognosis when compared to ovarian cancer but may over-represent the survival of advanced-stage disease that is attributed to advanced-stage ovarian cancer when the site of origin cannot be determined. Approximately 50% of patients have disease confined to the pelvis.[17] Higher stage disease correlates with worse prognosis.[20] The percentage of patients presenting with FIGO Stage I-IV disease and 5-year survival for the corresponding stage are as shown (Table 13.2).

When disease appears grossly confined to the Fallopian tube, surgical staging may uncover occult metastases. Fallopian tube carcinoma has a significant propensity for metastasis to the retroperitoneal lymph nodes with pelvic and para-aortic lymph node

Table 13.2. Stage distribution at presentation and 5-year survival of Fallopian tube carcinoma by FIGO stage

	Distribution (%)	5-Year Survival (%)
Stage I	27.0	95
Stage II	21.5	75
Stage III	34.5	69
Stage IV	11.5	45

involvement reported as high as 50%. A significant fraction of these patients (40%), have para-aortic lymph node involvement as their only evidence of metastatic disease.[21] In cases of advanced disease, optimal cytoreductive surgery appears to be therapeutic.

Traditionally, surgical staging of Fallopian tube and ovarian cancer involves exploratory laparotomy to ensure adequate visualization of the abdomen and pelvis. Recently, interest has focused on the feasibility of complete laparoscopic staging procedures in hopes of reducing operative morbidity. Small studies have shown that laparoscopic staging may be a safe and feasible option in selected patients.[22] This approach is more easily accomplished in women with a normal body mass index (BMI) and suspected Stage I disease. Port site metastases are reported in 1.7% of laparoscopic procedures on Fallopian tube carcinoma patients. This condition appears to be a risk primarily in patients with advanced-stage disease.[23]

Fertility-sparing surgery has been proposed for premenopausal patients with early-stage disease who want to preserve their fertility. In the presence of unilateral disease confined to the tubal lumen, unilateral salpingo-oophorectomy, lymphadenectomy, omentectomy and staging biopsies with close clinical follow-up has been utilized with success. In patients who no longer desire future fertility or who have extensive involvement of the pelvis, total hysterectomy and bilateral salpingo-oophorectomy may represent the best surgical option.

Although substantial investigations evaluating the success of such secondary cytoreductive surgery in Fallopian tube cancer are lacking, several studies have demonstrated procedure improved progression-free and overall survival in patients with ovarian carcinoma.[24,25] Secondary cytoreduction may be considered in select patients with recurrent Fallopian tube cancer. Factors such as location and number of implants and time since completion of chemotherapy should influence the decision to proceed with surgery.

Adjuvant Therapy

Adjuvant treatment is often suggested for Fallopian tube carcinoma. In cases of advanced disease, therapy is given to treat residual microscopic or macroscopic disease. Given the propensity for occult metastatic disease, especially in the para-aortic lymph nodes, adjuvant therapy is typically recommended even in Stage I disease. Both chemotherapy and radiation have been utilized as adjuvant therapies. Support for these treatments has largely been translated from investigations involving patients with ovarian malignancies.

No large prospective trials assessing adjuvant chemotherapy have been performed due to the small number of patients affected. As with epithelial ovarian cancer, platinum-based chemotherapy remains the primary agent, whether as a single agent or part of a multiagent regimen. In recent years, intravenous carboplatin and pacli-taxel has been evaluated given its excellent response in epithelial ovarian cancer. A retrospective review of the use of carboplatin and paclitaxel in patients with optimally debulked predominantly Stage III and IV Fallopian tube carcinoma demonstrated a progression-free survival of 28 months. These results were suggestive of an improve-ment in overall survival when compared to patients receiving only platinum-based chemotherapy at the same institution in the preceding years. Optimally cytoreduction (<1 cm residual disease) was associated with a 3-year progression-free survival of 67% compared with 45% after suboptimally debulking (≥1 cm residual disease). This dif-ference was statistically significant.[26]

Postoperative radiation therapy has been evaluated as adjuvant treatment of Fallopian tube carcinoma. This frequently includes whole pelvic external-beam radiation to a dose of 50 Gray with or without additional treatment of the para-aortic lymph node chains. The lack of prospective trials and inconsistent regimens in the limited literature that exists hinders establishing a standard of care for radiation therapy.

Several approaches have been utilized in evaluation of disease response and monitor-ing after adjuvant therapy. The value of a second-look laparotomy to assess for clinical response has not been determined. Serum CA125 levels can be used to assess clinical response to therapy with a decrease suggestive of chemotherapy-sensitive disease. In addition, an increase in CA125 levels precedes the clinical or radiographic onset of recurrent disease in 90% of patients.[27]

BRCA Mutations and Fallopian Tube Carcinoma

A strong association exists between deleterious *BRCA* germ line mutations and Fallopian tube carcinoma. *BRCA* mutation testing is often performed in women with a strong family and/or personal history of breast and/or ovarian cancer. If a *BRCA1* mutation is discovered, the subsequent risk of developing ovarian cancer is estimated at 20-40%.[28] Prophylactic bilateral oophorectomy in women with *BRCA1* or 2 mutations significantly reduces the risk of coelomic epithelial cancer (primary peritoneal carcinoma) and breast cancer.[29] Similarly, in a pathology review of *BRCA*-positive breast cancer patients undergoing bilateral salpingo-oophorec-tomy for ovarian cancer prophylaxis, a significant minority of specimens (15%) evaluated by the pathologist contain a small focus of Fallopian tube carcinoma.[30] Therefore, the Fallopian tubes should also be removed at the time of prophylactic oophorectomy. Given the persistence of an interstitial portion of the Fallopian tube following salpingo-oophorectomy, without removal of the uterus, some investigators have recommended hysterectomy at the time of prophylactic surgery although this hypothesis has not been tested.

Future Directions

Improvements in the outcomes of patients with Fallopian tube carcinoma are likely to mirror advances in the treatment of ovarian cancer. Screening biomarkers are likely to lead to significant improvements in survival if such screening tests can lead to diagnosis at an early stage. For patients with advanced disease, recent evidence suggests a survival benefit for patients with Stage IIIC epithelial ovarian cancer receiving adjuvant intra-peritoneal cisplatin and paclitaxel instead of intravenous alone.[31] Such improvements are likely to translate to patients with advanced Fallopian tube carcinoma. Therapies targeting the immune response to solid tumors as well as angiogenesis inhibitors may also hold promise for Fallopian tube cancer patients.

Suggested Reading

1. Alvarado-Cabrero I, Young RH, Vamvakas EC et al. Carcinoma of the Fallopian tube: a clinicopathological study of 105 cases with observations on staging and prognostic factors. Gynecol Oncol 1999; 72(3):367-79.
2. Rebbeck TR, Lynch HT, Neuhausen SL et al. Prophylactic oopherectomy in carriers of BRCA1 or BRCA2 mutations. NEJM 2002; 346(21):1616-22.
3. Gemignani ML, Hensley ML, Cohen R et al. Paclitaxeol-based chemotherapy in carcinoma of the Fallopian tube. Gynecol Oncol 2001; 80(1):16-20.

References

1. Nordin AJ. Primary carcinoma of the Fallopian tube: a 20-year literature review. Obstet Gynecol Surv 1994; 49(5):349–61.
2. Rosenblat KA, Weiss NS, Schwartz SM. Incidence of malignant Fallopian tube tumors. Gynecol Oncol 1989; 35(2):236-9.
3. Riska A, Leminen A, Pukkala E. Sociodemographic determinants of incidence of primary Fallopian tube carcinoma, Finland 1953-97. Int J Cancer 2003; 104(5): 643-5.
4. Demopoulos RI, Aronov R, Mesia A. Clues to the pathogenesis of Fallopian tube carcinoma: a morphological and immunohistochemical case control study. Int J Gynecol Pathol 2001; 20(2):128-32.
5. Aziz S, Kuperstein G, Rosen B et al. A genetic epidemiological study of carcinoma of the Fallopian tube. Gynecol Oncol 2001; 80(3):341-5.
6. Whittemore AS, Gong G, John EM et al. Prevalence of BRCA1 mutation carriers among U.S. non-Hispanic Whites. Cancer Epidemiol Biomarkers Prev 2004; 13(12):2078-83.
7. King A, Seraj IM, Thrasher T et al. Fallopian tube carcinoma: a clinicopathological study of 17 cases. Gynecol Oncol 1989; 33(3):351–5.
8. Obermair A, Taylor KH, Janda M et al. Primary Fallopian tube carcinoma: the Queensland experience. Int J Gynecol Cancer 2001; 11(1):69-72.
9. Ajithkumar TV, Minimole AL, John MM et al. Primary Fallopian tube carcinoma. Obstet Gynecol Surv 2005; 60(4):247-52.
10. Sasagawa M, Nishino K, Honma S et al. Origin of adenocarcinoma cells observed on cervical cytology. Acta Cytol 2003; 47(3):410-4.
11. Kurjak A, Kupesic S, Ilijas M et al. Preoperative diagnosis of primary Fallopian tube carcinoma. Gynecol Oncol 1998; 68(1):29-34.
12. Slanetz PJ, Whitman GJ, Halpern EF et al. Imaging of Fallopian tube tumors. Am J Roentgenol 1997; 169(5):1321-4.
13. Podratz KC, Schray MF, Gaffey TA et al. Primary carcinoma of the Fallopian tube. Am J Obstet Gynecol 1986; 154(6):1319-26.
14. Lee Y, Medeiros F, Kindelberger D et al. Advances in the recognition of tubal intraepithelial carcinoma: applications to cancer screening and the pathogenesis of ovarian cancer. Adv Anat Pathol 2006; 13(1):1-7.
15. Sedlis A. Carcinoma of the Fallopian tube. Surg Clin North Am 1978; 58(1):121-9.
16. Baekelandt M, Jorunn Nesbakken A, Kristensen GB et al. Carcinoma of the Fallopian tube. Cancer 2000; 89(10):2076-84.
17. Medeiros F, Muto MG, Lee Y et al. The tubal fimbria is a preferred site for early adenocarcinoma in women with familial ovarian cancer syndrome. Am J Surg Pathol 2006; 30(2):230-6.
18. Alvarado-Cabrero I, Young RH, Vamvakas EC et al. Carcinoma of the Fallopian tube: a clinicopathological study of 105 cases with observations on staging and prognostic factors. Gynecol Oncol 1999; 72(3):367-79.

13

19. Alvarado-Cabrero I, Navani SS, Young RH et al. Tumors of the fimbriated end of the Fallopian tube: a clinicopathologic analysis of 20 cases, including nine carcinomas. Int J Gynecol Pathol 1997; 16(3):189-96.

20. Kosary C, Trimble EL. Treatment and survival for women with Fallopian tube carcinoma: a population-based study. Gynecol Oncol 2002; 86(2):190-1.

21. Tamimi HK, Figge DC. Adenocarcinoma of the uterine tube: potential for lymph node metastasis. Am J Obstet Gynecol 1981; 141(2):132-7.

22. Chi DS, Abu-Rustum NR, Sonoda Y et al. The safety and efficacy of laparoscopic surgical staging of apparent stage I ovarian and Fallopian tube cancers. Am J Obstet Gynecol 2005; 192(5):1614-9.

23. Abu-Rustum NR, Rhee EH, Chi DS et al. Subcutaneous tumor implantation after laparoscopic procedures in women with metastatic disease. Obstet Gynecol 2004; 103(3):480-7.

24. Gungor M, Ortac F, Arvas M et al. The role of secondary cytoreductive surgery for recurrent ovarian cancer. Gynecol Oncol 2005; 97(1):74-9.

25. Eisenkop SM, Friedman RL, Wang HJ. Secondary cytoreductive surgery for recurrent ovarian cancer. A prospecitive study. Cancer 1995; 76(9):1606-14.

26. Gemignani ML, Hensley ML, Cohen R et al. Paclitaxeol-based chemotherapy in carcinoma of the Fallopian tube. Gynecol Oncol 2001; 80(1):16-20.

27. Hefler LA, Rosen AC, Graf AH et al. The clinical value of serum concentrations of cancer antigen 125 in patients with parimary Fallopian tube carcinoma. Cancer 2000; 89(7):1555-60.

28. Ford D, Easton DF, Stratton M et al. Genetic heterogeneity and penetrance analysis of the BRCA1 and BRCA2 genes in breast cancer families. Am J Hum Genet 1998; 62(3):676-89.

29. Rebbeck TR, Lynch HT, Neuhausen SL et al. Prophylactic oopherectomy in carriers of BRCA1 or BRCA2 mutations. NEJM 2002; 346(21):1616-22.

30. Coglan TJ. Challenges in the early diagnosis and staging of Fallopian-tube carcinomas associated with BRCA mutations. Int J Gynecol Pathol 2003; 22(2):109-20.

31. Armstrong DK, Bundy B, Wenzel L et al. Intraperitoneal cisplatin and paclitaxel for ovarian cancer. NEJM 2006; 354(1):34-43.

CHAPTER 14

Ovarian Sex Cord-Stromal Tumors

Lynne M. Knowles and John O. Schorge

Abstract

Ovarian sex cord-stromal tumors (SCSTs) are a heterogeneous group of neoplasms that develop from the intraovarian matrix. These tumors account for less than 5% of all ovarian malignancies and may develop at any age. SCSTs account for nearly 90% of all functioning ovarian neoplasms. Patients often present with clinical manifestations of excessive estrogen or androgen production. The scarcity of these tumors limits the understanding of their natural history, management and prognosis. Ovarian SCSTs exhibit indolent growth and either behave in a clinically benign fashion or have low malignant potential. Complete surgical resection is the mainstay of treatment. The vast majority of SCSTs are confined to one ovary at presentation, and few patients will require postoperative therapy. Recurrent disease is relatively insensitive to chemotherapy, but patients often live for many years due to slow tumor progression. The overall prognosis of ovarian SCSTs is excellent—primarily due to early diagnosis and curative surgery.

Ovarian neoplasms may originate from the surface epithelium, primitive germ cells or intraovarian matrix. Sex cord-stromal tumors are a heterogeneous group of neoplasms that develop from the intraovarian matrix supporting the germ cells. Granulosa cells and Sertoli cells are derived from the sex cord cells. Theca cells, Leydig cells and fibroblasts are derived from the pluripotential mesenchymal cells. Ovarian sex cord-stromal tumors represent the neoplastic transformation of any of these cellular constituents.

Epidemiology

Ovarian SCSTs account for 5% of all ovarian malignances. Their relative scarcity constitutes a limitation in our understanding of their natural history, management and prognosis. These tumors can occur at any age, but the peak incidence occurs in postmenopausal women just beyond 50 years. SCST subtypes display a bimodal age distribution. Juvenile granulosa cell tumors, Sertoli-Leydig cell tumors and other variants occur predominantly during the first three decades of life. Adult granulosa cell tumors develop thereafter—most commonly in the fifth, sixth and seventh decades. Black women have the highest overall incidence of developing SCSTs. Although there are no proven risk factors, exposure of the gonad to persistently high levels of pituitary gonadotropins is thought to potentially facilitate transformation. Ovarian SCSTs have been reported more commonly in oral contraceptive users and those women treated by ovulation induction for infertility.

SCSTs account for nearly 90% of all functioning ovarian neoplasms. The clinical presentation of patients is frequently governed by the clinical manifestations resulting from endocrinologic abnormalities. One-third of tumors produce estrogen, progesterone, testosterone or other androgens. Excessive tumor-induced estrogen production

Gynecologic Oncology, edited by Paola Gehrig and Angeles Secord.
©2009 Landes Bioscience.

Table 14.1. World Health Organization classification of ovarian sex cord-stromal tumors

Granulosa-stromal cell tumors
    Granulosa cell tumor
        Adult type
        Juvenile type
    Thecoma-fibroma group
        Thecoma
        Fibroma-fibrosarcoma
        Sclerosing stromal tumor
Sertoli-stromal cell tumors
        Sertoli-cell tumor
        Leydig-cell tumor
        Sertoli-Leydig cell tumor
Sex cord tumor with annular tubules
Steroid-cell tumors
        Stromal luteoma
        Leydig-cell tumor
        Steroid-cell tumor not otherwise specified
Unclassified
Gynandroblastoma

14

may result in a variety of age-dependent signs and symptoms (i.e., precocious puberty, postmenopausal bleeding). Peripheral aromatization may result in a hyperandrogenic state with associated sequelae (i.e., hirsutism, frank virilization). Many patients will seek medical attention for hormonally-induced symptoms rather than pain or other symptoms related to the ovarian tumor.

Classification

The intraovarian matrix consists of cells originating from the sex cords and mesenchyme of the embryonic gonad. Granulosa cells and Sertoli cells are derived from the sex cord cells, whereas the mesenchymal cells are the precursors of the theca cells, Leydig cells and fibroblasts. The primitive gonadal stroma possesses sexual bipotential, and therefore tumors that develop may be of a male-directed cell type (Sertoli or Leydig cell) or female-directed cell type (granulosa or theca cells). Table 14.1 shows the World Health Organization classification of ovarian SCSTs. Mixed tumors are also occasionally observed. Ovarian granulosa cell tumors may have admixed Sertoli components, just as tumors that are predominantly Sertoli or Sertoli-Leydig cells may contain minor granulosa elements. These mixed tumors are believed to arise from a common lineage with variable differentiation and do not represent two separate entities in apposition.

Clinical Features

Ovarian granulosa cell tumors are universally considered to have malignant potential. Most other SCST subtypes do not have definitive criteria for clearly making the distinction between benign and malignant. Histologic grading of these tumors using nuclear characteristics or mitotic activity counts has also produced inconsistent results.

Table 14.2. Tumor markers for ovarian sex cord-stromal tumors with malignant potential

Granulosa cell tumors (adult and juvenile)	inhibin, estradiol (not as reliable)
Sertoli-Leydig cell tumors	inhibin, alpha-fetoprotein (occasionally)
Sex cord tumor with annular tubules	inhibin
Steroid-cell tumors not otherwise specified	steroid hormones elevated pretreatment

Granulosa Cell Tumors

Granulosa cell tumors comprise 70% of ovarian SCSTs. There are two clinically and histologically distinct forms: adult and juvenile.

Adult Form

Adult-type tumors account for 95% of all granulosa cell tumors. The average age at presentation is 52 years and most patients are diagnosed after age 30. These tumors typically present with abdominal pain, abdominal distension and/or abnormal vaginal bleeding. Pain and distension are related to the size of a tumor that often exceeds 10 to 15 cm in diameter. Abnormal menses or postmenopausal bleeding most frequently prompt women to seek medical attention. These tumors can produce estrogen as well as progesterone. Up to one-quarter of these patients have coexisting endometrial pathology related to estrogen excess—such as hyperplasia or adenocarcinoma. Occasionally, tumor rupture with hemoperitoneum is the presentation. Tumor markers may include inhibin A and B or serum estradiol levels (Table 14.2). Inhibins have been demonstrated to be elevated months before clinical detection of disease and are more reliable.

Grossly, adult granulosa cell tumors are large and multicystic (Fig. 14.1A). The ovarian surface is often unusually adherent to other pelvic organs and requires more extensive dissection than epithelial ovarian cancers or malignant germ cell tumors. Inadvertent rupture and intraoperative bleeding are also more common. The interior of the tumor has a variable solid and cystic appearance with hemorrhagic areas (Fig. 14.1B). Microscopic examination shows predominately granulosa cells that have pale, grooved, "coffee bean" nuclei (Fig. 14.2). The characteristic microscopic feature is the Call-Exner body: a rosette arrangement of cells around an eosinophilic fluid space (Fig. 14.3).

Adult granulosa cell tumors are low grade malignancies that demonstrate indolent growth. The majority (95%) are unilateral and 90% are Stage I (confined to the ovary) at diagnosis. The 10-year survival for Stage I disease is approximately 90%. Fifteen to 25% of Stage I tumors will eventually relapse. Recurrences may occur years or even decades after diagnosis. The median time to documented relapse is 6 years. These indolent tumors usually progress slowly thereafter, and the median length of survival after recurrence is another 6 years.

Juvenile Form

Juvenile granulosa cell tumors are very rare and occur primarily in children and young adults. They account for 90% of granulosa cell tumors in prepubertal girls and women younger than 30 years of age. Patients may present with a pelvic mass, abdominal pain, increasing abdominal girth or endocrine manifestations. Clinical evidence of estrogen production in the prepubertal female is isosexual precocious puberty. Breast enlargement, development of pubic hair, vaginal secretions and other secondary sexual

14

14

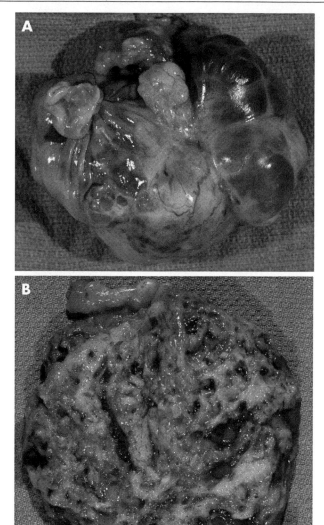

Figure 14.1. Ovarian granulosa cell tumor. A) Gross specimen; B) Sectioned surface consisting of solid and cystic areas with hemorrhage.

characteristics are common. Estrogen, progesterone and testosterone levels may be elevated and gonadotropins are suppressed. Juvenile granulosa cell tumors are infrequently androgen-secreting but may induce virilization in such patients.

Grossly, the juvenile tumor is similar to the adult form with solid and cystic features and areas of hemorrhage. Microscopic examination shows a predominantly solid cellular tumor with focal follicle formation. Call-Exner bodies are rarely encountered. The cytologic features that distinguish juvenile granulosa cell tumors from the adult type are

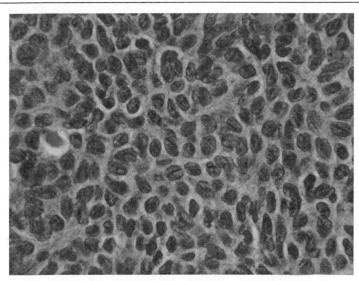

Figure 14.2. "Coffee bean" nuclei with grooves in adult granulosa cell tumor. (hematoxylin and eosin stain, magnification ×40). Provided courtesy of Kelley Carrick, MD, Department of Pathology, University of Texas Southwestern Medical Center, Dallas, TX.

14

their rounded, hyperchromatic nuclei without "coffee-bean" grooves. In addition, they exhibit moderate to abundant eosinophilic or vacuolated (luteinized) cytoplasm.

Juvenile granulosa cell tumors are sometimes associated with Ollier's disease (enchondromas) and Mafucci's syndrome (endochondromas and hemangiomas). Similar to adult type tumors, the majority (95%) are unilateral and 90% are Stage I at diagnosis. The 5-year survival rate is 95%. In contrast to adult types, juvenile granulosa cell tumors are more aggressive in advanced stages and the time to relapse and death is much shorter.

Thecoma

These clinically benign tumors are composed of lipid-laden stromal cells that resemble the theca cells normally surrounding the ovarian follicles. They account for approximately 1% of ovarian neoplasms and usually occur among patients in their 60s and 70s. They are rarely bilateral and virtually never present with extraovarian spread. The most common presenting signs and symptoms are abnormal vaginal bleeding and a pelvic mass. Thecomas are among the most hormonally active of the SCSTs and usually produce estrogen. Many women also present with concurrent endometrial hyperplasia or adenocarcinoma. Occasionally, thecomas demonstrate luteinization. Half of luteinized thecomas are either hormonally inactive or androgenic.

Fibromas

These solid ovarian neoplasms arise from the spindled stromal cells that form collagen. Fibromas are hormonally inactive. They can occur at any age, but are most common during middle age and rare before 30 years of age. Benign ascitic fluid is detected in 10% to 15% of fibromas larger than 10 cm. This results from transudate through

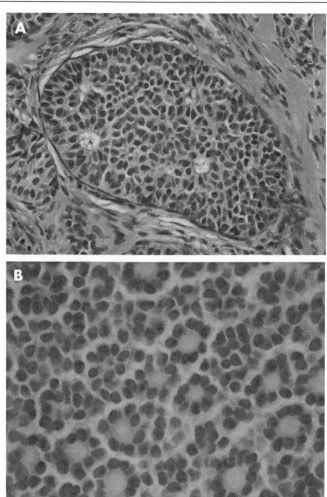

Figure 14.3. A) Nest of granulosa cells with Call-Exner bodies (hematoxylin and eosin stain, magnification ×20); B) Call-Exner bodies characterized by rosette arrangement of cells around a fluid space in adult granulosa cell tumor (hematoxylin and eosin stain, magnification ×40). Provided courtesy of Kelley Carrick, MD, Department of Pathology, University of Texas Southwestern Medical Center, Dallas, TX.

the surface of the enlarging and edematous tumor. Meig's syndrome occurs in 1% of patients and includes findings of hydrothorax, ovarian fibroma and ascites. Fibromas are generally considered benign tumors. However, 10% will demonstrate increased cellularity, varying degrees of pleomorphism and mitotic activity. These features indicate a tumor better characterized as having low malignant potential—especially if rupture has occurred or adhesions are encountered. Fibrosarcomas are very rare, highly malignant tumors that are distinguished by their greater cellular density and moderate to marked pleomorphorism.

Sclerosing Stromal Tumors

These tumors are histologically distinct from thecoma or fibromas. Eighty percent occur before age 30—which is unique among ovarian stromal tumors. Sclerosing stromal tumors are always benign and unilateral. Tumor size may be up to 20 cm, but ascites is seldom encountered; this further contrasts with fibromas.

Sertoli-Cell Tumors

These tumors are formed by cell proliferations that resemble rete ovarii and rete testis and usually have tubular differentiation as the predominant feature. Sertoli-cell tumors are rare and account for less than 5% of all SCSTs. The average age at diagnosis is 27 years, but tumors can occur at any age. Two-thirds of cases produce estrogen. Grossly, these tumors are solid, yellow and lobulated. Microscopic examination demonstrates hollow or solid tubules lined by cytologically bland cells. The majority of these tumors are well-differentiated and cured with surgery alone.

14

Sertoli-Leydig Cell Tumors

These morphologically diverse tumors have variable proportions of cells resembling epithelial and stromal testicular cells. Tumors can be solid, partially cystic, or completely cystic, and they may or may not have polypoid or vesicular structures in their interior. These rare tumors have an average age at diagnosis of 25 years and generally occur during the reproductive years. Fewer than 10% are diagnosed in premenarchal or postmenopausal women. Patients with well-differentiated tumors present around 35 years, approximately 10 years later than the intermediate or poorly differentiated tumors. Extraovarian spread at the time of diagnosis is 2 to 3%.

Patients usually present with complaints of menstrual disorders, virilization or nonspecific symptoms resulting from an abdominal mass. Frank virilization occurs in 35% of patients and another 10% to 15% have some clinical manifestations related to androgen excess. The most common androgenic symptoms include amenorrhea, voice deepening and hirsutism. Other androgenic symptoms that may occur include breast atrophy, clitoromegaly, loss of female contour and temporal hair recession. Most patients with signs of virilization will have elevated plasma testosterone levels. Plasma androstenedione is only occasionally elevated and the urinary 17-ketosteroids are usually normal. An elevated testosterone/androstenedione ratio suggests the presence of an androgen-secreting ovarian tumor, usually a Sertoli-Leydig cell tumor. Surgical removal results in an immediate drop in androgen levels and over time, partial to complete resolution of signs of virilization.

Sertoli-Leydig cell tumors vary in size, but most are 5 to 10 cm in diameter. They are solid yellow masses that are often lobulated. Well-differentiated tumors are characterized by a predominantly tubular pattern (Fig. 14.4A). Intermediate and poorly differentiated tumors have a variety of patterns and cell types (Fig. 14.4B).

The majority (97%) of Sertoli-Leydig cell tumors are Stage I at diagnosis. Poorly differentiated tumors and those with heterologous elements (i.e., muscle, cartilage) are more likely to recur. The mortality rate for patients presenting with greater than Stage I disease approaches 100%.

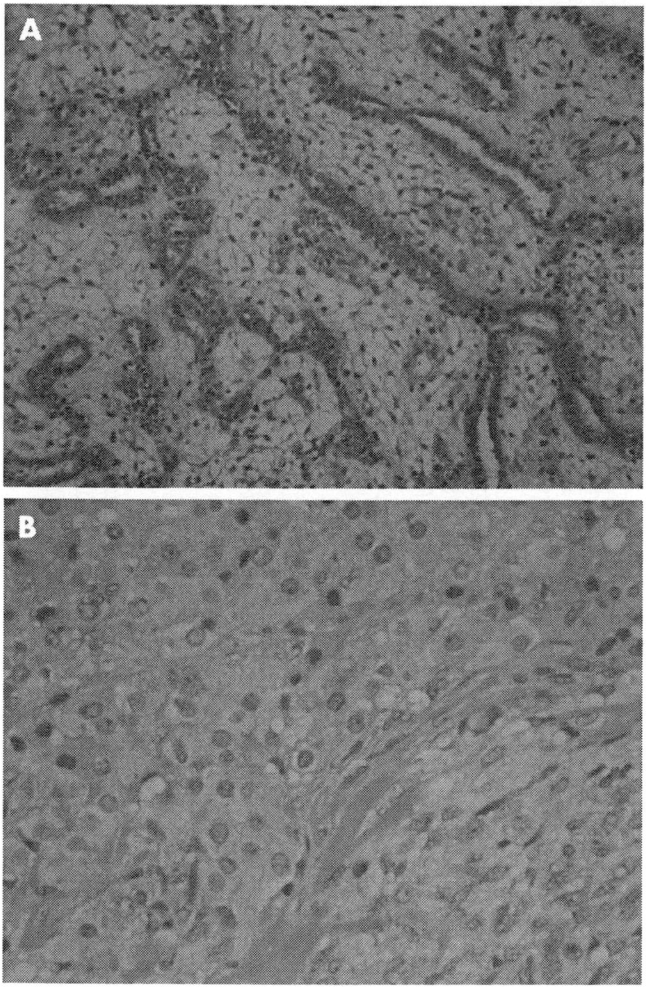

Figure 14.4. Sertoli-Leydig cell tumor (SLCT). A) Well-differentiated SLCT with Sertoli cells arranged in a tubular pattern (hematoxylin and eosin stain, magnification ×10); B) Leydig cells are present in top/left portion of picture and Sertoli cells with spindled morphology in bottom/right (hematoxylin and eosin stain, magnification ×20). Provided courtesy of Kelley Carrick, MD, Department of Pathology, University of Texas Southwestern Medical Center, Dallas, TX.

Leydig-Cell Tumors

These tumors occur in postmenopausal women, are usually benign, unilateral and secrete testosterone. They are usually small and solid and the cut surface has a yellow-orange appearance. These tumors must be differentiated from lipid (adrenal rest) cell tumors which behave malignantly in 20% of cases. Cellular atypia, mitosis, large size and the absence of Reinke crystals are features suggestive of more malignant behavior.

Other Sex Cord-Stromal Tumors

a. Sex cord tumors with annular tubules (SCTAT) have distinctive cellular elements that are histologically intermediate between Sertoli-cell and granulosa cell tumors. One-third of SCTATs occur in patients with Peutz-Jehgers syndrome (PJS). When associated with PJS they are usually small, bilateral and clinically benign. Fifteen percent of patients with PJS-associated SCTAT will also develop adenoma malignum of the cervix. SCTAT without PJS has a 20% clinical malignancy rate.

b. Steroid-cell tumors are rare tumors comprised of steroid hormone-secreting cells including lutein cells, Leydig cells and adrenocortical cells. Stromal luteomas and Leydig-cell tumors are benign tumors typically seen in postmenopausal women. Steroid-cell tumors not otherwise classified are more common and occur in younger women. These tumors can be associated with androgenic, estrogenic and/or cortisol excess (i.e., Cushing's syndrome). The rate of clinical malignancy is high and the prognosis is dismal.

c. Unclassified SCSTs are an ill-defined group comprising tumors without a predominant pattern of testicular or ovarian differentiation. This diagnosis is most common when a SCST is removed during pregnancy due to alterations in their usual clinical and pathologic features.

d. Gynandroblastomas are extremely rare SCSTs with intermingling granulosa cells and tubules of Sertoli cells. They are considered tumors of low malignant potential.

14

Treatment

The optimal management of ovarian SCSTs has been difficult to elucidate due to their low incidence and variable histology. Treatment recommendations have developed from observations of small groups and derived from the management principles of other ovarian neoplasms. The recent establishment of a Rare Tumor Committee by the Gynecologic Oncology Group will hopefully facilitate continued progress in research of SCSTs.

Surgery

The mainstay of treatment for patients with an ovarian SCST is surgical resection. The goals of surgery are to establish a definitive tissue diagnosis, determine the extent of disease by performing appropriate staging procedures and removal of all grossly visible disease. The type of surgery should take into account the patient's age and desire for future fertility. Hysterectomy with bilateral salpingo-oophorectomy (BSO) is most commonly performed because the majority of patients will have either completed childbearing or be postmenopausal. Fertility-sparing unilateral salpingo-oophorectomy (USO) with preservation of the uterus may also be appropriate in the absence of obvious disease spread to these organs. Endometrial sampling should be performed when fertility-sparing surgery is desired since many tumors will have co-existing hyperplasia or adenocarcinoma that might affect the decision to remove the uterus. Routine ovarian cancer surgical staging then proceeds with partial omentectomy, random peritoneal biopsies and pelvic/para-aortic nodal dissection.

Minimally-invasive laparoscopic surgery has a variety of relevant applications. Many patients ultimately diagnosed with ovarian SCST are taken to the operating

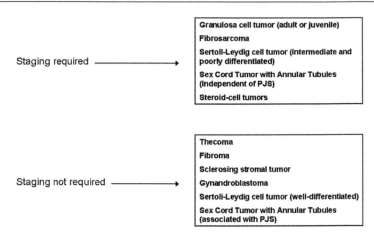

Figure 14.5. Recommendation for surgical staging of sex cord-stromal tumors.

room for evaluation of a presumed benign complex adnexal mass in the setting of a normal serum CA125 tumor marker. The diagnosis of SCST may not be discovered until the mass is sent for frozen section or perhaps when the final pathology report is confirmed. Laparoscopic staging may be safely performed in women with apparent early-stage disease detected intraoperatively. Laparoscopic staging may also decrease morbidity for unstaged patients referred after USO or hysterectomy and BSO. Staging laparotomy or laparoscopy is essential to determine the extent of disease and need for adjuvant therapy in most patients (Fig. 14.5).

Chemotherapy

Ovarian SCSTs display less sensitivity to chemotherapy than other types of ovarian malignancies. The decision to administer postoperative therapy depends on a variety of factors (Fig. 14.6). In general, women with surgical Stage I disease have an excellent prognosis with surgery alone and do not benefit from postoperative chemotherapy. Management should be individualized in the presence of large tumor size, high mitotic index, capsular excrescences, tumor rupture, incomplete staging or equivocal pathology. Patients with one or more of these suspicious features are thought to be at higher risk of relapse and should be considered for postoperative platinum-based chemotherapy.

Bleomycin, etoposide and cisplatin (BEP) is the most widely used regimen. Three cycles is sufficient for completely resected disease. Four cycles is recommended for patients with incompletely resected tumor. The relative rarity of ovarian SCST patients receiving chemotherapy makes it difficult to conduct randomized studies to evaluate the value of postoperative adjuvant chemotherapy. Taxanes have recently been shown to have significant activity against ovarian SCSTs. Future studies of paclitaxel and carboplatin chemotherapy are warranted to reduce the toxicity associated with BEP.

Radiation

The use of postoperative radiation therapy currently has a limited role in the management of ovarian SCSTs. There is some evidence for a prolonged survival in at least some

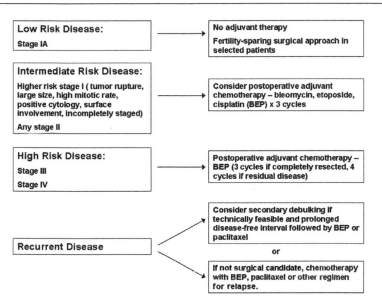

Figure 14.6. Postoperative management for patients with malignant ovarian sex cord-stromal tumors.

patients with newly diagnosed disease who received whole-abdominal radiotherapy. However, chemotherapy is usually the postoperative treatment of choice because it is generally better tolerated, more widely accessible and easier to administer.

Recurrent Disease

The optimal treatment for recurrent disease is less well-defined and must be individualized. Repeat surgical debulking is an appropriate first consideration given the indolent growth and long disease-free intervals seen with many SCSTs. Four cycles of BEP is a suitable option for patients who did not receive earlier treatment with BEP or have had a prolonged treatment-free interval. Paclitaxel is another promising second-line agent that is being evaluated in a Phase II Gynecologic Oncology Group trial. Vincristine, actinomycin D and cyclophosphamide (VAC) is a combination regimen with known efficacy. Hormonal therapy should be reserved for patients with progressive disease that have failed chemotherapy. Medroxyprogesterone acetate and leuoprolide acetate are both active in halting the growth of recurrent ovarian SCSTs. Radiation is best reserved for palliation of local symptoms.

Prognosis

Ovarian SCSTs portend a much better prognosis overall than epithelial ovarian carcinomas. This is chiefly because the majority of women with SCSTs are diagnosed with Stage I disease. Patients with Stage I granulosa cell tumors or Sertoli-Leydig cell tumors have a 10-year survival that exceeds 90%. Stage III-IV tumors are rare, but have a poor prognosis similar to their epithelial counterparts. Other poor prognostic clinical variables include age greater than 50 years, large tumor size and residual disease at the time of initial surgery.

Future Directions

A better understanding and determination of the optimal management of ovarian SCSTs will hopefully be achieved through the research conducted by the Rare Tumor Committee in the Gynecologic Oncology Group.

Suggested Reading

1. Gershenson DM, Hartmann LC, Young RH. Ovarian sex cord-stromal tumors. Practice of Gynecologic Oncology 4th ed. Philadelphia Lippincott Williams and Wilkins 2005; 1011-53.
2. Matei DE, Russell AH, Horowitz CJ et al. Ovarian germ-cell tumors. Principles and practice of gynecologic oncology 4th ed. Philadelphia Lippincott Williams and Wilkins 2005:989-1009.

References

1. Brewer M, Gershenson DM, Herzog CE et al. Outcome and reproductive function after chemotherapy for ovarian dysgerminoma. J Clin Oncol 1999; 17:2670.
2. Brown J, Shvartsman HS, Deavers MT et al. The activity of taxanes in the treatment of sex core-stromal ovarian tumors. J Clin Oncol 2004; 22:3517.
3. Chan JK, Zhang M, Kaleb V et al. Prognostic factors responsible for survival in sex cord stromal tumors of the ovary—a multivariate analysis. Gynecol Oncolo 2005; 96:204.
4. Homesley HD, Bundy BN, Hurteau JA et al. Bleomycin, etoposide and cisplatin combination therapy of ovarian granulosa cell tumors and other stromal malignancies: A gynecologic oncology group study. Gynecol Oncol 1999; 72:131.
5. Lai CH, Chang TC, Hsueh S et al. Outcome and prognostic factors in ovarian germ cell malignancies. Gynecol Oncol 2005; 96:784.
6. Low JJ, Perrin LC, Crandon AJ et al. Conservative surgery to preserve ovarian function in patients with malignant ovarian germ cell tumors. A review of 74 cases. Cancer 2000; 89:391.
7. Peccatori F, Bonazzi C, Chiari S et al. Surgical management of malignant ovarian germ-cell tumors: 10 years' experience of 129 patients. Obstet Gynecol 1995; 86:367.
8. Schneider DT, Calaminus G, Wessalowski R et al. Ovarian sex cord-stromal tumors in children and adolescents. J Clin Oncol 2003a; 21:2357.
9. Young RH. Sex cord-stromal tumors of the ovary and testis: their similarities and differences with consideration of selected problems. Mod Pathol 2005; 18(Suppl 2):S81.
10. Zanetta G, Bonazzi C, Cantu M et al. Survival and reproductive function after treatment of malignant germ cell ovarian tumors. J Clin Oncol 2001; 19:1015.

14

CHAPTER 15

Germ Cell and Metastatic Tumors of the Ovary

William P. Irvin, Jr. and Christopher Darus

Epithelial tumors account for approximately 90% of ovarian malignancies and one-half of all ovarian neoplasms. Primary nonepithelial ovarian tumors account for the remaining 10% of ovarian malignancies and 50% of the remaining nonmalignant ovarian neoplasms. Primary nonepithelial ovarian tumors can be divided into two major subgroups, germ cell tumors and sex cord-stromal tumors (please refer to Chapter 14). Additionally, tumors metastatic to the ovary and exceptionally rare primary ovarian malignancies (such as sarcomas) may be discovered at time of exploration for an adnexal mass. Table 15.1 shows the approximate percentages of ovarian tumors by cell of origin. The frequency of ovarian tumors varies by patient age and geography. For example, germ cell tumors are more prevalent in younger women and epithelial tumors are more common in western countries.

Germ Cell Tumors

Treatment and cure of malignant ovarian germ cell tumors (GCT) is a success story of modern oncology. Malignant germ cell neoplasms, like analogous tumors of the testis, are derived from immature ovarian germ cells. In western countries, germ cell tumors comprise approximately 20% of ovarian neoplasms overall and 2-5% of ovarian malignancies. In areas with a low incidence of epithelial ovarian cancer, such as Japan and other Asian countries, a much higher percentage of ovarian neoplasms are of germ cell origin. Most ovarian neoplasms diagnosed in children and adolescents are GCT. Approximately two-thirds of these tumors will be malignant at the time of diagnosis.

Pathology

The World Health Organization classification of germ cell tumors is shown in Table 15.2. Although GCT are the most common group of ovarian neoplasms, malignant GCT are rare. The most common GCT (95%), as well as the most common ovarian neoplasm overall, is the mature cystic teratoma.

Dysgerminoma

Dysgerminoma is the most common primitive GCT. Eighty percent are diagnosed in patients under the age of 30. It is also the most commonly bilateral of the GCT, grossly evident in approximately 10% of cases and microscopically in another 10%. Dysgerminomas have been reported in pregnancy. Lactate dehydrogenase (LDH) is frequently elevated in dysgerminomas. hCG is also elevated in a minority of the cases due to the presence of hCG-secreting syncytiotrophoblasts (Table 15.3). Characteristic

Gynecologic Oncology, edited by Paola Gehrig and Angeles Secord.
©2009 Landes Bioscience.

Table 15.1. Percentage of ovarian tumors by cell of origin

Cell Type	Total	Malignant Neoplasms
Epithelial	50	85-90
Germ cell	20	2-5
Sex cord-stromal	10-20	7
Metastatic	5-10	5-6

Table 15.2. World Health Organization classification of germ cell tumors[1]

Dysgerminoma
Endodermal sinus tumor
Embryonal carcinoma
Polyembryoma
Choriocarcinoma
Teratoma
    Immature
    Mature
        Solid
        Cystic
            Dermoid cyst with malignant transformation
        Monodermal
Mixed
Tumors composed of germ cells and sex cord-stromal derivative
    Gonadoblastoma
    Mixed germ cell-sex cord-stromal tumor

Table 15.3. Tumor markers in germ cell tumors

	hCG	AFP	LDH
Dysgerminoma	+/−	−	+
Endodermal sinus tumor	−	+	
Embryonal carcinoma	+	+	
Choriocarcinoma	+	−	
Immature teratoma	−	+/−	

radiographic appearance is that of a complex unilateral or bilateral ovarian mass, predominantly solid in appearance (Fig 15.1).

Endodermal Sinus Tumors

Endodermal sinus tumors, also known as yolk sac tumors, secrete alpha-fetoprotein (AFP). Endodermal sinus tumors show a similar age distribution as dysgerminomas. On hematoxylin and eosin staining, they show the characteristic Schiller-Duvall body, a perivascular structure similar in appearance to a glomerulus. These are aggressive tumors which tend to grow rapidly and metastasize early.

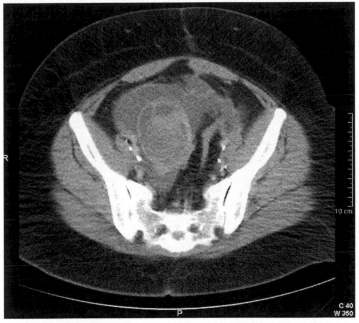

Figure 15.1. Dysgerminoma: Characteristic radiographic appearance per CT scan of enlarged right ovarian dysgerminoma.

Embryonal Carcinoma

Embryonal carcinomas are very rare malignancies that tend to occur in younger patients than GCT overall. They generally secrete both hCG and AFP (Table 15.3). They often produce estrogen which in turn may lead to precocious pseudopuberty or abnormal bleeding patterns.

Nongestational Choriocarcinoma

Nongestational choriocarcinoma is extremely rare GCT in its pure form; it is more commonly found as a constituent of a mixed germ cell tumor. Nongestational choriocarcinomas reliably secrete beta-hCG (Table 15.3).

Teratoma

Teratomas are derived from a solitary germ cell that has undergone defective meiosis. Malignant potential is related to the presence of immature elements. Mature teratoma, the most common of all ovarian neoplasm, is a benign tumor found in patients of varying age. Mature cystic teratoma, or dermoid cyst, consists of differentiated elements of all three germ layers. The cyst wall is usually lined with skin and dermal appendages. The cyst itself is commonly filled with hair and sebaceous material; teeth, sometimes associated with a jawbone, long bones, intestinal loops and other macroscopic structures, may be encountered inside the cyst as well. Very rarely a structure recapitulating a fetus, known as a homunculus, may be present. Radiographically, mature cystic teratomas have a characteristic sonographic appearance in which fat may be readily recognized within

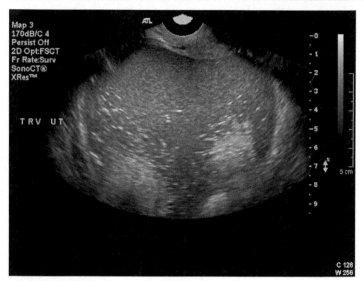

Figure 15.2. Teratoma: Characteristic ultrasound appearance of ovarian mature cystic teratoma with internal fat echogenicity.

the cyst wall. This is a finding that is considered to be pathognomonic for a mature cystic teratoma (Fig. 15.2). As opposed to its cystic counterpart, mature solid teratomas are rare and must be examined carefully to exclude immature elements. Teratomas are often discovered incidentally on physical exam or imaging, but patients may present with discomfort or ovarian torsion.

1-2% of mature teratomas contain a malignancy of one of its constituents. Although this is most commonly a squamous carcinoma, other carcinomas, sarcomas and melanomas have been reported. Careful histologic examination of the putative mature teratoma is essential in patients over the age of 40 to rule out the presence of a malignant component.

Immature Teratoma

As the name suggests, these neoplasms are noteworthy for the presence of immature (embryonic) tissue from the three germ layers. The immature tissue is often admixed with mature, differentiated tissue and can therefore be a difficult diagnosis to make if only a small amount of immature elements are present. Grade is determined by the amount of immature neural elements. A minority of immature teratomas produce AFP (Table 15.3). A mature teratoma is present in the contralateral ovary in 10-15% of cases.

Monodermal Teratomas

Struma ovarii is an uncommon teratoma consisting of mature thyroid tissue. Approximately 30% of patients will be clinically hyperthyroid. Most of these are benign lesions and are treated by simple excision. Primary ovarian carcinoids can also occur as a monodermal teratoma.

Mixed Germ Cell Tumors

Approximately 10% of GCTs are mixed. This is important, as the combination(s) of histologic subtypes may influence surgical and medical management. A large percentage contains dysgerminoma. Management should be based on the nondysgerminomatous element.

Gonadoblastoma

Gonadoblastoma is a rare entity that consists of germ cells and ovarian stroma. Patients usually have sex chromosome abnormalities, typically 46XY or mosaic 45X/46 XY. Seventy-five percent are phenotypically female but have pure gonadal dysgenesis, androgen insensitivity, or Turner's syndrome. Although gonadoblastomas are benign, they frequently produce malignant germ cell tumors. In patients with a Y chromosome, prepubertal bilateral gonadal excision is recommended.

Clinical Features

Malignant GCT are clinically distinctive malignancies. They occur in young women, with a median age of 16-20 at the time of diagnosis. Unlike epithelial ovarian cancer in which 75% of women will have advanced stage disease at the time of diagnosis, approximately 60-75% of malignant germ cell tumors will be Stage I when initially diagnosed.

GCT are rapidly growing malignancies and commonly present with abdominal pain or distention. Approximately 10% of patients present acutely with intraperitoneal tumor rupture, bleeding, or adnexal torsion. These patients may be initially misdiagnosed with pregnancy or appendicitis.

Imaging studies typically show a large unilateral solid or solid and cystic adnexal mass. Consideration should be given to karyotype assessment of premenarchal girls with a pelvic mass to exclude gonadal dysgenesis. GCT frequently have reliable tumor markers (Table 15.3) that aid not only in following patients through treatment, but also for initial diagnosis.

15

Treatment

Ovarian germ cell tumors are rare neoplasms. Management schema have been borrowed from epithelial ovarian cancer (surgical) and testicular germ cell tumors (medical). Although prospective collaborative data do exist, studies of ovarian GCT are generally small, retrospective and single-institution. Moreover, patients in prior studies commonly had not undergone comprehensive surgical staging. Nevertheless, GCT are frequently cured using multi-modality therapy based on available data.

Surgery

Surgical excision, with or without adjuvant chemotherapy, is the cornerstone of the management of ovarian GCT. Germ cell tumors are staged the same as are epithelial ovarian cancers (please refer to Chapter 12). Surgical management consists of excision of at least the affected ovary, abdominopelvic washings (or the removal of ascites), excision of any gross tumor, pelvic and para-aortic lymph node evaluation, peritoneal biopsies and omental sampling. GCT metastasize to regional lymph nodes more commonly than do epithelial ovarian malignancies, and therefore the lymph node dissection is a critical part of the surgical staging procedure for GCT.

The extent and utility of surgical debulking is controversial. In small studies of the Gynecologic Oncology Group (GOG), patients who were cytoreduced completely had better outcomes in terms of response and overall survival than did those patients with residual disease following surgical exploration.[2]

Table 15.4. BEP regimen for ovarian germ cell tumors[3]

Drug	Dose and Schedule (q 21 days for 3-4 cycles)
Cisplatin	20 mg/m^2 days 1-5
Etoposide	100 mg/m^2 days 1-5
Bleomycin	30 units IV weekly

The majority of patients diagnosed with GCT are young and of reproductive age. Given the fact that 60-75% of GCT will be Stage I at the time of diagnosis, there is rarely an indication for total hysterectomy and bilateral salpingo-oophorectomy. Intraoperative decisions regarding extent of surgery should be based on patient's age and fertility desires, preoperative tumor markers, intraoperative frozen section and intraoperative findings. The only GCT with any predilection for bilateralism is dysgerminoma. Some authors advocate biopsy of the contralateral ovary in patients with dysgerminoma (10-15%). This practice, which may lead to adnexal adhesions, may not be preferable in the patient who desires to maintain fertility. In the interest of preserving ovarian function, it may in some cases be appropriate to leave small-volume disease behind as most GCT are highly chemosensitive.

Chemotherapy

Malignant GCT are generally very chemosensitive tumors. Because of their rarity, much of the chemotherapeutic management has been borrowed from analogous testicular GCT. Treatment of GCT has evolved as newer prospective data from the management of testicular tumors has emerged. Initial therapy consisted of vinblastine, adriamycin (doxorubicin) and cyclophosphamide (VAC). With the development of platinum agents, clinical trials confirmed cisplatin, vinblastine and bleomycin (PVB) to be superior to VAC as primary chemotherapy for the management of GCT. Substitution of etoposide for vinblastine was subsequently shown in testicular cancer to have lower toxicity. Moreover, comparison of testicular germ cell cancer patients treated with and without etoposide-containing regimens showed improved survival with the addition of etoposide. Currently, the standard chemotherapeutic regimen for the treatment of patients with malignant germ cell tumors is bleomycin, etoposide and cisplatin (BEP) as depicted in Table 15.4.[3]

Management of Early Stage Disease

Only those patients with surgical Stage I dysgerminoma or low grade Stage I immature teratoma can be observed following surgical staging without further adjuvant chemotherapy. All other patients with early stage GCT require three to four cycles of BEP chemotherapy following their surgical staging procedure (Table 15.4).

Dysgerminoma is unique among the ovarian germ cell malignancies in that it is a radiosensitive tumor. Consequently, radiation therapy is an effective treatment modality for those patients not felt to be candidates for chemotherapy. Radiation therapy will result in infertility however, and therefore it is rarely given as primary treatment for dysgerminoma today.

Management of Advanced Stage Disease

After initial treatment, patients are followed closely at regular surveillance intervals to detect recurrences. Most recurrences will be detected within the first 2 years following the patient's initial diagnosis and treatment. A history and physical exam is performed at each visit, with regular assessment of tumor marker(s) and imaging studies as indicated. Patients diagnosed with recurrent disease who did not receive adjuvant chemotherapy initially (Stage I Grade 1 immature teratoma and Stage I dysgerminoma) are treated with BEP at the time of recurrence.

In patients previously treated with chemotherapy, clinical judgment is guided by small series and the testicular germ cell tumor data. In testicular cancer patients, 'platinum sensitivity,' defined by greater than 6 months of remission following initial treatment with platinum compounds, portends an improved course and retreatment with a platinum-containing regimen is indicated. As many as 50% of platinum-sensitive patients can be successfully retreated with platinum. Secondary cytoreductive surgery may be beneficial in some cases, but data are lacking.

Because of the youth of most patients with GCT, long-term sequelae of treatment must be considered. Patients are at risk for infertility with surgical and/or chemotherapeutic treatment. Pelvic surgery with resultant adhesion formation is a known risk factor for infertility. Premature menopause is an additional risk of therapy. Etoposide carries the rare but devastating risk of secondary poor-prognosis leukemia. This is a dose-related phenomenon. In addition, bleomycin is associated with pulmonary fibrosis. Assessment of diffusion capacity of O_2 (DLCO) may be performed if clinically indicated.

Fertility Following Treatment of Germ Cell Tumors

Cure rates are high with modern management of GCT. Given that most patients diagnosed with GCT are of childbearing age, fertility preservation is generally a major goal in the management of GCT. A number of small series have demonstrated a high percentage of women conceiving following the surgical and chemotherapeutic management of GCT without increased pregnancy complications or teratogenicity.

15

Metastatic Tumors

Five to six percent of ovarian tumors are metastatic. Metastatic tumors are commonly bilateral and solid, although cystic areas may be present.

Krukenberg tumors account for a high proportion of metastatic ovarian tumors. These tumors are characterized by mucinous signet-ring cells within ovarian stroma. Historically, this histology was associated with tumors arising in the stomach. With the decreasing incidence of gastric cancer in western countries, colon, breast and biliary signet-ring metastatic tumors have been described. Breast cancer and colorectal cancer are the most common primary tumors associated with ovarian metastasis. Nonovarian gynecologic tumors, including Fallopian tube, endometrial and rarely cervical carcinomas may directly extend to the ovaries as well. Primary peritoneal cancer, which clinically mimics papillary serous ovarian cancer, often presents with ovarian surface implants. Other malignancies may spread to the ovaries including pancreatic carcinoma, lymphoma and melanoma.

Tumors metastatic to the ovary generally carry a poor prognosis, as patients generally have advanced disease by the time the adnexa are involved. Should additional surgical evaluation be required upon discovery of a metastatic tumor, consideration should be given to an intraoperative consultation to a surgeon who specializes in the management of the primary tumor. In some cases, adnexectomy of a metastatic tumor is indicated for palliative or diagnostic reasons. Preoperative evaluation with appropriate diagnostic studies, such as colonoscopy and mammogram, can help exclude metastatic disease in the patient presenting with adnexal masses.

Suggested Reading

1. Williams SD, Blessing JA, Moore DH et al. Cisplatin, vinblastine and bleomycin in advanced and recurrent ovarian germ-cell tumors. Ann Intern Med 1989; 111:22-7.

References

1. Serov SF, Sully RE, Sobin LJ. Histological typing of ovarian tumors. World Health Organization International Histological Classifications of Tumors. Geneva: World Health Organization 1973.
2. Williams SD, Blessing JA, Liao S et al. Adjuvant therapy of ovarian germ cell tumors with cisplatin, etoposide and bleomycin: a trial of the gynecologic oncology group. J Clin Oncol 1994; 12:701-6.
3. Gershenson DM, Morris M, Cangir A et al. Treatment of malignant germ cell tumors of the ovary with bleomycin, etoposide and cisplatin. J Clin Oncol 1990; 8:715-20.

15

CHAPTER 16

Vulvar Cancer

Victoria Bae-Jump, Elizabeth N. Skinner, Bradley Sakaguchi and Michael E. Carney

Epidemiology

Vulvar cancer is the fourth most common gynecologic cancer following cancers of the uterus, ovary and cervix, representing 3-5% of all cancers in the female genital tract. According to National Cancer Institute statistics, there will be an estimated 3,460 new vulvar cancer cases in 2008 resulting in approximately 870 deaths in the United States.[1] However, the incidence of this cancer has been steadily increasing in younger women over the past several decades. This rise in invasive cancer of the vulva among this younger population is thought to be associated with a greater number of women exposed to the human papillomavirus (HPV), leading to a 2-3 fold increase in vulvar intra-epithelial neoplasia (VIN) and in situ disease over the past 20 years.[2] It is estimated that up to 80% of women with untreated VIN will develop a carcinoma. Despite this phenomenon observed in younger woman, it is important to note that there has been no significant overall increase in the incidence of invasive vulvar cancer. One study noted that in women under age 50, the incidence has risen from 2 to 21 percent in the last 20 years. Thirty percent of women with vulvar cancer present at age 70 or older, and the rate increases with age, peaking at 20 per 100,000 by 75 years of age. Neoplasms in older women may not be as closely related to a HPV infection as they are in younger women.

The most common histology of vulvar cancer is squamous representing approximately 90% of cases. Other less common histological subtypes include adenocarcinoma, melanoma, basal cell and sarcoma. Vulvar dysplasia which is often multifocal is easily curable with local resection, laser or even cryotherapy. Early invasive vulvar cancer has an excellent prognosis, even though the treatments can have significant associated morbidity. Surgery, chemotherapy and radiation are all treatments for vulvar cancer, but they can lead to significant disfigurement and psychological as well a sexual morbidity. Decreasing this morbidity relies on early detection and adequate surveillance by providers. Recently, therapeutic trends have moved toward more conservative approaches.

Etiology

No specific causative factor has been identified in vulvar cancer. Historically the disease has been associated with hypertension, diabetes and obesity in up to 25% of patients; however none of these are independent risk factors. Sexually transmitted diseases and granulomatous conditions have also been associated. Identified risk factors include: cigarette smoking, vulvar dystrophy, VIN, HPV infection, immune deficiencies, prior cervical cancers and northern European ancestry.

There seem to be two distinct types of vulvar cancer. Type I is characterized by younger patients, HPV infection, cervical dysplasia, smoking, previous VIN, history

Gynecologic Oncology, edited by Paola Gehrig and Angeles Secord.
©2009 Landes Bioscience.

Table 16.1. Vulvar disease

1. Nonneoplastic epithelial disorders of skin and mucosa (Fig. 16.1)
   a. lichen sclerosis
   b. squamous hyperplasia
   c. other dermatoses
2. Mixed nonneoplastic and neoplastic disorders
3. Intra-epithelial neoplasia
   a. VIN 1
   b. VIN 2
   c. VIN 3
   d. Paget's disease
4. Invasive cancer

of STD and condyloma. Type II is characterized by older patients, vulvar atypias and dystrophies, lower HPV infection rates and only rarely associated with smoking, STDs or condylomas.

The etiology for the Type II disease is less well-defined but is associated with the group of hypertrophic disorders of the vulva formerly known as chronic vulvar dystrophies. The new preferred terminology is Nonneoplastic Epithelial Disorders of Skin and Mucosa. Of these types of disorders, lichen sclerosis is the classic example of a hypertrophic condition causing pruritis leading to an itch-scratch-itch cycle. This cycle theoretically compounds the hypertrophy and eventually leads to atypia, VIN and finally cancer. The spectrum of vulvar diseases can be summarized as per Table 16.1.

Paget's disease of the vulva typically affects postmenopausal Caucasian women and usually presents with pruritis and vulvar pain. Initially, the lesions appear in the hair bearing portion of the vulva and may spread to involve the mons, pubis, thighs and anus. As the lesion progresses it may take on a raised and eczematoid appearance. Microscopically, lesions typically extend beyond the gross lesion and account for the high local recurrence rate even after primary excision. Approximately 10% of patients with Paget's will have invasive disease and only 4-8% will have an underlying adenocarcinoma.

Anatomy

The vulva is comprised of the external genital organs including the mons pubis, labia majora, labia minora, clitoris, vaginal vestibule and perineal body (Fig. 16.1). The major blood supply to the vulva is from the internal pudendal artery which arises from the anterior branch of the internal iliac artery. Additional blood supply to the vulva is achieved through the superficial and deep pudendal arteries, both of which arise as branches from the femoral artery. Innervation to the superior vulva is through the ilioinguinal and genitofemoral nerve while the pudendal nerve innervates the lower vagina, labia, clitoris and perineal body. Lymphatic drainage from the vulva is primarily to the ipsilateral superficial inguinal lymph nodes, located within the femoral triangle. Contralateral drainage is seen for the more midline structures such as those occurring in the clitoris and perineal body. The superficial inguinal lymph nodes subsequently drain into the deep inguinal lymph nodes, found beneath the cribriform fascia and midline to the femoral vein

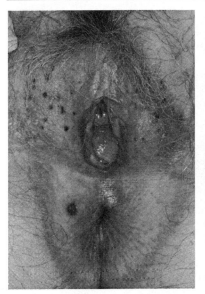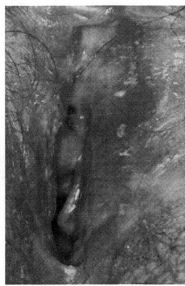

Figure 16.1. Vulvar dystrophies (courtesy of Wesley C. Fowler Jr, MD, Division of Gynecologic Oncology, University of North Carolina at Chapel Hill, Chapel Hill, North Carolina).

16

and ultimately into the pelvic lymph nodes. The pelvic lymph nodes are considered minor drainage for the vulva.

Clinical Manifestations

Vulvar cancer most commonly presents as a single plaque, ulcer or mass but can be multifocal in approximately 5% of cases.[3] These lesions can be nodular, fleshy or warty and occur on the labia majora in an estimated 40% of cases. Other common sites include the labia minora (20%), perineal body (15%), mons pubis (10%) or the clitoris (10%).[3] Often, these lesions are surrounded by co-exisitng VIN or vulvar dystrophies such as lichen sclerosis and squamous hyperplasia. Twenty-two percent of cases present concurrent to another primary cancer. Typically these are also HPV- or smoking-related and associated with younger patients.

The presenting symptoms can be varied and initially rather benign. Complaints include nodularity, persistent ulcerations, itching, bleeding, pain, vaginal discharge and dysuria. The most common presenting symptom is long-standing pruritis. Findings at the time of initial presentation may also vary from subtle to obvious and vary with the age of the patient, anatomy, presence of irritation, skin condition, underlying infection and how long the condition has existed. A small percentage will present with inguinal lymphadenopathy. The older patient with coexisting Nonneoplastic Epithelial Disorder may present with a lesion and an underlying lichen sclerosis. Providers should also be vigilant for the possibility of vulvar melanoma which may present with melanosis or altered pigmentation.

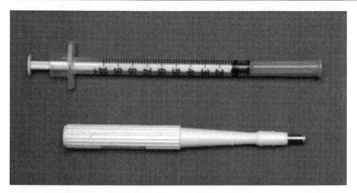

Figure 16.2. Keyes punch biopsy (courtesy of Wesley C. Fowler Jr, MD, Division of Gynecologic Oncology, University of North Carolina at Chapel Hill, Chapel Hill, North Carolina).

Diagnostic Evaluation

The initial evaluation of the patient with vulvar cancer should include a complete history and physical exam, including a thorough survey of all lymphatic areas (i.e., inguinal, axillary and supraclavicular). Given that neoplasias of the female genital tract are often multifocal, a thorough exam of the vagina and cervix, including a Pap smear, should also be performed. Differential diagnosis includes epidermal inclusions cysts, lentigos, benign Bartholin's gland lesions, seborrheic keratoses, hidradenomas, lichen sclerosis, dermatoses and condyloma accuminata. All concerning lesions should be biopsied, with the biopsy centered over the middle or most worrisome portion of the lesion. Local anesthesia injected directly around the area to be biopsied generally provides sufficient anesthesia. Post biopsy hemostasis can be obtained with epinephrine in the local anesthetic, application of Monsel's solution, silver nitrate or suturing. Ulcerating lesions should be biopsied from the center of the lesion where invasive disease is most likely to be found. Biopsy can easily be performed in the clinic with local anesthesia and a punch biopsy (Fig. 16.2). Colposcopy with application of 5% acetic acid may be a useful aid in determining the best site for biopsy. In addition, colposcopy can be used to follow lesions that have already been biopsied. If acetic acid is used as an adjunct to directed biopsy, the solution should be applied liberally and for several minutes due to the thicker keratin layer on the vulvar skin.

Women with small lesions and clinically negative nodes have limited benefit from undergoing extensive radiologic workup. However, those women with large, bulky tumors or clinical evidence of metastatic disease may need further evaluation with such modalities as barium enema, proctosigmoidoscopy, cystourethroscopy, computed tomographic (CT) scan and intravenous pyelography. Fine needle aspirate of suspicious nodes or metastatic lesions may also be helpful in delineating the extent of disease.

Staging

The International Federation of Gynecology and Obstetrics (FIGO) in 1989 adopted a modified surgical staging system for vulvar cancer. This system was revised in 1995 with few changes. Prior to 1989, a clinical staging system was used; however, this system missed 20-30% of nodal metastases when clinical diagnosis was correlated with surgical-pathologic diagnosis.[3] Thus, the surgical staging system was implemented which

Table 16.2. American joint committee on vulvar cancer staging

TNM	FIGO	
TX		Primary tumor cannot be assessed.
T0		No evidence of primary tumor.
Tis	0	Carcinoma in situ (pre-invasive carcinoma).
T1	I	Tumor confined to the vulva or to the vulva and perineum, 2 cm or less in greatest dimension.
T1a	Ia	Tumor confined to the vulva or to the vulva and perineum, 2 cm or less in greatest dimension and with stromal invasion no greater than 1 mm.
T1b	Ib	Tumor confined to the vulva or to the vulva and perineum, 2 cm or less in greatest dimension and with stromal invasion greater than 1 mm.
T2	II	Tumor confined to the vulva or to the vulva and perineum, more than 2 cm in greatest dimension.
T3	III	Tumor of any size with contiguous spread to the lower urethra and/or vagina or anus.
T4	IV	Tumor invades any of the following: upper urethra, bladder mucosa, rectal mucosa, or is fixed to the pubic bone.

Regional Lymph Nodes (N)

NX		Regional lymph nodes cannot be assessed.
N0		No regional lymph node metastases.
N1	III	Unilateral regional lymph node metastases.
N2	IVA	Bilateral regional lymph node metastases.

Distant Metastases (M)

MX		Distant metastases cannot be assessed.
M0		No distant metastases.
M1	IVB	Distant metastases (including pelvic lymph nodes).

Stage Grouping

Stage 0	Tis	N0	M0
Stage I	T1	N0	M0
Stage Ia	T1a	N0	M0
Stage Ib	T1b	N0	M0
Stage II	T2	N0	M0
Stage III	T1	N1	M0
	T2	N1	M0
	T3	N0	M0
	T3	N1	M0
Stage IVa	T1	N2	M0
	T2	N2	M0
	T3	N2	M0
	T4	Any N	M0
Stage IVb	Any T	Any N	M1

16

assesses nodal involvement, size and spread of primary lesions and depth of invasion. The American Joint Committee on Cancer Staging (AJCC) developed a TNM classification scheme that correlates with the FIGO staging (Table 16.2).

Table 16.3. Frequency of groin node metastases relative to depth of invasion

Depth of Invasion (mm)	% Positive
≤ 1	2.6
2	8.9
3	18.6
4	30.9
5	33.3
>5	47.9

Patterns of Spread

Vulvar cancers spread in three different ways including (1) direct extension into local organs, including the vagina, bladder and rectum, (2) lymphatic embolization and (3) hematologic spread. These tumor cells usually embolize first to the "sentinel" node located on the ipsilateral groin before moving to the deeper nodes within the pelvis.

Risk factors for lymphatic spread include tumor size ≥2 cm, poor tumor differentiation, depth of invasion >1 mm and lymphovascular space involvement.[4] Depth of invasion directly correlates with risk of positive inguinal nodes (Table 16.3).[4] Hematological spread results in distant metastases to areas such as the lung and bone. This is rare to see at initial presentation, but is a more common phenomenon seen with recurrent or advanced disease.

Pathology

Ninety percent of vulvar cancers are squamous cell carcinoma. Table 16.4 lists all the histologic types and subtypes.

Squamous Cell Carcinomas (Fig. 16.3)

Most vulvar cancers arise within the squamous epithelium; and thus, squamous cell carcinomas are the most common histologic type, comprising over 90% of vulvar

Table 16.4. Histologic types of vulvar cancer

Squamous cell carcinoma
Melanoma
Bartholin's gland carcinoma
Adenocarcinoma
Basal cell carcinoma
Verrucous carcinoma
Sarcomas
    Leiomyosarcoma
    Epithelioid sarcoma
    Rhabdomyosarcoma
Lymphoma
Endodermal sinus tumor
Merkel's cell carcinoma
Dermatofibrosarcoma protuberans
Malignant schwannoma

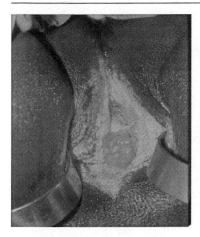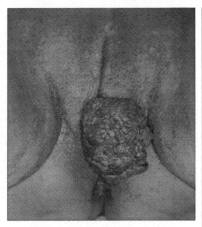

Figure 16.3. Squamous cell carcinoma of the vulva (courtesy of Wesley C. Fowler Jr, MD, Division of Gynecologic Oncology, University of North Carolina at Chapel Hill, Chapel Hill, North Carolina).

malignancies. The two most common histologic subtypes of squamous cell tumors are the (1) keratinizing, differentiated, simplex type and the (2) warty/basaloid, classic type. Keratinizing tumors are the more prevalent of the two subtypes and comprise 65% of squamous cell carcinomas of the vulva and tend to occur in older women.[3] These are seen primarily in older women and are not associated with HPV. In contrast, warty/basaloid or Bowenoid tumors comprise the other 35% of squamous cell carcinomas, are generally seen in younger women and are associated with HPV, HIV, early age at first intercourse, multiple sexual partners, smoking and the presence of coexisting vulvar/vaginal/cervical dysplasias.[3] Two different grading systems are used to describe these tumors with no definitive accepted scheme. The Kurman system categorizes these tumors as well, moderate or poorly differentiated based on nests of tumor cells versus diffuse infiltration and the presence or absence of keratinization and nuclear atypia.[3] The Gynecologic Oncology Group (GOG) system classifies these tumors based on the percentage of the tumor that is undifferentiated.[4]

Verrucous Carcinomas

Verrucous carcinomas of the vulva are a rare, highly differentiated variant of squamous cell carcinomas. These are typically very slow growing tumors with an exophytic growth pattern that is cauliflower-like in appearance. Many times, these tumors are mistaken for condylomas or squamous papillomas, reiterating the necessity for biopsy of any suspicious vulvar lesion to confirm the diagnosis. These tumors rarely metastasize to the lymph nodes but can be very destructive locally.

Melanomas

Melanomas of the vulva comprise 5-10% of vulvar neoplasms and are the second most common histologic type.[3] These lesions are most commonly seen in postmenopausal, Caucasian women and can arise de novo or at the site of a preexisting compound or junctional nevi. There are three subtypes: (1) superficial spreading malignant melanoma, (2) nodular melanoma and (3) acral lentiginous melanoma.[3]

Table 16.5. Staging systems for vulvar melanomas

Level	Clark's	Chung	Breslow
I	Intra-epithelial	Intra-epithelial	<0.76 mm
II	Into papillary dermis	<1 mm from granular layer	0.76-1.50 mm
III	Filling dermal papillae	1.1-2 mm from granular layer	1.51-2.25 mm
IV	Into reticular dermis	>2 mm from granular layer	2.26-3 mm
V	Into subcuta- neous fat	Into subcuta- neous fat	>3 mm

The nodular variant is typically raised, deeply invasive and metastasizes aggressively. The level of invasion and tumor thickness are essential measurements in evaluating vulvar melanomas. Depth of invasion is assessed similarly to other melanomas, utilizing either the Breslow's method or the modified Clark's levels (Table 16.5). The Clark level definitions were subsequently modified by Chung and colleagues to better describe vulvar melanomas (Table 16.5). These tumors can metastasize to other sites in the female genital tract including the cervix, vagina, urethra and rectum. Distant metastases are common, especially in the setting of recurrent disease and long-term survival is poor.

Vulvar Paget's Disease

Extramammary Paget's disease is an intra-epithelial adenocarcinoma that accounts for <1% of all vulvar malignancies.[3] In general, these lesions arise in Caucasian women in their 60s and 70s with puritus as the most common complaint on presentation. These lesions are similar to those seen on the breast and appear as well-demarcated, erythematous lesions. Most often, these lesions are multi-focal in nature and may be found anywhere on the vulva, mons, perineum or inner thigh. Interestingly, 20-30% of women with Paget's disease of the vulva will have a synchronous neoplasm; and thus, need to be evaluated for a noncontiguous tumor of the breast, bladder, colon, ovary or endometrium. Therefore, one may consider screening women with Paget's disease for its associated malignancies. The treament for these lesions is a simple vulvectomy with the goal being negative margins.

Adenocarcinomas

Most adenocarcinomas of the vulva arise in the Bartholin's glands, but these tumors can also occur in the skin appendages and Skene's glands. On rare occasions, squamous cell carcinomas may also arise in the the Bartholin's glands. The mean age at diagnosis is 57 years. Enlargement of the Bartholin's glands in women over the age of 40 is concerning because of the increased risk of malignancy in this age group as opposed to a benign process such as a simple abscess.[3] Thus, women in this older age group should all undergo biopsy of the Bartholin's gland if a mass is noted, even if the lesion appears to be cystic in nature. Because of the rich vascular and lymphatic supply to the Bartholin's glands, metastatic disease is quite common.

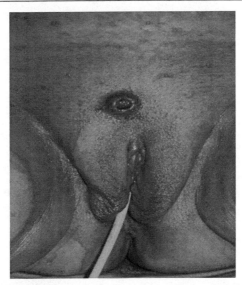

Figure 16.4. Basal cell carcinoma of the vulva (courtesy of Wesley C. Fowler Jr, MD, Division of Gynecologic Oncology, University of North Carolina at Chapel Hill, Chapel Hill, North Carolina).

16

Basal Cell Carcinomas (Fig. 16.4)

Two percent of basal cell carcinomas occur on the vulva and two percent of vulvar cancers are of basal cell histology.[3] These are typically seen in postmenopausal women and are described as "rodent" ulcers, with rolled edges surrounding a central area of ulceration. Although these lesions can be locally invasive, metastatic disease is rare. Basal cell carcinomas are often associated with a concomitant malignancy; and thus, women with this diagnosis should be thoroughly evaluated for evidence of another primary cancer.

Sarcomas

Vulvar sarcomas represent only 1-2% of vulvar cancers and as a group are heterogenous. The most common type is leiomyosarcomas, but other types include: fibrosarcomas, neurofibrosarcomas, liposarcomas, rhabdomyosarcomas, angiosarcomas, epitheliod sarcomas and malignant schwannomas. Primary treatment is wide local excision. Radiation therapy may be helpful in aggressive forms. Leiomyosarcomas typically present as a painful mass in the labia majora; treatment is radical local excision, and they rarely metastasize. Epithelioid sarcomas are generally soft tissue sarcomas that develop in the extremities of young adults may may be found in the vulva. These may mimic a Bartholin cyst and therefore may go undiagnosed. Treatment is wide local excision. Rhabdomyosarcomas are typically seen in childhood, and 20% may involve the urogenital tract. A multimodality treatment approach using multiple chemotherapeutic regimens following wide local excision has proven beneficial over radical surgery. Risk factors associated with an increased rate of recurrence include high grade lesions, size greater than 5 cm in diameter, infiltrating margins, positive margins and a high mitotic count.[3]

Other

Other much less common vulvar malignancies include merkel-cell tumors, transitional cell carcinomas, sebaceous carcinomas, adenosquamous carcinomas, adenoid cystic carcinomas, malignant schwannomas, yolk sac tumors and metastatic tumors.[3] A list of the most common histologic types of vulvar carcinoma is summarized in Table 16.4 .

Metastatic Disease

As previously stated, vulvar cancer may spread in three ways: direct extension, lymphatic metastasis and hematogenous metastasis. Direct extension may affect any adjacent organs to include the clitoris, vagina, anus and urethra. Lymphatic metastasis may occur early and one report showed nodal involvement in approximately 10% of superficially invasive vulvar lesion (Table 16.3). The lymphatic drainage of the vulva generally occurs sequentially from the superficial inguinal nodes to the deep femoral nodes below the cribiform fascia and then to the pelvic lymph nodes. Cases of "skipped" metastasis directly to the femoral nodes are rare but have been reported. This predictable lymphatic spread provides the basis for sentinel lymph node dissection (SN) and staging. Hematogenous metastasis is generally found later in the disease course and is rare in the absence of nodal metastatic disease. The risk for hematogenous spread is only 4% in those patients with three or fewer positive nodes at the time of initial diagnosis. This risk increases susbstantially to 66% if more than three nodes are found to be positive at the time of diagnosis.

Metastasis to pelvic nodes is relatively uncommon with an overall frequency of approximately 9%. In the presence of positive inguinal nodes, the risk of finding positive pelvic nodes is approximately 20% and more than three positive inguinal nodes is a risk factor.[4,5]

Presence of lymph node metastasis is related to size of primary lesion, stage at the time of diagnosis and the depth of invasion (Table 16.3). As the size of the primary lesion increases, the percentage of patients found to have nodal disease present also increases (Table 16.6). Nodal disease is also related to stage of disease at the time of diagnosis (Table 16.7).

Table 16.6. Influence of lesion size on regional lymph node mestastasis

Size	% of patients with + nodes
1 cm or less	5
1-2 cm	16
2-4 cm	33
More than 4 cm	53

Table 16.7. Influence of stage on regional lymph node mestastasis

Stage	% of patients with + nodes
I	10
II	26
III	64
IV	84

Similarly, depth of invasion predicts lymph node metastasis (Table 16.3). There is little risk of metastatic disease with less than 1 mm of invasion, but the risk increases quickly as up to 30% of patients with 4 mm of invasion will have lymph node metastases.

A 1993 prospective GOG sponsored study by Homesley et al involving 588 patients studied prognostic factors for lymph node metastasis. They concluded that clinically positive nodes, poor histologic grade, capillary/lymphatic involvement, tumor depth and increased patient age were all associated with lymph node metastatic disease.[4]

Treatment by Stage

Stage 0 Vulvar Cancer

Simple vulvectomy offers a 5-year survival rate of 100% but is seldom necessary. Treatment with less deforming procedures is preferred. VIN in nonhairy areas can be considered an epithelial disease, whereas VIN in hair bearing areas requires a greater depth of excision. Recurrence is common and usually occurs around the clitoral hood, perianal area and presacral area.

Standard treatment options are:

1. Wide local excision or laser beam therapy or a combination of both
2. Skinning vulvectomy with or without grafting
3. 5% fluorouracil cream (only a 50-60% response rate and should not be first line of treatment).

Stage IA (Microinvasive) Tumors

Historically Stage I vulvar cancer has been treated with radical vulvectomy with 5-year survival rates over 90%.[4-8]

As a result these and other studies two conclusions can be made regarding Stage I vulvar cancer:

1. Patients with tumor depth greater than 1 mm are at risk for lymph node involvement.[9-12]
2. Patients who have recurrent disease in an undissected groin have a very high mortality rate.

Standard treatment options:

1. Radical vulvectomy with unilateral or bilateral groin dissections depending on the site of the primary lesion. Separate groin incisions are recommended to reduce morbidity. Margins should be at least 8 mm to reduce incidence of recurrence.
2. For microinvasive lesions (<1 mm invasion) and no severe dystrophy, a wide local excision is indicated. For all other lateralized Stage I lesions, a radical local excision with complete unilateral lymphadenectomy should be performed. Candidates for this procedure should have lesions 2 cm or less in diameter with 5 mm or less invasion, no capillary or lymphatic invasion and clinically uninvolved nodes. Local recurrence rate with radical local excision is 7.2% compared with 6.3% after radical vulvectomy.
3. Radical vulvectomy and sentinel lymph node dissection.
4. Radical radiation therapy can achieve reasonable long-term survival rates for those few patients unable to tolerate surgery.

A prospective GOG trial evaluating groin radiation alone compared to groin dissection and adjuvant XRT for positive nodes was undertaken, but due to poor technical design the results were not able to be interpreted.[13] Retrospective studies show no improvement in outcome with groin dissection over XRT alone. Groin radiation is an alternative treatment for women with clinical N0 nodes and do not wish dissection or have medical contraindications to surgery.

16

Stage II Tumors

Historically, these tumors were treated by radical vulvectomy with bilateral inguinofemoral lymphadenctomy via an en bloc resection through a "butterfly" incision. This operation required radical removal of the tumor with a wide margin of normal skin, removal of the remaining vulva, dermal lymphatics and regional lymph nodes, all through a single incision. This approach had excellent cure rates (90% long-term survival); however, this was at the cost of significant postoperative morbidity, including wound breakdown (50% incidence at the vulva, 30% at the groin), lymphocyst formation, lower extremity lymphedema (10-15%) and lymphangitis.[3] The dramatic loss of normal appearing vulvar tissue also had a negative impact on sexual function for these women. In addition, this radical surgical approach did not negate the need for adjuvant postoperative radiation in high risk populations but did increase the risk of further complications, especially in regards to lymphedema.[3]

Several modifications have been made on this surgical procedure to allow for adequate treatment of the lesion but sparing the morbidity. A three incision technique or modified radical vulvectomy allows for radical excision of the vulvar lesion and bilateral inguinofemoral lymph nodes while retaining the skin over the groin. This technique decreased the incidence of wound breakdown to approximately 20%.[6,7] Survival and disease-free interval have been shown to be similar in women undergoing a modified radical vulvectomy versus an en bloc resection.[6]

A more conservative approach to the inguinal lymph node dissection is also currently employed to further spare morbidity. As opposed to removing both the superficial and deep lymph nodes in the inguinal region, removal of the superficial nodes acts as a "sentinel" biopsy to assess the probability of the deeper nodes being positive for disease. If the superficial lymph nodes are negative, there is a very small probability of the deeper nodes and/or contralateral nodes containing tumor cells.[9-12] Unanticipated groin failure in women with a negative superficial lymphadenectomy is approximated at 3-5%.[3] However, if the superficial lymph nodes are positive, additional treatment is needed due to the risk of deeper nodal involvement. This can entail either excision of the deeper nodes and the nodes of the contralateral groin and/or irradiation of the bilateral groins.

An ipsilateral lymph node dissection is sufficient for unilateral lesions that are ≥2 cm from the midline. For these lesions, the risk of contralateral lymph node involvement is minimal (2.8%).[3,8] In the case of lesions that are central (within 1 cm of the midline or involving midline structures), bilateral lymph node dissection is indicated.

Standard treatment options:

1. Modified radical vulvectomy with bilateral inguinal and femoral node dissection. Surgical margins should be clear by 10 mm. Separate groin incisions are recommended to decrease postoperative morbidity.

2. Adjuvant local radiation therapy may be indicated for surgical margins less than 8 mm, capillary-lymphatic space invasion and thickness greater than 5 mm particularly if the patient has positive nodes.[14,15]

Despite the trend towards less radical surgery for vulvar cancer, there is often the need for reconstructive surgery of the vulva, especially given the location and size of some of these lesions. Although the majority of the time the vulvar incision can be closed primarily, there are also cases where the area of excision is too large. Under these circumstances, tissue flaps may need to be created to provide normal, healthy tissue to reconstruct the wound (Chapter 20A).

In addition, depth of invasion and the presence of lymphovascular space invasion may be other risk factors for recurrent disease and should also be taken in to consideration when evaluating the need for adjuvant radiation therapy.[3]

Stage III

Advanced vulvar carcinomas involving the anus, rectum, rectovaginal septum or proximal urethra can be potentially curatively resected by radical operations, such as a radical vulvectomy or some combination of pelvic exenteration and vulvectomy. However, the current paradigm shift has been to a combined modality approach, involving sequenced chemoradiation and radical surgery. Generally, chemoradiation precedes surgery in an effort to decrease tumor bulk and permit easier resection. The basis for this therapeutic approach is that vulvar cancers tend to be very radioresponsive and that this may be a means to spare the high morbidity of radical surgery. Neoadjuvant chemotherapy as been shown to be inferior to chemoradiation for reduction of tumor burden.[16,17]

Standard treatment options:

1. Modified radical vulvectomy with inguinal and femoral node dissection. Pelvic and groin irradiation should be performed if inguinal nodes are positive.
2. Radical vulvectomy with inguinal and femoral node dissection followed by radiation therapy to the vulva with large primary lesions and narrow margins. Localized adjuvant radiation (45-50 Gy) may be indicated when there is capillary or lymphatic space invasion and a thickness greater than 5 mm; particularly if nodes are involved. Pelvic and groin XRT if two or more groin nodes are involved.
3. Preoperative radiation may be used in selected cases to improve operability or to decrease the extent of surgery.
4. For patients unsuitable for surgery, radical radiation therapy may result in long-term survival. Phase II trials studying concurrent radiation and 5-FU with or without cisplatin showed complete response rate of 53% to 89%.[18]

Chemotherapy and Radiation Treatment in Stage III and Stage IV Vulvar Cancer

Both chemotherapy and radiation therapy have been used pre- and postoperatively in patients with advanced vulvar cancer. Studies have shown that radiation therapy improves outcomes particularly in locally advanced disease to improve operability. A cisplatin and 5-fluorouracil combination has also demonstrated improved outcomes.

Benedett-Panici et al (1993) treated 21 patients with Stage III or IV disease with 2-3 cycles of cisplatin/bleomycin/methotrexate combination followed by surgery. They achieved 90% operability, but a only a 24% 3-year survival and 70% recurrence rate within 17 months.[17] A Phase II GOG trial by Moore et al in 1998, studied the use of preoperative chemoradiation to avert the need for more radical surgery in patients with T3 lesions or pelvic exenteration in patients with T4 lesions.[19] A cisplatin/5-fluorouracil regimen with concurrent radiation was given to 71 patients followed by resection of the residual tumor and bilateral groin dissections. 46.5% of the patients had no visible cancer at the time of surgery and 53.5% had gross residual cancer at the time of surgery. Only 2 of the 71 patients had unresectable residual tumors following this regimen.

Stage IV Vulvar Cancer

Standard treatment options:

1. Radical vulvectomy and pelvic exenteration.
2. Surgery followed by radiation therapy to the vulva for large resected lesions with narrow margins. Local adjuvant therapy may also be indicated if there is capillary-lymphatic space invasion and thickness greater than 5 mm, particularly in nodes are positive. Pelvic and groin irradiation should be performed if two or more groin nodes are involved.

16

3. Radiation therapy to primary lesions to improve operability followed by radical surgery. A radiation dose up to 55 Gy with concomitant 5-FU has been suggested.
4. Radical radiation therapy may result in long-term survival for those patients deemed unsuitable for surgical resection.

Sentinel Node Biopsy

Sentinel lymph node dissection has been of considerable interest in order to determine if SN assessment is an appropriate alternative to inguinofemoral lymph node dissection and minimizes morbidity from the lymphatic assessment. SN biopsy is already the accepted standard for patients with cutaneous melanoma and breast cancer. The feasibility of intraoperative lymphatic mapping in patients with vulvar cancer has been demonstrated in several studies.[20-22] Van der Zee and colleagues conducted a multicenter observational study on SN detection using a radioactive tracer and blue dye in women with stage I and II squamous cell cancers of the vulva that were <4 cm. Intraoperative lymphatic mapping can be performed using either the blue dye or lymphoscintigraphy with the radioactive tracer, but typically the combination is preferred to provide more accurate results. Six (2.3%) groin recurrences occurred in 259 patients with unifocal vulvar disease and a negative SN, and the 3-year survival rate was 97%. Morbidity was decreased in patients who underwent the SN compared to those who had a positive SN and underwent inguinofemoral lymphadenectomy. Specifically, wound separation (11.7% vs 34.0%), cellulitis (4.5% vs 21.3%), and lymphedema (1.9% vs 25.2%) were significantly decreased in the SN group.[23] In January 2009 a GOG study on intraoperative lymphatic mapping and SN identification in patients with squamous cell carcinoma of the vulva was completed after 9 years of enrollment. The objective of this study was to determine the negative predictive value of a negative SN and its location. This study highlights the challenges of conducting an assessment of SN in vulvar cancer which is a relatively rare disease. Currently no randomized trials have been performed comparing SN to inguinofemoral lymphadenectomy. In addition there is a steep learning curve in performing the SN procedure and adequate experience is needed prior to implementation in practice. Van der Zee and colleagues have recommended exposure of at least 5 to 10 cases per year per surgeon as a minimum figure in order to have the necessary expertise to perform SN.[23]

Follow-Up

The risk of recurrence of vulvar cancer is greatest during the first 2 years after treatment. In general, patients should be closely followed by a gynecologic examination every 3 months for the first 2 years after diagnosis and then every 6 months for the next 3

Table 16.8. Five-year survival for patients with squamous cell carcinomas of the vulva

FIGO Surgical Stage	% Survival
I	87
II	67
III	40
IV	22
Node Status	
Negative	91
Positive	52

years. At this point, patients can then return to annual exams. The physical examination should focus on careful inspection of the vulva and a complete lymph node survey and ideally be performed by a trained gynecologic oncologist. Vulvar biopsy and colposcopy are employed if abnormalities are found. Some reports have shown the risk of recurrence to be as high as 10% even when patients are >5 years out from their original diagnosis, demonstrating the need for long-term close follow-up.[24]

Outcomes

Overall survival for squamous cell carcinoma of the vulva is excellent, mainly because over two-thirds of patients are diagnosed with early stage disease. The 5-year survival for Stage I and II disease is 60-90%, Stage III disease is 40% and Stage IV disease is 20% (Table 16.8).[25]

Recurrent Disease

Vulvar tumor recurrences can be classified as local, inguinal or distant. In one series of over 502 patients, 187 patients (37%) recurred following primary surgical management. The distribution of recurrences was 53% on the vulva, 19% in the inguinal region, 6% in the pelvis, 8% distant sites and 14% with multiple sites.[26] Local recurrences can be treated successfully by re-excision in approximately 75% of cases. For those patients with inguinal or distant recurrences, resection may be an option, but most will need combination modality treatment, consisting of radiation and chemotherapy. In the same series of 502 patients, the 5-year survival rates according to site of recurrence were 60% for local recurrences, 27% for inguinal and pelvic disease and 15% for those with distant metastases.[26] There is no standard chemotherapy treatment regimen for patients with metastatic disease. These patients should be considered for treatment under clinical trials.

Standard treatment options:

1. Wide local excision with or without radiation in those patients with local recurrence.
2. Radical vulvectomy and pelvic exenteration.
3. Synchronous radiation and cytotoxic chemotherapy with or without surgery.

16

References

1. Jemal A, Murray T, Ward E et al. Cancer statistics. CA Cancer J Clin 2 2005; 55(1):10-30.
2. Messing MJ, Gallup DG. Carcinoma of the vulva in young women. Obstet Gynecol 1995; 86(1):51-4.
3. Moore DH K, WJ, McGuire WP et al. Vulva. In: Hoskins W, Perez CA, Young RC et al, eds. Fourth ed. Principles and Practice of Gynecologic Oncology. Philadelphia: Lippincott Williams and Wilkins, 2005:665-705.
4. Homesley HD, Bundy BN, Sedlis A et al. Prognostic factors for groin node metastasis in squamous cell carcinoma of the vulva (a Gynecologic Oncology Group study). Gynecol Oncol 1993; 49(3):279-83.
5. Drew PA, al-Abbadi MA, Orlando CA et al. Prognostic factors in carcinoma of the vulva: a clinicopathologic and DNA flow cytometric study. Int J Gynecol Pathol 1996; 15(3):235-41.
6. Farias-Eisner R, Cirisano FD, Grouse D et al. Conservative and individualized surgery for early squamous carcinoma of the vulva: the treatment of choice for stage I and II (T1-2 N0-1M0) disease. Gynecol Oncol 1994; 53(1):55-8.
7. Hacker NF, Berek JS, Lagasse LD et al. Individualization of treatment for stage I squamous cell vulvar carcinoma. Obstet Gynecol 1984; 63(2):155-62.

8. Homesley HD, Bundy BN, Sedlis A et al. Assessment of current International Federation of Gynecology and Obstetrics staging of vulvar carcinoma relative to prognostic factors for survival (a gynecologic oncology group study). Am J Obstet Gynecol 1991; 164(4):997-1003; discussion 1003-4.

9. Moore RG, DePasquale SE, Steinhoff MM et al. Sentinel node identification and the ability to detect metastatic tumor to inguinal lymph nodes in squamous cell cancer of the vulva. Gynecol Oncol 2003; 89(3):475-9.

10. Sliutz G, Reinthaller A, Lantzsch TM et al. Lymphatic mapping of sentinel nodes in early vulvar cancer. Gynecol Oncol 2002; 84(3):449-52.

11. Levenback C, Coleman RL, Burke TW et al. Intraoperative lymphatic mapping and sentinel node identification with blue dye in patients with vulvar cancer. Gynecol Oncol 2001; 83(2):276-81.

12. De Cicco C, Sideri M, Bartolomei M et al. Sentinel node biopsy in early vulvar cancer. Br J Cancer 2000; 82(2):295-9.

13. Stehman FB, Bundy BN, Thomas G et al. Groin dissection versus groin radiation in carcinoma of the vulva: a Gynecologic Oncology Group study. Int J Radiat Oncol Biol Phys 1992; 24(2):389-96.

14. Faul CM, Mirmow D, Huang Q et al. Adjuvant radiation for vulvar carcinoma: improved local control. Int J Radiat Oncol Biol Phys 1997; 38(2):381-9.

15. Homesley HD, Bundy BN, Sedlis A et al. Radiation therapy versus pelvic node resection for carcinoma of the vulva with positive groin nodes. Obstet Gynecol 1986; 68(6):733-40.

16. Wagenaar HC, Colombo N, Vergote I et al. Bleomycin, methotrexate and CCNU in locally advanced or recurrent, inoperable, squamous-cell carcinoma of the vulva: an EORTC gynaecological cancer cooperative group study. European Organization for Research and Treatment of Cancer. Gynecol Oncol 2001; 81(3):348-54.

17. Benedetti-Panici P, Greggi S, Scambia G et al. Cisplatin (P), bleomycin (B) and methotrexate (M) preoperative chemotherapy in locally advanced vulvar carcinoma. Gynecol Oncol 1993; 50(1):49-53.

18. http://www.cancer.gov/cancertopics/pdq/treatment/vulvar/HealthProfessional/page9

19. Moore DH, Thomas GM, Montana GS et al. Olt G. Preoperative chemoradiation for advanced vulvar cancer: a phase II study of the Gynecologic Oncology Group. Int J Radiat Oncol Biol Phys 1998; 42(1):79-85.

20. Levenback C, Coleman RL, Burke TW et al. Intraoperative lymphatic mapping and sentinel node identification with blue dye in patients with vulvar cancer. Gynecol Oncol 2001; 83(2):276-81.

21. Terada KY, Coel MN, Ko P, Wong JH. Combined use of intraoperative lymphatic mapping and lymphoscintigraphy in the management of squamous cell cancer of the vulva. Gynecol Oncol 1998; 70(1):65-9.

22. Van der Zee AG, Oonk MH, De Hullu JA et al. Sentinel node dissection is safe in the treatment of early-stage vulvar cancer. J Clin Oncol 2008; 26(6):884-9.

23. http://www.cancer.gov/clinicaltrials/results/vulvar-SNB0308.

24. Gonzalez Bosquet J, Magrina JF, Gaffey TA et al. Long-term survival and disease recurrence in patients with primary squamous cell carcinoma of the vulva. Gynecol Oncol 2005; 97(3):828-33.

25. Beller U, Sideri M, Maisonneuve P et al. Carcinoma of the vulva. J Epidemiol Biostat 2001; 6(1):155-73.

26. Maggino T, Landoni F, Sartori et al. Patterns of recurrence in patients with squamous cell carcinoma of the vulva. A multicenter CTF Study. Cancer 2000; 89(1):116-22.

16

CHAPTER 17

Vaginal Cancer

Thanasak Sueblinvong and Michael E. Carney

The most common carcinoma found in the vagina is actually metastatic from other primary gynecologic and nongynecologic sites including cervix (32%), endometrial (18%), colon and rectum (9%), ovary (6%) and vulva (6%), Metastatic carcinoma accounts for 84% of all vaginal carcinomas. Primary cancer of the vagina is considered rare and constitutes 2% of all female genital tract malignancies.

Squamous cell carcinomas account for the majority of primary vaginal carcinomas. Other histologic subtypes include adenocarcinoma, melanoma and sarcoma (Table 17.1). Most patients are in their sixth and seventh decades of life with only 10% of cases occurring in patients 40 years of age or younger. However, vaginal cancer is increasingly being seen in younger women, possibly due to human papillomavirus (HPV) infection or other sexually transmitted infections. Similar to cervical cancer, most vaginal cancers are thought to arise in a premalignant vaginal intra-epithelial dysplastic lesion (VAIN).

Anatomy

The vagina is a tubular structure lined by nonkeratinized squamous epithelium extending from the vestibule to the uterus. It lies dorsal to the bladder and ventral to the rectum. The average vaginal length is 7.5 cm. Beneath the mucosa lies a submucosal layer of elastin and a double muscularis layer, which is highly vascularized, includes lymphatics and is richly innervated.

The lymphatic drainage of the vagina is complex, consisting of an extensive intercommunicating network. Fine lymphatic vessels coursing through the submucosa and muscularis coalesce into small trunks running laterally along the walls of the vagina. The upper anterior vagina drains along cervical channels to the interilliac chain; the posterior vagina drains into the inferior gluteal, presacral and anorectal

Table 17.1. Primary vaginal tumors

Histologic Types	Percentage
Squamous cell carcinoma	81.5
Adenocarcinoma (including clear cell)	10.5
Sarcoma	3.0
Melanoma	3.2
Small cell	0.9
Lymphoma	0.6
Carcinoid	0.1

Gynecologic Oncology, edited by Paola Gehrig and Angeles Secord.
©2009 Landes Bioscience.

nodes. The distal vaginal lymphatics drain into the inguinal and femoral nodes and subsequently to the pelvic nodes. Lymphatic flow from lesions in the midvagina may drain either to the inguinal nodes or to the iliac nodal system. However, because of the presence of intercommunicating lymphatics along the terminal branches of the vaginal artery and near the vaginal wall, the external iliac nodes are at high risk even in lesions of the lower third of the vagina. Such a complex lymphatic drainage pattern has significant implications for therapeutics planning. Therefore, bilateral pelvic nodes should be considered to be at risk in any invasive vaginal carcinoma, and bilateral groin nodes should be considered to be at risk in those lesions involving the distal third of the vagina.

Etiology

Squamous Cell Carcinoma (SCC)

Potential risk factors for squamous cell carcinoma of the vagina include HPV infection especially HPV 16.[1] HPV DNA has been recovered from 80% of VAIN lesions and 60% of invasive SCC of the vagina.

VAIN is far less common that CIN representing only 1% of all gynecologic intra-epithelial neoplasias. The VAIN process most commonly occurs in the upper third of vagina (75%), most commonly in the posterior wall and it is frequently multifocal. However, it may also occur in the lower third of the vagina in approximately a third of the cases. Other risk factors for SCC of vagina and VAIN include five or more sexual partners, sexual debut before age 17 years, smoking, low socioeconomic status, a history of genital warts, prior abnormal cytology, immunosuppression, history of radiotherapy, history of sexually transmissible diseases and/or human papillomavirus infection and prior hysterectomy.[2]

About 30% of patients with primary vaginal carcinoma have a history of in situ or invasive cervical cancer treated at least 5 years earlier. There are three possible mechanisms for the occurrence of vaginal cancer after cervical neoplasia: (1) occult residual cervical disease; (2) new primary disease arising in an "at-risk" lower genital tract and (3) radiation carcinogenicity. However, a recent study identifying similar molecular clonality leaned towards occult residual disease from cervical neoplasia as the most likely explanation.[3]

Adenocarcinoma

Approximately 10% of primary vaginal carcinomas are adenocarcinomas, and they affect a younger population of women, regardless of whether exposure to diethylstilbestrol has occurred in utero.

In 1971, Herbst et al first reported the association of in utero exposure to DES during the first 16 weeks of pregnancy and clear cell adenocarcinoma of the vagina.[4] DES (diethylstilbesterol) is a synthetic estrogen previously used in women with recurrent pregnancy loss. More than 500 cases of clear cell carcinoma of the vagina or cervix have been reported to the registry, although only two-thirds of the completely investigated cases have an actual documented history of prenatal exposure to DES. These tumors become most frequent after the age of 14 years and peak age at diagnosis is 19 years. The incidence of clear cell carcinoma in the exposed female population from birth to 34 years is estimated to be between 0.14 and 1.4 per 1,000. Approximately 90% of the patients had Stage I-II disease at diagnosis.

Vaginal adenosis occurs in about 45% of DES-exposed patients. The lining of the vagina of young girls is made of cells that give it a red and rough look. As the young women mature, these areas of vaginal and cervical ectropion are progressively covered with metaplastic squamous epithelium looking flat and pale. This maturation does not occur in vaginal mucosa of women with in-utero DES exposure. There are two types of cells described in vaginal adenosis: the mucinous cell, which resembles the endocervical epithelium and the tuboendometrial cell which is believed to be the origin of carcinogenic process. It is recommended that DES-exposed daughters have twice yearly Pap smears beginning at age 14. Colposcopy is not essential if clinical and cytologic evaluations are negative, but staining with half-strength Lugol's iodine may help delineate areas of adenosis. There are also a number of genital abnormalities associated with DES exposure which include a T-shaped uterus, a small uterine cavity, constriction rings, uterine filling defects, synechae, diverticulae and unicornuate or bicornuate uteri.

Melanoma

Vaginal melanomas account for approximately 0.3% of all melanomas, with an incidence of 0.026 per 100,000 women per year and a median age at diagnosis of 66.3 years. Vaginal melanoma represents 4% of primary vaginal cancers and most commonly occurs in Caucasian women.

Sarcoma

Sarcomas represent 3% of primary vaginal cancers and are more common in adults, with leiomyosarcoma representing 50-65% of vaginal sarcomas. Malignant mixed mullerian tumor (MMMT, carcinosarcoma), endometrial stromal sarcoma and angiosarcoma are less common. MMMT and vaginal angiosarcoma are associated with a history of pelvic radiation. Embryonal rhabdomyosarcoma (sarcoma botryoides) is a rare pediatric tumor.

Germ Cell Tumor

Malignant vaginal germ cell tumors (GCT) account for less than 3% of all cancers in children. Endodermal sinus tumor (EST) of the vagina is a rare, highly malignant GCT that occurs in children less than 3 years of age. It is difficult to explain the etiology of EST arising in the vagina as they are thought to arise from germ cells that have failed to complete their migration to the gonad. The most common vaginal cancers in infancy are embryonal rhabdomyosarcomas and endodermal sinus tumors.

Screening

For screening to be cost-effective, the incidence of the disease must be sufficient to justify the cost of screening. This makes routine screening for vaginal cancer inappropriate. When age and prior cervical disease are controlled for, there is no increased risk of vaginal cancer in women who have had hysterectomy for benign disease. However, it is recommended that patients previously treated for pre-invasive cervical disease or cervical carcinoma continue to undergo lifelong surveillance with vaginal cytologic evaluation—even after hysterectomy.

Symptoms and Signs

Most patients with vaginal cancer present with vaginal bleeding or vaginal discharge. This is most common after menopause. Approximately 5% of patients present with pelvic pain because of extension of disease beyond the vagina and approximately 5-10% of the patients are asymptomatic. Occassionally, vaginal cancer can present with a vaginal mass protruding through the introitus.

Table 17.2.   Carcinoma of the vagina: FIGO nomenclature

Stage 0	Carcinoma in situ; intra-epithelial neoplasia Grade 3.
Stage 1	The carcinoma is limited to the vaginal wall.
Stage II	The carcinoma has involved the subvaginal tissue but has not extended to the pelvic wall.
Stage III	The carcinoma has extended to the pelvic wall.
Stage IV	The carcinoma has extended beyond the true pelvis or has involved the mucosa of the bladder or rectum; bullous edema as such does not permit a case to be allotted to Stage IV.
Stage IVA	Tumor invades bladder and/or rectal mucosa and/or direct extension beyond the true pelvis.
Stage IVB	Spread to distant organs.

Diagnosis

The diagnosis of vaginal carcinoma is often missed on examination, particularly if the lesion is small and situated in the lower two-thirds of the vagina where it may be covered by the speculum blade. The definitive diagnosis is made by biopsy of a lesion which may be detected through physical examination, digital palpation, colposcopic and cytologic evaluation. Examination under anesthesia may be recommended if cytology suggests a lesion that is not seen on clinical exam. If a cervix is present, this represents the most likely source of any abnormal cytology or discharge and needs to be evaluated completely. Biopsy of the cervix, if present, is necessary to rule out primary cervical cancer.

Staging

At present, primary malignancies of the vagina are all staged clinically. The International Federation of Gynecology and Obstetrics (FIGO) staging for vaginal carcinoma is shown in Table 17.2.

Surgical staging for vaginal cancer has been used less commonly than for cervical cancer, but in selected premenopausal patients a pretreatment laparotomy may allow better definition of the extent of disease, excision of any grossly enlarged lymph nodes and placement of an ovary into the paracolic glutter beyond the radiation field. Positron emission tomography (PET) may also be increasingly used in the future to determine the extent of metastatic disease.

Pattern of Spread/Metastasis

Vaginal cancer spreads by the following routes:

1. Direct extension to the pelvic soft tissues, pelvic bones and adjacent organs (bladder and rectum).
2. Lymphatic dissemination to the pelvic and later the para-aortic lymph nodes in lesions occurring in the upper vagina. Lesions in the lower one-third of the vagina metastasize directly to the inguinofemoral lymph nodes, with the pelvic nodes being involved secondarily.
3. Hematogenous dissemination to distant organs, including lungs, liver and bone. Hematogenous dissemination is a late phenomenon in vaginal cancer and the disease usually remains confined to the pelvis for most of its course.

General Management:
Treatment options and outcome:

1. **Surgery**: Surgery is an option in very selected patients with vaginal cancer. The approach can range from a partial vaginectomy for a small Stage I lesion that is in the distal vagina to a complete pelvic exenteration for certain histologic subtypes and recurrent disease.

2. **Radiation Therapy**: Radiation therapy is considered the preferred primary treatment for most women with vaginal cancer. Radiation achieves excellent tumor control and good posttreatment results for most early and locally advanced disease. The radiation technique usually combines external beam radiation therapy (EBRT) and brachytherapy but in patients with small Stage I tumors intracavitary brachytherapy (ICB) can be used alone. As with cervical carcinoma, chemosensitization should be considered when treating women with vaginal cancer with primary radiotherapy.

   Intracavitary brachytherapy uses a vaginal or uterine applicator/tandem devices to deliver a radioactive source to the tumor, and interstitial brachytherapy (ITB) uses a template with a series of needles implanted through the perineum and loaded with radiation sources. Interstitial brachytherapy is important in more regionally advanced and irregular tumors where dose-rate distributions are calculated by means of computer-assisted optimization of the source placement and strength.

3. **Chemotherapy**: Chemotherapy theoretically works as a radio-sensitizer in vaginal squamous cell cancers as it does in cervical cancers although there are no large studies documenting its efficacy. Chemotherapy is more important in the control of nonsquamous vaginal cancers—especially sarcomas and germ cell tumors. Multimodal treatment with radiation, chemotherapy and radical surgery has helped to improve survival rates in children with embryonal rhabdomyosarcoma from 20% to 70%.

Recurrent of the Disease
The treatment of recurrent vaginal cancer can be considered either potentially curative or palliative. For cure, in most cases, a total pelvic exenteration is offered in patients who recur centrally after radical radiotherapy. Cure rates are improved when there is a long time period since initial treatment and the extensive surgery is generally not indicated in patients with significant unilateral or sciatic pain, hydronephrosis, lymph node involvement, or lower extremity edema. Most patients with vaginal cancer treated with chemoradiation will not recur in the pelvis. For these patients, palliation with radiation and chemotherapy is used for symptom control. Chemotherapy in vaginal cancer is poorly studied, but most would use a cervical cancer regimen including cisplatin.

Future Directions
Vaginal cancer is a rare tumor. Large trials examining newer treatments for this cancer are unlikely given the rarity, but newer treatments for the more common cervical cancer can often be extrapolated to this disease. Most recently, treatments incorporating radiosensitizers such as cisplatin for vaginal cancer based on cervical cancer data is being employed. As newer chemotherapies and targeted agents including vascular endothelial growth factor inhibitors, kinase inhibitors and epidermal growth factor inhibitors are examined in cervical cancer, it is likely that we will see more agents used in the treatment of vaginal cancer.

Suggested Reading

1. Hoskins WJ, Perez CA, Young RC et al. The Principles and Practice of Gynecologic Oncology "Vaginal Cancer", Fourth Edition. Lippincot, Williams and Wilkins, 2004.
2. Griffiths CT, Fuller AR. Gynecologic Oncology (Cancer Treatment and Research), "Vaginal Cancer", First Edition. Springer, 2001.
3. Beller U, Maisonneuve P, Benedet JL et al. Carcinoma of the vagina. Int J Gynaecol Obstet 2003; 83:27-39.

References

1. Hacker NF. Vaginal cancer. In: Berek JS, Hacker NF, eds. Practical Gynecologic Oncology. 4th Ed. Philadelphia: Lippincott Williams and Wilkins 2005:585-601.
2. Gardens HR, Roth LM, McGuire WP et al. Vagina. In: Hoskins WJ, Perez CA, Young RC et al, eds. Principles and Practice of Gynecologic Oncology. 4th Ed. Philadelphia: Lippincott Williams and Wilkins 2005:707-42.
3. Vinokurova S, Wentzensen N, Einenkel J et al. Clonal history of papillomavirus-induced dysplasia in the female lower genital tract. J Natl Cancer Inst 2005; 97:1816-21.
4. Herbst AL, Ulfelder H, Poskanzer DC. Adenocarcinoma of the vagina: association of maternal stibestrol therapy with tumor appearance in young women. N Engl J Med 1971; 284:878-81.
5. Hampl M, Sarajuuri H, Wentzensen N et al. Effect of human papillomavirus vaccines on vulvar, vaginal and anal intraepithelial lesions and vulvar cancer. Obstet Gynecol 2006; 108:1361-8.
6. Haidopoulos D, Diakomanolis E, Rodolakis A et al. Can local application of imiquimod cream be an alternative mode of therapy for patients with high-grade intraepithelial lesions of the vagina? Int J Gynecol Cancer 2005; 15:898-902.

17

CHAPTER 18

Gestational Trophoblastic Disease

John T. Soper

Introduction

Gestational trophoblastic disease (GTD) and malignant forms of gestational trophoblastic neoplasia (GTN) form a spectrum of interrelated disease processes originating from the placenta. This general terminology comprises many clinical and histological entities including hydatidiform moles, invasive moles, gestational choriocarcinomas and placental site trophoblastic tumors (PSTT). Before the development of sensitive assays for human chorionic gonadotropin (hCG) and effective chemotherapy, mortality from all forms of malignant GTN was substantial. Currently, most women with malignant GTN can be cured, and their reproductive function can be preserved.

Estimates for the incidence of various forms of GTD vary. In the United States, hydatidiform moles are observed in approximately 1/600 therapeutic abortions and 1/1,000-1,200 pregnancies.[1] Approximately 20% of patients will be treated for malignant sequelae after evacuation of a hydatidiform mole. The majority of patients with postmolar GTN will have nonmetastatic molar proliferation or invasive moles, but gestational choriocarcinoma is present in approximately 30-50% of postmolar GTN. Gestational choriocarcinoma occurs in approximately 1/20,000-40,000 pregnancies; approximately 50% present after term pregnancies, 25% after molar pregnancies and the remainder after other gestational events.[1] Although much more rare than hydatidiform mole or gestational choriocarcinoma, PSTT can develop after any type of pregnancy.

Hydatidiform Mole

Partial and complete hydatidiform moles are distinct diseases with characteristic cytogenetic, histologic and clinical features.[1,2] These do not represent a "transition" or degeneration from normal gestation to hydatidiform mole (Table 18.1). In both conditions, the placental villi become edematous, forming small grape-like (hydatidiform) structures to a variable degree (Fig. 18.1). The increasing use of early obstetrical ultrasounds and greater availability of sensitive serum hCG assays has shifted many of the clinical features of complete hydatidiform moles to resemble those traditionally associated with partial moles, largely due to earlier diagnosis. However, a reduction in the incidence of postmolar GTN has not been observed. Despite the cytogenetic, pathologic and clinical differences, the management of patients with complete and partial moles is similar.[1,2]

Partial hydatidiform moles (Fig. 18.2) usually have complete trisomy derived from two paternal and one maternal haploid sets of chromosomes (Table 18.1).[1,2] Most have a 69, XXX or 69, XXY genotype derived from a haploid ovum with either reduplication of the paternal haploid set from a single sperm, or less frequently, from dispermic fertilization. Gross or histologic evidence of fetal development, such as amnion or vessels with fetal red blood cells (Fig. 18.3), is a prominent pathologic feature.

Gynecologic Oncology, edited by Paola Gehrig and Angeles Secord.
©2009 Landes Bioscience.

Table 18.1. Features of partial and complete hydatidiform moles

Feature	Partial Mole	Complete Mole
Karyotype	Most commonly 69, XXX or -, XXY	Most commonly 46, XX or -, XY
Pathology		
Fetus	Often present	Absent
Amnion, fetal rbcs	Usually present	Absent
Villous edema	Variable, focal	Diffuse
Trophoblastic Proliferation	Focal, slight-moderate	Diffuse, slight-severe
Clinical presentation		
Diagnosis	Missed abortion	Molar gestation
Uterine size	Small for dates	50% large for dates
Theca lutein cysts	Rare	25-30%
Medical complications	Rare	10-25%
Postmolar GTN*	2.5-7.5%	7.8-30%

*Gestational Trophoblastic Neoplasia, Modified from Soper, 1996.

In contrast, complete moles (Fig. 18.4) almost always have a chromosomal complement totally derived from the paternal genome.[1,2] The 46, XX karyotype is most common (Table 18.1), representing reduplication of the haploid genome of the sperm and exclusion of the maternal chromosomal complement. A smaller portion of complete moles have a 46, XY karyotype consistent with dispermic fertilization. Fetal development is not observed in complete moles because the fetus resorbs before the development of fetal circulatory system. Therefore fetal red blood cells are absent.

Figure 18.1. Gross photograph of an evacuated complete mole. The specimen is an aggregate comprised of diffuse hydropic villi.

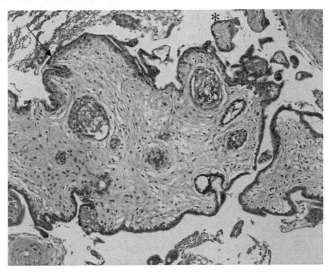

Figure 18.2. Photomicrograph of a partial mole with characteristic features of slightly edematous villi with scalloped borders, trophoblastic inclusions and slight trophoblastic proliferation (H and E, original magnification 20X).

18

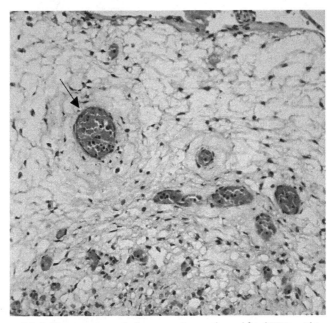

Figure 18.3. Photomicrograph demonstrating nucleated fetal RBCs within vessels in the villus of a partial mole (H and E, original magnification 40X).

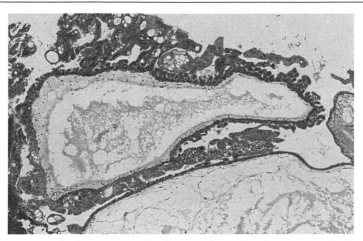

Figure 18.4. Photomicrograph of a complete mole illustrating markedly edematous villi with absence of vessels in the villi and marked trophoblastic proliferation (H and E, original magnification 20X).

Among complete moles the amount of hydropic villi and trophoblastic proliferation generally exceeds that observed in partial moles and is reflected in the different clinical presentations (Table 18.1). Serum hCG levels are usually higher in patients with complete moles. Although an increasing proportion of complete moles are diagnosed as missed abortion on the basis of an early ultrasound, the majority of patients with complete moles have a clinical or sonographic diagnosis of hydatidiform mole (Fig. 18.5).[1] Uterine enlargement beyond the expected gestational age is observed in up to 50% of patients with complete moles. These patients may present with vaginal bleeding or expulsion of molar vesicles. Medical complications of molar pregnancy, including pregnancy-induced hypertension, hyperthyroidism, anemia and hyperemesis are more frequently seen among patients with complete moles. Approximately 15-25% of patients with complete moles will have theca lutein cysts with ovarian enlargement >6 cm (Fig. 18.5). Malignant sequelae occur in less than 5-10% of patients with partial moles compared to approximately 20% after evacuation of a complete hydatidiform mole (Table 18.1).

Hydatidiform moles are usually diagnosed during the first half of pregnancy. The most common presenting symptom is abnormal bleeding, occasionally with passage of hydropic villi. Other signs and symptoms include uterine enlargement greater than expected for gestational dates, absent fetal heart tones, theca lutein cysts, hyperemesis and an abnormally high level of hCG for gestational dates. Pregnancy-induced hypertension in the first half of pregnancy is virtually diagnostic of molar pregnancy. Ultrasound has replaced all other noninvasive means of establishing the diagnosis. Molar tissue typically is identified as a diffuse mixed echogenic pattern replacing the placenta (Fig. 18.5). The combination of ultrasound findings with an elevation of hCG above expected for the duration of pregnancy is highly suggestive of molar pregnancy and should alert the clinician to the possibility of a molar gestation.[1,2]

The diagnosis of complete or partial mole will sometimes be made only after histologic evaluation after dilatation and evacuation (D and E) is performed for a suspected

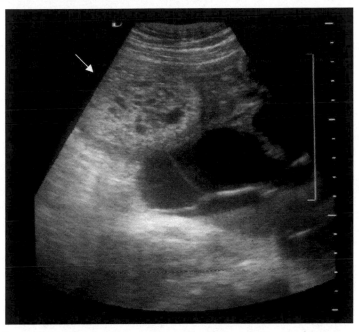

Figure 18.5. Ultrasound of complete mole with theca lutein cyst of the ovary. The uterus is filled with a characteristic amorphous mass of mixed echogenic material, corresponding to the diffusely hydropic villi. The ovarian cyst has thin-walled septae; theca lutein cysts usually resolve without intervention as hCG levels decline following evacuation of the mole.

incomplete spontaneous abortion. In these cases, patients should be monitored with serum quantitative hCG values. For patients in whom hydatidiform mole is suspected prior to evacuation, a complete blood count with platelet determination, clotting function studies, renal and liver function studies, blood type with antibody screen, preevacuation hCG level and chest X-ray should be obtained.

Medical complications of hydatidiform mole are observed in approximately 25% of patients with uterine enlargement >14-16 weeks gestational size.[1,2] Common medical complications include anemia, infection, hyperthyroidism, pregnancy-induced hypertension, respiratory distress and coagulopathy. The mole should be evacuated as soon as possible after stabilization of any medical complications. The preferred method of evacuation for most patients is suction D and E.[1-4] Medical induction of labor with oxytocin or prostaglandin and hysterotomy are not recommended for evacuation because they increase blood loss and increase the risk for malignant sequelae compared with suction D and E.[4] Furthermore, patients most often require D and E to complete the evacuation of the mole after medical induction of labor.

Postoperative pulmonary complications are frequently observed around the time of molar evacuation among patients with uterine enlargement >14-16 weeks' size.[1-3] Although the syndrome of trophoblastic embolization has been emphasized as an underlying cause for respiratory distress syndrome following molar evacuation, there are many other potential causes of pulmonary complications in these women. Respiratory distress syndrome can be caused by high-output congestive heart failure due to anemia, hyperthyroidism, preeclampsia, or iatrogenic fluid overload. In general, these complications should be treated aggressively with therapy directed by intravenous monitoring and ventilator support as required.

Hysterectomy is an alternative to suction D and E for molar evacuation in selected patients who do not wish to preserve childbearing.[5] Usually the ovaries may be preserved. Hysterectomy reduces the risk of malignant postmolar sequelae compared to evacuation by D and E. However, the risk of postmolar GTN after hysterectomy remains approximately 3-5%; these patients should be monitored postoperatively with serial hCG levels.

Serial quantitative serum hCG determinations should be performed after molar evacuation using one of several commercially available assays capable of detecting beta hCG to baseline values (<5 mIU/mL). Ideally, serum hCG levels should be obtained within 48 hours of evacuation, every 1-2 weeks while elevated and then at 1-2 month intervals for an additional 6-12 months. Reliable contraception is mandatory during monitoring of hCG values.

Although rare instances of long latent periods between evacuation and postmolar GTN have been reported, the vast majority of episodes of malignant sequelae occur within approximately 6 months of evacuation;[1-3] traditionally hCG surveillance has been recommended for up to 12 months after evacuation of a complete mole. Although early pregnancies after molar evacuation are usually normal gestations, an early pregnancy obscures the value of monitoring hCG values during this interval and may result in a delayed diagnosis of postmolar malignant GTN. After completion of surveillance documenting remission, pregnancy can be permitted and hCG monitoring discontinued. Patients with prior partial or complete mole have a 10-fold increased risk (1-2% incidence) of a second mole in subsequent pregnancy.[1-3] Therefore, all future pregnancies should be evaluated by an early obstetrical ultrasound.

Prophylactic chemotherapy with methotrexate or dactinomycin will reduce the incidence of postmolar GTN in women with high-risk moles and usually has very little toxicity.[1,2] However, there are anecdotal cases of fatalities caused by prophylactic chemotherapy, and it does not eliminate the need for postevacuation followup. In compliant patients the low morbidity and mortality achieved by monitoring patients with serial hCG determinations and instituting chemotherapy only in patients with postmolar GTN outweighs the potential risk and small benefit for routine prophylactic chemotherapy.

Coexistence of a fetus with molar change of the placenta (both complete and partial) is relatively rare, occurring in 1:22,000—1:100,000 pregnancies.[1,2] Most of these twin pregnancies are diagnosed antepartum by ultrasound findings of a complex, cystic placental component distinct from the fetoplacental unit; however, in a few cases the diagnosis is not suspected until examination of the placenta following delivery. Compared to singleton hydatidiform moles, twin pregnancy with fetus and mole has an increased risk for postmolar GTN[1,2] with a higher proportion of patients having metastatic disease and requiring multiagent chemotherapy. Among patients with coexistent mole and fetus who continue pregnancy beyond the first trimester, there is a subset with early complications leading to termination of the pregnancy before fetal viability. This group of patients has a markedly increased risk of postmolar GTN compared to patients whose pregnancy continues into the third trimester.

For patients with coexistent hydatidiform mole and fetus suspected by ultrasound, there are no clear guidelines for management. Ultrasound monitoring of the pregnancy should be repeated to exclude retroplacental hematoma, other placental abnormalities, or degenerating myoma and to fully evaluate the fetoplacental unit for evidence of a partial mole or gross fetal malformations. If the diagnosis is still suspected and continuation of pregnancy is desired, fetal karyotype should be obtained, a chest X-ray performed to screen for metastases and serial serum hCG values should be followed. If fetal karyotype is normal, major fetal malformations are excluded by ultrasound and there is no evidence of metastatic disease, it is reasonable to allow the pregnancy to continue unless pregnancy-related complications force delivery. Patients should be counseled about the increase in obstetrical risks and the increased risk of postmolar GTN after evacuation or delivery. After delivery, the placenta should be histologically evaluated, and the patient followed closely with serial hCG values, similar to a followup after a singleton hydatidiform mole.

As long as hCG values are declining after molar evacuation, there is no role for chemotherapy. However, if hCG levels rise or plateau over several weeks, immediate evaluation and treatment for malignant postmolar GTN is indicated.[1-3] Other criteria for postmolar GTN include diagnosis of choriocarcinoma or invasive mole on uterine curettings, or identification of metastases. The role of repeat curettage in the setting of an hCG rise or plateau is controversial. Most investigators have reported that repeat curettage does not often induce remission or influence treatment.[5]

A variety of hCG criteria have been used to diagnose postmolar GTN. In 2000 the International Federation of Gynecologists and Obstetricians (FIGO) standardized hCG criteria used to make the diagnosis of postmolar GTN:[7]

1. An hCG level plateau of 4 values +/– 10% recorded over a 3 week duration (days 1, 7, 14, 21),
2. An hCG level rise >10% of 3 values recorded over a 2 week duration (days 1, 7, 14), and
3. Persistence of detectable hCG for >6 months after molar evacuation.

Malignant GTN

The clinical presentation of the patient with malignant GTN is more important for determining treatment and outcome than the precise histologic diagnosis.[1-3] Postmolar GTN comprises noninvasive postmolar trophoblastic proliferation, invasive moles, gestational choriocarcinoma and placental site trophoblastic tumors (PSTTs). Invasive moles are characterized by presence of edematous chorionic villi with trophoblastic proliferation that invade directly into the myometrium. They rarely metastasize, but are treated with chemotherapy to prevent morbidity and mortality caused by uterine perforation, hemorrhage, or infection. Gestational choriocarcinoma is a pure epithelial malignancy, composed of both neoplastic syncytiotrophoblast and cytotrophoblast elements without chorionic villi (Fig. 18.6). Gestational choriocarcinoma tends to develop early systemic metastasis; chemotherapy is clearly indicated when choriocarcinoma is histologically diagnosed.[1-3] PSTTs are relatively rare and these tumors are characterized by absence of villi with proliferation of intermediate trophoblast cells (Fig. 18.7).[1,8] In general, PSTTs are not as sensitive to simple chemotherapy as other forms of malignant GTN, and surgery is typically required.[8] Fortunately, the majority of patients present with disease confined to the uterus and can be treated by hysterectomy.

Postmolar GTN is most frequently diagnosed on the basis of rising or plateauing hCG values as discussed previously.[7] Women with malignant GTN following nonmolar

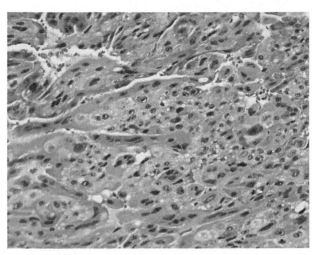

Figure 18.6. Photomicrograph of gestational choriocarcinoma demonstrating the characteristic dimorphic cell population. Multinucleate syncytiotrophoblastic cells with basophilic cytoplasm produce hCG, while a significant proportion of the cell population is composed of polygonal cytotrophoblastic cells (H and E, original magnification 100X).

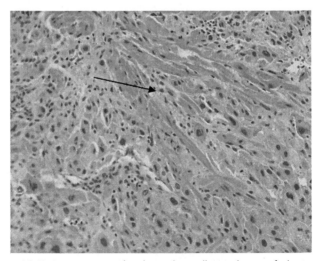

Figure 18.7. In contrast to the dimorphic cell population of choriocarcinoma, this photomicrograph illustrates a placental site tumor invading into the myometrium with a more uniform cell population. Although some of the polygonal intermediate trophoblastic cells are multinucleate, they lack the basophilic cytoplasm of syncytiotrophoblastic cells and secrete lower levels of intact hCG (H and E, original magnification 100X).

Table 18.2. Distribution of metastatic sites from GTN

Metastatic Site	Number	(%)	(% metastatic)
Nonmetastatic	195	(54)	
Metastatic	136	(46)	
Lung only	110		(81.0)
Vagina only	7		(5.0)
Central nervous system*	9		(7.0)
Gastrointestinal*	5		(4.0)
Liver*	2		(1.5)
Kidney*	1		(0.7)
Unknown**	4		(3.0)

*Concurrent lung metastases, highest risk site recorded. **Rising hCG after hysterectomy for hydatidiform mole, no identifiable metastases.

pregnancies may present with subtle signs and symptoms of disease, making the diagnosis difficult.[1] Abnormal bleeding following any pregnancy should be evaluated with hCG testing to exclude the diagnosis of malignant GTN. Metastases of gestational choriocarcinoma have been reported in virtually every body site. Gestational choriocarcinoma should be considered in any woman of reproductive age presenting with metastatic disease from an unknown primary site.[1,2] A serum hCG determination and exclusion of normal pregnancy are all that are required to diagnose metastatic GTN in these circumstances.

Once the diagnosis of malignant GTN is suspected or established, patients should be immediately evaluated for metastases and risk factors.[1,2,7] Along with history and physical examination, the following laboratory studies should be performed: complete blood count with platelet determinations, clotting function studies, renal and liver function studies, blood type and antibody screen and determination of baseline (pretherapy) hCG level. Recommended radiographic studies include chest X-ray or computerized tomography (CT) scan, pelvic ultrasound, brain magnetic resonance imaging (MRI) or CT scans and abdominopelvic CT or MRI scans.[7] Distribution of metastatic lesions encountered among patients presenting for primary therapy of GTN is presented in Table 18.2. Examples of pulmonary and brain metastases from GTN are illustrated in Figures 18.8 and 18.9.

For the minimum evaluation of a patient with postmolar GTN, a chest X-ray should be performed. If positive, further imaging of the abdomen and brain should be performed. However, if the chest X-ray is negative, CT scan of the chest should be considered before forgoing the remainder of the metastatic evaluation. Up to 40% of patients treated for postmolar GTN with negative chest X-rays have metastatic lesions detected on chest CT scan;[1] these patients would be at risk for systemic metastases. It is important to complete a full radiographic evaluation for women with GTN developing after nonmolar pregnancies. Virtually all large series of patients with high-risk sites of metastases include patients who initially presented with negative chest X-rays.

Rarely women present with persistently elevated hCG levels but are subsequently found to have a false positive hCG assay result known as "phantom hCG".[2,9] Most patients with phantom hCG present with low-level hCG elevations, but occasionally values above 500 mIU/mL have been recorded.[9] False positive hCG values are most often caused by nonspecific heterophile antibodies in the patients' sera. Phantom

18

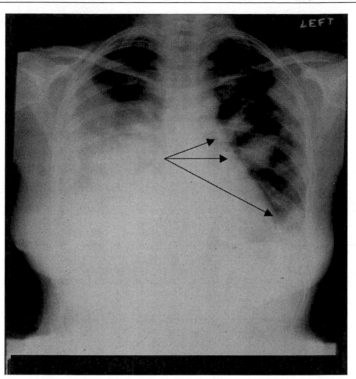

Figure 18.8. This pretherapy chest X-ray of a patient with metastatic GTN has diffuse large pulmonary metastases, with partial collapse of the right lower lobe and hemothorax. The patient has been in sustained remission for more than 5 years after completion of chemotherapy.

hCG should be suspected if hCG values plateau at relatively low levels and do not respond to therapeutic maneuvers such as methotrexate given for a presumed missed abortion or ectopic pregnancy. Work-up should include evaluation of serum hCG using a variety of assay techniques at different dilutions of patient serum, combined with a urinary hCG level.[9] Heterophile antibodies and other proteins producing false elevations of hCG are not excreted in the urine;[8] therefore urinary hCG values will not be detectable. Also false positive hCG assays usually will not be affected by serial dilution of patient sera and will have marked variability using different assay techniques, with the majority of assays reflecting undetectable hCG.[9] It is important to exclude the possibility of phantom hCG before subjecting patients with persistent low levels of hCG without clinical evidence of GTN to hysterectomy or chemotherapy for GTN.

Three systems have been used to categorize patients with malignant GTN:[1,2,7] the World Health Organization (WHO) prognostic index score, the Clinical Classification system developed from early experience with chemotherapy for patients treated at the National Institutes of Health and the FIGO staging system, which was most recently

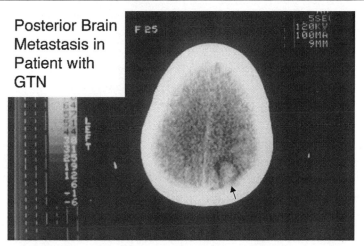

Posterior Brain Metastasis in Patient with GTN

Figure 18.9. This CT image is of a brain metastasis from GTN. The patient presented with seizures and vision loss within 4 months of a term delivery. She has remained in remission >10 years after completing whole brain radiation and combination chemotherapy.

revised in 2000. The original anatomic FIGO staging system (Table 18.3) did not take into account other factors which might reflect disease outcome, such as hCG level, duration of disease, or type of antecedent pregnancy. The latest FIGO revision includes a modification of the WHO prognostic index score for risk assessment (Table 18.4).[7] All systems correlate with clinical outcomes of patients treated for malignant GTN and identify patients at risk for failure of treatment.

The WHO prognostic index score assigned a weighted value to several individual clinical variables. The total prognostic index score used a sum of the individual component scores to generate three risk categories. In the 2000 FIGO risk index, blood type analysis was eliminated as a risk factor and risk categories were consolidated into low-risk (total score <7) and high-risk (total score ≥7) categories. The new FIGO risk index also standardized the radiological studies to be used for determining the number and size of metastases.[7] Several studies have correlated outcome of treatment for GTN with WHO prognostic index score. In one retrospective analysis, the new FIGO risk index correlated with outcome better than previous modifications of the WHO prognostic index score.

Table 18.3. Anatomic FIGO staging system for GTN*

Stage	Criteria
I	Disease confined to the uterus
II	Disease outside of uterus but is limited to the genital structures
III	Disease extends to the lungs with or without known genital tract involvement
IV	All other metastatic sites

*Gestational Trophoblastic Neoplasia, Modified from Kohorn, 2001.

18

Table 18.4. The FIGO risk scoring system for GTN*

FIGO SCORE	0	1	2	4
Age (years)	<39	>39		
Antecedent pregnancy	Hydatidiform Mole	Abortion	Term pregnancy	
Interval from index pregnancy (months)	<4	4-6	6-12	>12
Pretreatment hCG (mIU/mL)	<1000	1000-10,000	>10,000-100,000	>100,000
Largest tumor size including Uterus (cm)	3-4	≥5		
Site of metastases		Spleen Kidney	Gastro-intestinal	Brain Liver
Number of metastases identified	0	1-4	4-8	>8
Previous failed chemo- therapy			Single drug	Two or more drugs

The total score for a patient is obtained by adding the individual scores for each prognostic factor. Total score 0-6 = Low risk; ≥7 = High risk. *Gestational Trophoblastic Neoplasia. Modified from Kohorn, 2001.

The original analyses of patients treated for metastatic GTN at the NIH led to the current Clinical Classification system[1] that is frequently used in the United States (Table 18.5). This system segregates patients with nonmetastatic disease from those with metastatic disease because virtually all patients who present with nonmetastatic disease can be cured using initial single-agent chemotherapy, regardless of other risk factors. Patients with metastatic disease are further subdivided depending on the presence or absence of factors that correlated with response to initial single-agent chemotherapy. Those who lack any of the high-risk clinical factors are highly likely to respond to initial single-agent therapy and are classified as good prognosis metastatic GTN. Patients who have any single high-risk clinical factor are classified as poor prognosis disease. These patients are not only at an increased risk of failure of single-agent chemotherapy but also have an increased risk of death if treated with single-agent therapy followed by multiagent regimens compared to patients receiving initial multiagent regimens.[1]

The FIGO staging system is currently the standard for reporting results of treatment for patients with malignant GTN. However, the Clinical Classification system is less complicated and allows easy identification of patients who are likely to fail

Table 18.5. Clinical classification system for patients with malignant GTN*

Category	Criteria
Nonmetastatic GTN	No evidence of metastases; not assigned to prognostic category
Metastatic GTN	Any extrauterine metastases
Good prognosis metastatic GTN	No risk factors: Short duration (<4 months) Pretherapy hCG <40,000 mIU/mL No brain or liver metastases No antecedent term pregnancy No prior chemotherapy
Poor prognosis metastatic GTN	Any risk factor: Long duration (>4 months) Pretherapy hCG >40,000 mIU/mL Brain or liver metastases Antecedent term pregnancy Prior chemotherapy

*Gestational trophoblastic neoplasia. Modified from Soper, 1996.

initial single-agent chemotherapy. Virtually all deaths from malignant GTD occur among women who fall into the poor prognosis metastatic category, and these patients should be considered to have high-risk disease. All patients with high-risk malignant GTN should be referred for management in consultation with individuals who are experienced in the therapy of this relatively rare malignancy.

Because of the relative rarity of malignant GTN, there are very few randomized trials of therapy that have been attempted.[1] The majority of studies are retrospective analyses of single-institution experiences or prospective Phase II trials, but these confirm high activity for a variety of agents in the treatment of malignant GTD including methotrexate, dactinomycin, etoposide, 5-flurouracil and cisplatin.

Primary remission rates of patients treated for nonmetastatic GTN are similar with a variety of chemotherapy regimens (Table 18.6). Essentially all patients with this condition can be cured, usually without the need for hysterectomy.[1] The current regimen most often used to treat nonmetastatic GTN is the weekly methotrexate regimen. Chemotherapy is continued until hCG values have achieved normal levels and then an additional course is administered after the first normal hCG value has been recorded. Hematologic indices must be monitored carefully during chemotherapy, but significant hematologic toxicity is infrequent among patients treated with weekly methotrexate. Because methotrexate is excreted entirely by the kidney and can produce hepatic toxicity, patients must have normal renal and liver functions before each treatment.

Dactinomycin regimens may provide a slightly higher primary remission rate than methotrexate but generally are associated with more toxicity.[1] For this reason, they are usually used as second-line therapy or for initial therapy of patients who are not appropriate to be treated with methotrexate. Based on high primary remission rates and lower toxicity, bolus dactinomycin is used more frequently than the 5-day dactinomycin regimen.

18

Table 18.6. Chemotherapy regimens for nonmetastatic and low-risk metastatic GTN

Agent/schedule	Dosage
Methotrexate (1)	
Weekly	30 mg/m^2 IM
5 day/every 2 weeks	0.4 mg/kg IM (maximum 25 mg/d total dose)
Methotrexate/folinic acid rescue	methotrexate 1 mg/kg IM
Every 2 weeks	days 1, 3, 5, 7
	Folinic acid 0.1 mg/kg IM days 2, 4, 6, 8
Methotrexate infusion/folinic acid	Methotrexate 100 mg/m^2 IV bolus
Every 2 weeks	and 200 mg/m^2 12 hr infusion folinic acid 15 mg po every 6 hr for four doses
Dactinomycin (2)	
5 day/every 2 weeks	9-13 mcg/kg/d IV (maximum dose 500 mcg/day)
bolus, every 2 weeks	1.25 mg/m^2 IV bolus
Etoposide (3)	
5 day/every 2 weeks	200 mg/m^2/day po

1. Dose based on ideal body weight, maximum 2 m^2. 2. Potential extravasation injury, gastrointestinal toxicity common. 3. Alopecia, small leukemogenic risk. Abbreviations: GTN, gestational trophoblastic neoplasia; IM, intramuscular; IV, intravenous; po, oral. Modified from Soper, 1996.

In the group of patients with nonmetastatic disease, early hysterectomy will shorten the duration and amount of chemotherapy required to produce remission.[5] Therefore, each patient's desire for fertility should be evaluated at the onset of treatment. Hysterectomy may be performed during the first cycle of chemotherapy. However, further chemotherapy after hysterectomy is mandatory until hCG values are normal.

Patients whose levels of hCG have plateaued or are rising during therapy should be switched to an alternative single-agent regimen. Dosages for dactinomycin and etoposide regimens are shown in Table 18.6. If there is appearance of new metastases or failure of alternative single-agent chemotherapy, the patient should be treated with multiagent regimens. Hysterectomy should be considered for the treatment of nonmetastatic disease that is refractory to chemotherapy and remains confined to the uterus. The overall cure rate for patients with nonmetastatic disease approaches 100%.[1-3] When chemotherapy is given for an additional 1-2 cycles after the first normal hCG value, recurrence rates are <5%.

Patients with metastatic GTN who lack any of the clinical high-risk factors or a FIGO risk score of <7 have low-risk disease. They can be treated successfully with initial single-agent regimens and have essentially a 100% cure rate, similar to patients with nonmetastatic GTN.[1] Most often treatment has consisted of 5-day treatment regimens using intramuscular methotrexate or intravenous dactinomycin recycled at 14 day intervals (Table 18.6). Approximately 30-40% of patients with low-risk metastatic GTN will require alternative therapy to achieve remission; however, fewer than 10% of patients

Table 18.7. Triple agent (MAC) chemotherapy for high-risk GTN*

Day	Drug	Dose
Day 1-5	Methotrexate	15 mg IM
	Dactinomycin	500 mcg IV
	Cholrambucil	8-10 mg po
	OR	
	Cyclophosphamide	3 mg/kg IV
Day 15-22	Begin next cycle	

*Gestational Trophoblastic Neoplasia. Modified from Soper, 1996.

with low-risk metastatic GTN will require multiagent chemotherapy.[1] Primary hysterectomy in conjunction with chemotherapy may also decrease the amount of chemotherapy required to achieve remission in the group of patients with low-risk metastatic disease.[5] Similar to the treatment of women with nonmetastatic GTN, 1-2 cycles of maintenance chemotherapy should be given after the first normal hCG. Recurrence rates are <5% among patients successfully treated for low-risk metastatic disease.

Patients with one or more of the Clinical Classification System risk factors or a FIGO risk score of ≥7 have high-risk disease.[1-3,5,7] They should be initially treated with multiagent chemotherapy; surgery and/or radiation are often incorporated into treatment.[5] Survival reported by trophoblastic disease centers ranges up to 86%.[1,2] In contrast to patients with nonmetastatic or low-risk metastatic GTN, early hysterectomy does not appear to improve the outcome in women with high-risk metastatic disease,[5] probably because of the larger extrauterine tumor burden.

Aggressive treatment with multiagent chemotherapy is an important component for management of these patients. Triple therapy (Table 18.7) with methotrexate, dactinomycin and either chlorambucil or cyclophosphamide (MAC) was the standard regimen for many years in the United States.[1] However, some investigators have reported poor survival of patients with WHO risk scores >12 who were treated initially with MAC. More recent multiagent regimens have incorporated etoposide with or without cisplatin into cyclical combination chemotherapy with high rates of success and apparent lower acute toxicity, but with a potential increased risk of leukemia in survivors compared to MAC, due to the etoposide.[1] Currently, the most frequently used of these combinations is EMA/CO (Table 18.8).[1-3,11]

High-risk metastatic sites include brain, liver and kidney lesions. Many patients with high-risk metastatic sites of disease will require coordinated multimodality therapy for optimal treatment. Radiation therapy has been used concurrently with chemotherapy in an attempt to limit acute hemorrhagic complications from brain metastases (Fig. 18.9). Brain irradiation combined with systemic combination chemotherapy results in cure rates up to 75% among patients who initially present with brain metastases.[5] However, a similar primary remission rate has also been reported among patients treated with modifications of the EMA/CO regimen that incorporated high-dose systemic methotrexate combined with intrathecal methotrexate infusions (Table 18.8) and relies on early neurosurgical intervention rather than brain irradiation for control of metastatic sites in the brain.[11] The best treatment for liver or other high-risk sites of metastases has not been established; these patients are most often managed with highly individualized multimodality therapy,

18

Table 18.8. Alternating weekly EMA/CO chemotherapy for high-risk GTN*

Day	Drug	Dose
1	Etoposide	100 mg/m^2 IV
	Methotrexate**	100 mg/m^2 IV bolus
	Methotrexate	200 mg/m^2 IV infusion over 12 hr
	Dactinomycin	350 mcg/m^2 IV
2	Etoposide	100 mg/m^2 IV
	Dactinomycin	350 mcg/m^2 IV
	Folinic acid	15 mg po, IM or IV every 12 hr × 4 doses, begin 24 hr after methotrexate bolus
8+	Cyclophosphamide	600 mg/m^2 IV
	Vincristine	1.0 mg/m^2 IV
15	Begin next cycle	

**Methotrexate is sometimes given as a 12 hr IV infusion at a dose of 500-1,000 mg/m^2 for treatment of brain metastases, with folinic acid rescue increased to 15 mg every 6 hours × 48 hours or 30 mg every 12 hours × 48 hours. Some investigators give methotrexate 15 mg intrathecal injection for prophylaxis or treatment of brain metastases on day 8. *Gestational Trophoblastic Disease. Modified from Soper, 1996.

18

including surgical resection of isolated metastases, embolization of liver metastases and localized radiation therapy.[5] Even with intense chemotherapy, additional surgery may be necessary to control hemorrhage from metastases, remove chemoresistant disease or treat other complications in order to stabilize high-risk patients during therapy.[5] Chemotherapy is continued until hCG values have normalized, and this is followed by at least three courses of maintenance chemotherapy for the purpose of eradicating all viable tumor. Despite using sensitive hCG assays and maintenance chemotherapy, up to 12.5% of patients with high-risk disease will develop recurrence after achieving an initial remission.

After hCG remission has been achieved, patients with malignant GTN should be managed with serial determinations of hCG levels at 2-week intervals for the first 3 months of remission and then at 1 month intervals to complete a year of normal hCG monitoring.[1-3] The risk of recurrence after 1 year of remission is <1%, but late recurrences have been observed. Sustained remissions can be achieved in ≥50% of patients treated for an episode of recurrent GTN. Therefore, it is recommended that patients continue to receive hCG monitoring at 6-12 month intervals indefinitely.

Contraception, preferably with oral contraceptives, should be used during the first year of remission. Because of the 1-2% risk for a second mole in subsequent pregnancy, early ultrasound is recommended during all future pregnancies. There does not appear to be an increase in the risk of congenital malformations or other complications related to pregnancy.[1]

Future Directions

Although the current management of patients with hydatidiform moles and GTN has resulted in preservation of fertility and high cure rates for the majority of these patients, it is anticipated that future studies will help to further refine our understanding of this disease. Molecular genomic or proteomic profiling may allow earlier determination of women who are destined to develop postmolar GTN, allowing earlier treatment. A Gynecologic Oncology Group Phase III trial comparing single agent dactinomycin and methotrexate regimens in women with low-risk GTN is nearing completion and will establish the most efficacious and cost-effective treatment for these women. It is anticipated that treatment strategies for women with high-risk disease will also be refined to allow selection of appropriate women for less toxic therapy, while retaining the high cure rates recorded for this population.

NOTE: All images are from the personal collection of John Soper, MD.

Suggested Readings

1. Soper JT, Lewis JL Jr, Hammond CB. Gestational Trophoblastic Disease. In: Hoskins WJ, Perez CA, Young RC eds. Principals and practice of gynecologic oncology, 2nd edition. Philadelphia Lippincott-Raven 1996; 1039-77.
2. Soper JT, Mutch DG, Schink JC. Diagnosis and treatment of gestational trophoblastic disease: ACOG Practice Bulletin No. 53. Gynecol Oncol 2004; 93:575-85.
3. Soper JT. Gestational trophoblastic disease. Obstet Gynecol 2006; 108:176-87.

References

1. Soper JT, Lewis JL Jr, Hammond CB. Gestational Trophoblastic Disease. In: Hoskins WJ, Perez CA, Young RC, eds. Principals and Practice of Gynecologic Oncology, 2nd Ed. Philadelphia Lippincott-Raven, 1996:1039-77.
2. Soper JT. Gestational trophoblastic disease. Obstet Gynecol 2006; 108:176-87.
3. Kohler MF, Soper JT. Gestational trophoblastic disease. In: Seils A, Noujaim S, Davis K, eds. Operative Obstetrics, 2nd Ed. New York: McGraw-Hill, 2002:615-28.
4. Tidy JA, Gillespie AM, Bright N et al. Gestational trophoblastic disease: a study of the mode of evacuation and the subsequent need for treatment with chemotherapy. Gynecol Oncol 2000; 78:309-31.
5. Soper JT. Role of surgery and radiation therapy in gestational trophoblastic disease. Best Prac Res Clin Obstet Gynaecol 2003; 17:943-57.
6. Wolfberg AJ, Feltmate C, Goldstein DP et al. Low risk of relapse after achieving undetectable hCG levels in women with complete molar pregnancy. Obstet Gynecology 2004; 104:551-4.
7. Kohorn EI. The new FIGO 2000 staging and risk factor scoring system for gestational trophoblastic disease: description and clinical assessment. Int J Gynecol Cancer 2001; 11:73-7.
8. Papadopoulos AJ, Foskett M, Seckl MJ et al. Twenty-five years' clinical experience with placental site trophoblastic tumors. J Reprod Med 2002; 47:460-4.
9. Cole LA. Phantom hCG and phantom choriocarcinoma. Gynecol Oncol 1998; 71:325-9.
10. Soper JT, Evans AC, Conaway MR et al. Evaluation of prognostic factors and staging in gestational trophoblastic tumor. Obstet Gynecol 1994; 84:969-73.
11. Rustin GJ, Newlands ES, Begent RH et al. Weekly alternating etoposide, methotrexate and actinomycin/vincristine and cyclophosphamide chemotherapy for the treatment of CNS metastases of choriocarcinoma. J Clin Oncol 1989; 7:900-3.

18

CHAPTER 19

Supportive Care:
Nutrition and Growth Factors

Mildred Ridgway and Tarrik Zaid

Nutrition for the human body is imperative to sustain life. Nutrients, vitamins and minerals are provided primarily through food and allow us to function in our daily living. Appropriate nutrition is needed for metabolic and active processes such as breathing, digesting food, walking and more vigorous exercise. Patients who have a healthy balance of nutrition with proteins, sugars and electrolytes have a more stable foundation from which to overcome their disease and the medical interventions necessary to treat the disease, whether surgery, chemotherapy, or radiation. Women with malignancies of the female genital tract are susceptible to malnutrition secondary to their disease. Addressing optimal nutritional support is a vital part of providing care for these patients. This chapter is aimed to help the physician understand the importance of nutrition and how best to meet the nutritional goals of their patients.

Nutrition

Nutrition is the process by which organisms assimilate food, composed of protein, carbohydrates, fats, vitamins and minerals, into their blood for the essential activities of cell growth and repair for all organ systems. Nutrients serve to provide energy (usually measured in calories) through a catabolic process that produces ATP. They also serve as substrates for structural (e.g., collagen), functional (e.g., albumin) and biochemical (e.g., enzymes/co-enzymes) components of the body.

The gynecological oncology patient is at particular risk for nutritional problems from the disease process itself (metabolic and mechanical) and nutritional challenges afforded by treatment modalities (surgery, radiation and chemotherapy). The goal of nutritional management in these patients is to provide these nutrients in such away as to prevent catabolism, support wound healing, improve immune response and assist in the therapy of existing medical comorbidities.

Calories are units of energy producing potential that protein, fat and carbohydrates create in animals. Each gram of protein and carbohydrate provides four calories while each gram of fat provides nine calories.

Proteins are the major structural component of all living cells and provide multiple cell functions. Composed of single or multiple chains of amino acids, proteins function as enzymes, transport carriers and hormones. Nine amino acids are considered essential as they cannot be made de novo and must be provided in the diet. The remainder can be made by transamination reactions involving simple sugars. Amino acids can be burned for fuel after their nitrogenous component is removed in the urea cycle. This catabolic process usually contributes only 15% of the normal energy expenditure, with the remainder supplied by carbohydrates and fat. Each gram of protein (amino acids) can

Gynecologic Oncology, edited by Paola Gehrig and Angeles Secord.
©2009 Landes Bioscience.

be converted into 4 kcal energy. Measurement of urinary nitrogen excretion provides a useful way to estimate the degree of protein catabolism.

Digestion of proteins occurs mainly in the duodenum but can occur at any level of the small bowel. Hence, protein malabsorption is rare even after extensive intestinal resection. The main site of protein turnover is the GI tract, as mucosal cells are shed and digestive enzymes are secreted into enteric content the majority of which is re-absorbed. Excessive GI tract losses usually occur through abnormal communications of the GI tract through a fistula, an ileostomy, a draining gastrostomy, or short gut syndrome from extensive bowel resection. Total body protein is about 15% of body weight which is mainly stored in striated muscle. Protein requirements in healthy adults are approximately 1 g/kg body weight. Requirements in surgical/ICU patients increase to 1.5 to 2 g/kg.

Carbohydrates are mainly used as energy substrates but also serve important structural and functional roles. Carbohydrates are the energy substrate of choice in the human body. In the presence of carbohydrates, catabolism of proteins and fats for fuel is diminished. Absorption is generally completed within the jejunum in the form of mono- and disaccharides. Deficiencies in intrinsic carbohydrate digestion and absorption are rare in surgical patients.

Lipids are a diverse group of hydrophilic molecules that have high caloric density with 9 Kcal/g. They provide the remaining 25-45% of calories in the typical diet. Triglycerides (esters of glycerol plus three fatty acids) are the main lipids in human diets. Digestion and absorption of lipids is a complex process involving interplay of several enzymes and bile salts in the distal small bowel. Malabsorption of lipids is a more common pathologic process than with proteins or carbohydrates, which can be caused by primary pancreatic or liver disease or secondary to ileal resection. Two essential fatty acids, linoleic and linolenic acid are required in the diet. In addition to their use as fuel lipids, they also serve as precursors to steroids and eiconasoid production. Clinical deficiency results in a generalized scaling rash, poor wound healing and hepatic steatosis.

Excess carbohydrates can be converted to fat, but the reverse is not true. The end product of fatty acid anabolism is acetyl-CoA which is not a substrate for gluconeo-gensis. Instead, calories provided by lipids are in the form of ketone bodies converted by the liver from fatty acids.

Vitamins and minerals are nutrients that are required in very small amounts for essential metabolic reactions in the body. Humans need 13 vitamins divided into two groups: four fat-soluble vitamins (A, D, E and K) and nine water-soluble vitamins (eight B vitamins and vitamin C) in addition to several trace elements. Clinical deficiency is rare in the developed world and can be easily remedied orally or parenterally.

Malnutrition is the resulting process by which an organism has poor nutrition from poor intake, poor absorption, or poor utilization of food (calories and protein) in the enteral tract. Patients can become malnourished from all three of these causes both directly and indirectly from their malignancy.

Nutritional Assessment

Preoperative nutritional assessment aims to identify patients at risk of malnutrition or with special nutritional needs. The majority of patients will not need preoperative nutritional supplementation unless severe malnutrition is noted.

A history and physical is always the first step. A recent weight loss of 5% of total body weight in the past month or 10% in the past 6 months is significant. Several indices and laboratory tests can be measured directly or calculated to assess a patient's nutritional status.

- Weight—A person's weight in pounds or kilograms is the simplest measurement in nutritional assessment. Weight alone however is not adequate. Body mass index (BMI) is an assessment which takes a patient's weight and height into consideration. BMI is weight in kilograms per height in meters squared (kg/m^2). BMI provides a more accurate assessment of the patient's level of obesity and/or malnutrition. A normal BMI is 19-25 kg/m^2.
- Labs—Total protein and albumin—These are helpful but lack specificity. Normal values are >6.5 total protein and >3.5 albumin.

  Prealbumin and transferrin—These have a shorter half-life and are more specific. Normal levels are >1.5 g/dl and 150 mg/dl respectively.

  Nitrogen balance—This measurement provides the best qualitative indicator of nitrogen balance and requires a 24-hour urine specimen. Formula: (protein intake/6.25)—(24-hour urea nitrogen in urine + 4 g)
- Subjective Global Assessment (SGA)—a clinical assessment using the history and physical examination
- Prognostic Nutritional Index (PNI)—an objective assessment predicting the percent risk of complication from malnutrition: Low risk (PNI < 40%), intermediate (PNI of 40% to 49%) and high risk (PNI > 50%). It involves a calculation which is: 158%—16.6 (albumin in g/100 mL)—0.78 (triceps skinfold in mm)—0.2 serum transferrin in mg/dL—5.8 (delayed skin hypersensitivity).
- Base Energy Expenditure (BEE)—655 + (9.6 × weight in kg) + (1.8 × height in cm)—(4.7 × age in years). An additional 500-1000 calories per day may be needed in patients under metabolic stress such as surgery or sepsis.

In day to day clinical practice, objective measurements of nutritional status are infrequently used. The cumbersome techniques and additional lab work contributes to the under utilization. However, the use of a quantitative approach such as the prognostic nutritional index may better guide management, especially for the more complex patient.

Etiology of Malnutrition

The etiology of malnutrition in the cancer patient is diverse. Decreased oral intake due to anorexia is both directly related to tumor bulk (by increased intra-abdominal pressure) and indirectly related by factors such as depression. Malabsorption can result in malnutrition from therapeutic modalities such as radiation (radiation enteritis) or surgery (extensive bowel resection). Metabolic derangements resulting in increased proteolysis, gluconeogenesis and lipolysis also describe cancer cachexia and play a role in the malnourished cancer patient.

Complications with Malnutrition

Malnutrition is the leading cause of mortality and morbidity in cancer patients. These patients suffer from a significantly decreased quality of life in their final weeks from starvation and malnutrition. Postoperative complications are more common in the malnourished patient. In addition, malnutrition prolongs hospital stay.

Nutrition/Malnutrition in the Surgical Patient

Patients undergoing major gynecologic surgery are usually kept NPO the night before and usually do not return to a regular diet until the second or third postoperative day. After an overnight fast, liver glycogen is rapidly depleted as insulin falls and glucagon levels rise. If the starvation state continues, caloric needs are supplied by fat and protein degradation. Only protein can be converted back to glucose which the brain uses preferentially. Fatty acids are broken down in the liver to produce ketone bodies which are used by other organs as a fuel source. The brain can later switch to ketones as a fuel source after several days of starvation.

The majority of patients have adequate reserves to withstand postsurgical catabolic stresses and a state of partial starvation for 7-10 days. Five percent dextrose intravenous fluid solutions contain 200 Kcal per liter (50 g of glucose). This should minimize protein catabolism until the patients are able to tolerate a diet.

The physiologic stress afforded by surgery has been classically described to occur in three phases.

- Catabolic phase: After a surgical injury, metabolic demand increases. Protein catabolism inevitably occurs because patients are commonly prevented from eating despite having an increased metabolic demand. Serum levels of gluconeogenic and catabolic hormones such as glucagon, corticosteroids and catecholamines increase while levels of insulin decrease.
- Early anabolic phase: Once in the presence of adequate nutrition, the body shifts from catabolism to anabolism. This usually takes several days but may take longer.
- Late anabolic phase: Several weeks to months after surgery the body's adipose stores are restored and the equilibration of nitrogen balance is attained.

Therapy

Enteral Feeding

Nutrition via the gastrointestinal tract is the preferred administration if a patient can tolerate gut stimulation. Enteral intake maintains the GI tract integrity and normal flora. Patients who should be considered for enteral feeds who cannot tolerate oral intake include those with (1) an upper obstruction of the esophagus, stomach or proximal small bowel, or (2) anorexia due to oral or esophageal stomatitis or marked depression. There are also supplements for those who can tolerate oral intake (Table 16.1).

Perioperative support of malnourished patient with an intact and functioning gut who otherwise cannot voluntarily eat can be obtained through a variety of routes (Table 16.2):

PEG—percutaneous esophagogastrostomy tube

PEJ—percutaneous esophagojejunostomy tube

NG—nasogastric tube

Dubhoff—nasoduodenal tube

- Enteral feeding protocols

Bolus feeding should be used in patients with nasogastric or gastrostomy feeding tubes. Feedings are administered by gravity and begin at 50-100 mL every 4-6 hours and are increased incrementally until the intake goal is reached. Residual gastric volume should be measured before each feed and feeding withheld if the volume exceeds 50% of the last feed.

Continuous infusion utilizing pumps is required with jejunal feeding. The rate usually begins between 25-50 mL/hour and titrated according to nutritional requirement.

Complications with tube-feeds

- Metabolic

  Electrolyte abnormalities—These should be monitored and corrected accordingly.

Volume Overload—Cardiac patients are at particular risk and can result in pulmonary edema or anasarca.

Hyperglycemia—Diabetic patients are at particular risk and glucose should be followed at routine intervals.

Diarrhea

- Mechanical

  Clogging of the tube—The tube should be flushed regularly, but may need to be exchanged if clog cannot be dislodged.

  Aspiration—This results from reflux or overfilling the stomach/duodenum. This can be a severe morbidity and possible mortality

Total Paraenteral Nutrition (TPN)

TPN should be used only in a select group of cancer patients. Choosing the appropriate patient is often difficult for the oncologist and needs to be considered cautiously. Patients for whom TPN would only prolong suffering without giving them a realistic chance of overcoming their disease should NOT be offered this form of nutritional support. However, patients who are disease-free or have marked regression of disease and need temporary nutritional support are excellent candidates for TPN. Preoperative TPN in the cancer patient has not been proven to be routinely beneficial. This should be reserved for only a select few. On the other hand, postoperative paraenteral nutrition has been proven to be beneficial. The National Institutes of Health, the American Society of Clinical Nutrition and the American Society of Parenteral and Enteral Nutrition all agree that TPN should be administered by the fifth postoperative day if a patient is clinically unable to tolerate oral or enteral nutrition.

Administration

TPN needs to be administered via a central line or catheter. It is preferable is the line is dedicated to TPN alone, such as a PICC line. A chemotherapy port-a-cath may be used, but is not ideal. Sterile access must be strictly performed.

Ingredients/Contents

TPN is composed of protein, carbohydrates and water (30-40 mL/kg/day) all of which need to be carefully calculated for total calorie, amino acid, dextrose and total fluid volume. Total calorie goal is 20-30 Cal/kg and total protein goal is 1.5 gm/kg.

TPN solutions generally are administered as a 3-in-1 mixture of protein, as amino acids, carbohydrate as dextrose and fat, as a lipid emulsion of soybean or sunflower oil. Alternatively, the lipid emulsion can be administered as a separate intravenous "piggyback" infusion. Standard preparations of TPN are used for most patients and provide total calories that are comprised of 50-60% carbohydrate, 24-34% fat and 16% protein. Special solutions that contain low, intermediate, or high protein and nitrogen concentrations as well as varying amounts of fat and carbohydrate are available for some patients with diabetes, renal failure, hepatic dysfunction or other systemic diseases.

Other elements can be administered in conjunction with the basic caloric and protein solutions. Electrolytes (e.g., sodium, potassium, chloride, acetate, calcium, magnesium and phosphate) that are added to the TPN solution should be adjusted daily. For the patient whose current serum electrolytes and renal function are normal, suggested ranges for these additives include sodium 60-80 mEq per day; potassium 30-60 mEq per day; chloride 80-100 mEq per day; calcium 4.6-9.2 mEq per day; magnesium 8.1-20.0 mEq per day; and phosphate 12-24 mmol per day. Chloride and

Table 19.1. Supplement options for patients who can tolerate oral feeds

Product	Ensure/Boost	Ensure Plus	TwoCal HN	Glucerna	Nepro
Cal/240 mL	250	355	475	220	475
Protein gm/240 mL	9	13	20	10	16
Fiber	0	0	0	3	0
Indications	Patient in need of oral supplement	Patient with poor PO intake; moderately stressed	Patient in need of maximum nutrition with volume restriction	Diabetic patient in need of oral supplement	Renal patient or patient requiring electrolyte restriction

Table 19.2. Supplement options for patients who require tube feeds

Product	Jevity	Osmolite	Ensure Plus HN	Glucerna	Nepro	Suplena
Calories	1.2/mL	1.06/mL	1.5/mL	1.0/mL	2.0/mL	2.0/mL
Protein	55.5 gm/liter	44.3 gm/liter	62.7 gm/liter	50 gm/liter	70 gm/liter	30 gm/liter
Water volume	810 mL	842 mL	769 mL	839 mL	699 mL	713 mL
Indications	Contains fiber; for patients without medical comorbidities	Isotonic; fiber-free	High calorie and protein; fiber-free	Diabetic patients	Renal patients on dialysis and/or fluid restriction	For protein, fluid and electrolyte restricted patients

19

acetate are used to balance serum pH and should be adjusted accordingly. If the serum bicarbonate is low, the solution should contain more acetate.

To discontinue TPN, the infusion rate should be halved for 1 hour, halved again the next hour and then discontinued. It is not necessary to taper the rate of TPN infusion if the patient is receiving less than 1,000 kcal per day. This is done to prevent complications that are caused by hyperinsulinemia.

Complications of TPN
Catheter-related:
>   Thrombus
>   Infection

Metabolic:
>   Hyperglycemia
>   Metabolic acidosis
>   Re-feeding syndrome
>>   This occurs when an excessive carbohydrate load is administered to a mal-nourished patient, resulting in a dramatic shift of extracellular ions into the intracellular space. This causes a precipitous drop in serum phosphate. Phosphate levels need to be checked frequently in these patients.
>   Hepatic dysfunction—This is a common manifestation of long-term TPN.
>   Cholecystitis

Special Diet Considerations
Diabetes mellitus often complicates nutritional management. Complications that are associated with TPN administration (e.g., catheter-related sepsis) are more common with these patients and those with prolonged hyperglycemia. The goal in glucose-intolerant patients is to maintain the serum glucose level at 100-200 mg/dL. Hypoglycemia can result in shock, seizures, or vascular instability. Unopposed glycosuria may cause osmotic diuresis, loss of electrolytes in urine and nonketotic coma.

Renal failure often co-exists with other medical comorbidities and can be associated with glucose intolerance, negative nitrogen balance, loss of protein with decreased protein synthesis and diminished excretion of phosphorus. Dialysis should be adjusted accordingly and these patients should be nutritionally replenished according to their calculated needs. Patients who receive peritoneal dialysis absorb approximately 80% of the dextrose in the dialysate fluid (assuming a normal serum glucose level). These factors must be considered when designing a nutritional support strategy.

Hepatic failure may result in wasting of lean body mass, fluid retention, vitamin and trace metal deficiencies, anemia and encephalopathy. More than 70-80 g per day of amino acids is required to maintain nitrogen balance in these patients. It may be difficult or impossible to limit the amount of nitrogen that a patient receives each day yet still provide adequate nutritional support.

Cachexia and cancer are associated with lean muscle wasting. More than two-thirds of patients with cancer experience significant weight loss during their illness. Malnutrition is a contributing cause of mortality in 20-40% of these individuals. Reasons for this development include decreased nutrient intake and impaired nutrient use. Antineoplastic therapies, such as chemotherapy, radiation therapy, or surgery can worsen preexisting malnutrition. Although the addition of TPN to these modalities in clinical studies has shown improvement in weight, nitrogen balance and biochemical markers, there is little evidence to suggest better response rates or survival.

Short-bowel syndrome commonly occurs in patients with less than 200 cm of functional jejunum. It may result from mesenteric ischemia, primary colon disease, or extensive surgical resection. It is characterized by nutrient malabsorption, electrolyte imbalance, diarrhea and dehydration. Some of these patients require intravenous nutrition for life with frequent hospitalizations and severely morbid diarrhea. The estimated length of small bowel that is required for adult patients to become independent of TPN is greater than 120 cm without colon or greater than 60 cm with some colonic continuity. Salvage of the ileocecal valve improves outcome. Intestinal adaptation may occur in some patients, thereby allowing for the transition from intravenous to enteral feeding. Uniquely formulated diets (supplemented with glutamine and growth hormone) show promise for accelerating this process.

Appetite Stimulants

Dronabinol (Marinol®)
Dronabinol is an orally active cannabinoid which affects the central nervous system inducing sympathomimetic activity. Dronabinol is currently indicated for both anorexia and severe nausea and vomiting in the cancer and chemotherapy patient. Patients may take Dronabinol as a capsule at 2.5 mg bid and can be increased to a maximum of 20 mg daily divided bid.

Megestrol Acetate (Megace®)
Megestrol acetate, a synthetic derivative of progesterone, is available in oral suspension (40 mg/mL) and is indicated for anorexia and cachexia in cancer patients. The recommended dose is 80 mg bid, but may be increased to 400 mg daily.

Anti-Emetics
Chemotherapeutic agents are often associated with nausea, vomiting and anorexia. Different agents have markedly different effects; however, in general some degree of these symptoms will appear in most patients. There are three distinct categories of chemotherapy-induced nausea and vomiting.

Anticipatory Nausea and Vomiting
Anticipatory nausea and vomiting can occur prior to initiating a course of chemotherapy. It is stimulated by the patient's thoughts of receiving the therapy and the stimuli, such as the sights and sounds of the chemotherapy suite. It is a conditioned response typically after three or four previous administrations.

Acute Nausea and Vomiting
Nausea and vomiting experienced during the first 24 hours of administration of chemotherapy.

Delayed Nausea and Vomiting
Nausea and vomiting associated with the administration of chemotherapy greater than 24 hours after the initiation of chemotherapy. Cisplatin and ifosfamide are two common agents we use that result in delayed nausea and vomiting.

The anti-emetic agents used for chemotherapy-induced nausea and vomiting can be used in all three settings; however some agents are better suited for anticipatory or acute or delayed depending on their pharmacotherapeutic action.

Corticosteroids
Dexamethasone

5-HT3 antagonists—
    Granisetron (Kytril®)
    Odansetron (Zofran®)
    Palonosetron (Aloxi®)
    P450 Isoenzyme inhibitor
    Aprepitant (Emend®)
    Prochlorperazine (Compazine®)
    Promethazine (Phenergan®)

Growth Factors

Bone marrow suppression is the single most common complication from cytotoxic chemotherapy in gynecologic oncology patients. Neutropenia and anemia are more common than thrombocytopenia. Neutropenia is the most common myelosuppressive effect and is measured by absolute neutrophil count (ANC). ANC is the total WBC × (%PMN). Neutropenia is classified by grading criteria where Grade 3 and 4 are the most clinically concerning. Grade 3 and 4 neutropenia are an ANC <1,000-500 cells/mm³ and an ANC of <500 cells/mm³ respectively.

Neutropenia

Pegfilgrastim (*Neulasta®*)

Pegfilgrastim (FDA approved in 2002) is a covalent conjugate of recombinant human G-CSF and monomethoxypolyethylene glycol. It is a colony stimulating factor that stimulates proliferation, differentiation, commitment and end cell functional activation of hematopoietic cells. Pegfilgrastim has reduced renal clearance and prolonged persistence in vivo as compared to filgrastim.

Dose: 6 mg subcutaneously (SQ) on day #2 of therapy

Indication: To decrease the incidence of severe (Grade 3 and 4) neutropenia and febrile neutropenia

Filgrastim (*Neupogen®*)

Filgrastim (FDA approved in 1991) is a recombinant human G-CSF that similarly stimulates proliferation, differentiation, commitment and end cell functional activation of hematopoietic cells.

Dose: 5 mg/kg/day for 5 days

Indication: To decrease the incidence of severe neutropenia and febrile netropenia

Anemia

Epoetin Alfa (*Procrit®*)

Epoetin Alfa is a glycoprotein manufactured by recombinant DNA technology that has the same biological effects as endogenous erythropoietin. Erythropoietin stimulates bone marrow red blood cell production and is produced in the kidney. In chemotherapy-induced anemia, Epoetin Alfa stimulates the production of red blood cells decreasing the need for packed red blood cell transfusions.

Dose: 40,000 units SQ weekly or 60,000 units SQ weekly

Indication: hgb ≤ 11 gm/dl

Darbepoetin (*Aranesp®*)

Darbepoetin is a similar recombinant DNA technology glycoprotein that differs by having a 5 N-linked oligosaccharide chain. It stimulates erythropoiesis in the bone marrow by the same mechanism as endogenous erythropoietin and Epoetin Alfa. The two additional carbohydrate chains increase the molecular weight of the glycoprotein. Its advantage is less frequent administration.

Dose: 100 mcg SQ (or IV) weekly

200 mcg SQ (or IV) biweekly

300 mcg SQ or IV every 3 weeks

Indication—the same as Epoetin Alfa

Suggested Reading

1. Lin E. Lowry SF. Substrate metabolism in surgery. In: Norton JA, ed. Surgery: Scientific Basis and Clinical Evidence. New York: Springer-Verlag 2000:95.
2. Rombeau J, Rolandelli R, Wilmore D et al. Gastroenterology, XIII: enteral and parenteral nutritional support. In: Dale DC, Federman DD, eds. ACP Medicine. New York: Scientific America WebMD, 2004:1-16.
3. Shatnner MA, Shike M. Nutritional support in patients with gynecologic cancers. In: Hoskins WJ, Perez CA, Young RC, eds. Principles and Practice of Gynecologic Oncology. 4th Ed. Philadelphia: Lippincott Williams and Wilkins 2005:611-28.

References

1. Baker JP, Detsky AS, Wesson DE et al. Nutritional assessment: a comparison of clinical judgment and objective measurements. N Engl J Med 1982; 306:969-72.
2. Burrin DG, Davis TA. Proteins and amino acids in enteral nutrition Curr Opin Clin Nutr Metab Care 2004; 7:79-87.
3. Eisenberg P. An overview of diarrhea in the patient receiving enteral nutrition. Gastroenterol Nurs 2002; 25(3):95-104.
4. Gopalan S, Khanna S. Enteral nutrition delivery technique. Cun Clin Nutr Metab Care 2003; 6:313-7.
5. Heslin MJ, Brennan MF. Advances in perioperative nutrition: Cancer. World J Surg 2000; 24:l477.
6. Heyland DK, Macdonald S, Keefe L et al. Total parenteral nutrition in the critically ill patient: a meta-analysis. JAMA 1998; 280:2013-9.
7. Koretz RL, Lipman TO, Klein S. AGA technical review on parenteral nutrition. American Gastroenterological Association. Gastroenterology 2001; 121:970-1001.
8. Patton KM, Aranda-Michel J. Nutritional aspects in liver disease and liver transplantation. Nutr Clin Pract 2002; 17:332.
9. Perioperative total parenteral nutrition in surgical patients. The Veterans Affairs Total Parenteral Nutrition Cooperative Study Group. N Engl J Med 1991; 325:525-32.
10. Parrish CR. Enteral feeding: The art and the science. Nutr Clin Pract 2003; 18:76.
11. Souba WW. Nutritional support. N Engl J Med 1997; 336:41-8.
12. Spain DA. When is the seriously ill patient ready to be fed? J Parenter Enteral Nutr 2002; 26:862.

19

Grafts and Flaps in Gynecologic Oncology

John T. Soper

Introduction

This chapter will review common principles of reconstructive surgery, with special emphasis on application to pelvic reconstructions. Many of these principles can be applied to less radical pelvic procedures or are used for reconstruction after complications of benign procedures, treatment of hydradinitis suppuritiva or necrotizing fasciitis, pelvic trauma and in conjunction with radical general surgery or urological procedures involving the female pelvis. The alterations in anatomy produced by radical pelvic surgery and/or radiation therapy often results in disruption of functional pelvic anatomy, sexual response, body self-image and psychosocial interactions.[1,2] Recognition of these issues has led to recommendations for appropriate individualization of the radical extent of major resections for some diseases, such as vulvar cancer and an increasing awareness for the use of reconstructive techniques following radical pelvic surgery.

Direct Closure

As a general principle, the most simple repair that can adequately close a surgical defect and restore functional anatomy will be the most effective and least likely to result in wound-related morbidity. The relative complexity of various reconstructive techniques is presented in Table 20A.1. Direct closure of the defect is preferred if tissue loss is minimal, surrounding tissues have good mobility and if local factors favor wound healing.[1] Adverse factors for wound healing include infection, malnutrition, microvascular disease, diabetes and prior radiotherapy. Fortunately, the pelvis has an excellent anastomotic blood supply and adjacent tissues can often be mobilized to fill in a surgical defect.

Surgical techniques that will improve the success of primary closure include the following: (1) incise the vulva along "wrinkle lines", which indicate the native lines of tissue stress;[1] (2) close your incision parallel to the "wrinkle lines" to avoid contractures; (3) avoid circular incisions around the introitus or in the vagina which tend to produce progressive contracture during healing and can lead to progressive vaginal and introital stenosis. Finally, the important surgical principals of hemostasis, debridement of subviable or infected tissue, wound irrigation and wide mobilization of surrounding tissues to allow closure without tension and obliteration of dead space will improve the chances of healing for any defect. Specific reconstructive techniques are discussed in the following sections.

Skin Grafts

Skin grafts are often used for vulvar and vaginal reconstruction. Full-thickness skin grafts may be used, but split-thickness grafts, comprising the thin layer of epidermis

Gynecologic Oncology, edited by Paola Gehrig and Angeles Secord.

Table 20A.1. Relative complexity of reconstructive surgical techniques used in gynecologic oncology procedures

Complexity	Example	Technique
Simple	Vulvar incision	Simple closure
	Partial vulvectomy	Direct closure
Intermediate	Skinning vulvectomy	Split-thickness skin graft
	Radical vulvectomy	Local "random" rotational flap
	Vaginal defect	Fasciocutaneous flap
Complex	Vulvovaginal defect at exenteration	Myocutaneous flap
	Massive tissue loss at exenteration	Free flap with microvascular reanastomosis

and reticular dermis within the rete pegs, are used more frequently.[1] Thigh, buttocks, lower abdomen, or mons are usually selected as the donor site because these sites have skin that is a similar consistency to the vulvar skin and vaginal mucosa. A dermatome is used to harvest a 0.16-0.20 mm thickness of skin, approximately 4-5 cm in width (Fig. 20A.1A). More than one strip of skin can be harvested for grafting, or the skin graft can be expanded with a meshing device to fit virtually any size defect (Fig. 20A. 1B). The donor site is a raw surface consisting of reticular dermis, epidermal rete pegs and skin adnexal structures. It can be dressed with a variety of dressings, such as scarlet red or dermoplast, to allow healing. Re-epithelialization of the donor site proceeds outward from the rete pegs and skin adnexae over several weeks. Often the donor site is the major source of pain after the procedure. The mons and gluteus can also be used as the donor site and allows use of a location that is not usually visible.

The skin graft initially adheres to the recipient site by fibrin bonding. During the first 24-48 hours it is oxygenated and nourished by plasmatic imbibition from the underlying raw surface. Meshing of the graft or making small "pie crust" incisions through an intact graft prevents accumulation of serum or blood under the graft so that revascularization can take place. Revascularization takes place over the next 4-7 days; it is critical that the graft remain immobilized during this process to prevent shearing of the immature vascular bed.[1]

Local factors can influence the chances of a successful split-thickness skin graft. Fascia, muscle, vascularized adipose tissue, perineurium, pericardium and intact periostium have sufficient vascularity to support a skin graft, but denuded tendon, nerve, cartilage, or bone lack sufficient vascularity to allow revascularization.[1] Likewise, irradiated tissues and adipose tissue in patients with microvascular disease are poor recipient tissues for a split-thickness skin graft. Heavy bacterial contamination of the recipient site may result in infection of the graft recipient site and loss of the graft. In these situations, delayed skin grafting is probably preferable to immediate skin grafting, allowing a bed of vascularized noninfected granulation tissue to become established at the recipient site. Finally, the importance of immobilization of the graft during the critical phase of revascularization cannot be over-emphasized.

20A

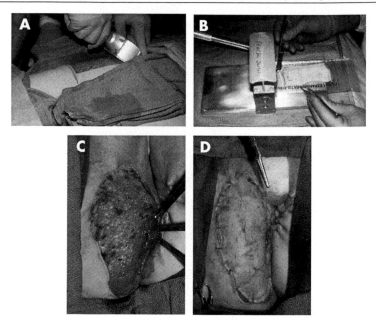

Figure 20A.1. Use of split-thickness skin graft during resection for multifocal vulvar and perianal dysplasia. A) A dermatome is used to harvest split-thickness (0.16-0.20 mm) skin graft from the donor site. B) A mesher can be used to expand the potential area of coverage for the skin graft and allows escape of serum from the slits produced by the mesher. C) An extensive vulvar defect produced during skinning vulvectomy for resection of multifocal vulvar carcinoma in situ. The larger Foley catheter is in the rectum. D) The skin graft has been placed over the operative defect and sutured into place. "Pie crust" incisions have been made through the graft to allow escape of serum from under the graft. Sutures at the margins of the graft are left long, so that packing material can be tied down over the graft to immobilize it during early healing. The larger Foley catheter is in the rectum.

Split-thickness skin grafts are often used to cover large vulvar, perineal, or perianal defects produced by surgical resections of extensive vulvar intra-epithelial neoplasia, hydradenitis suppuritiva, or after radical resections of vulvar cancer. Skinning vulvectomy with immediate split-thickness skin grafting has been used since the 1960s to treat women with extensive multifocal intra-epithelial neoplasia.[1] It results in better cosmetic results and a more normal functional anatomy than simple vulvectomy. A single surgical procedure may be preferable to multiple laser treatments, which are often required to treat extensive multifocal disease. Disease in the vagina, perianal region and anal canal can be resected and grafted at the time of vulvectomy (Fig. 20A.1C). Furthermore, in contrast to laser vaporization, skinning vulvectomy produces a definitive surgical specimen for pathologic review; approximately 5-10% of women with multifocal high-grade intra-epithelial neoplasia will have an undetected invasive malignancy.[1]

If the graft is harvested before the resection procedure, it is kept moist by saturating it in normal saline until use. The graft is trimmed to fit the defect and anchored in place, dermal surface down, with absorbable sutures (Fig. 20A.1D). Excess skin is trimmed so that there are no wrinkles or tension that elevates the graft from the recipient bed. If a meshed graft is not used, small full-thickness "pie crust" incisions are made at 2-3 cm intervals to allow escape of blood and serum during the first few days of healing. A bolus dressing is tied down over the graft to provide even and constant light pressure to immobilize the operative site. A Foley catheter is placed in the bladder, with a large Foley catheter in the anus if the anal canal has been grafted. The patient is kept at bed rest for 5-7 days to prevent the graft from shearing off of the immature vascular bed. Appropriate therapeutic measures should be carried out during this time of immobility to prevent the development of thromboembolic disease and pulmonary sequelae. In the absence of infection, more than 95% of the graft will take. The graft is cleansed with frequent sitz baths or irrigation of the operative site, followed by gentle drying with a hair drier on a cool or low heat setting.

Split-thickness skin grafting can also be used for vaginal reconstruction in women with congenital absence of the vagina (McIndoe procedure), repair of vaginal stenosis following surgery or radiation therapy and for a component of neovaginal reconstruction after exenteration or other radical vaginal resections.[1,2] In McIndoe vaginal reconstructions and repairs of vaginal stenosis, the split-thickness graft serves as vaginal mucosa to line the vaginal tube that has been created surgically. Because the skin graft does not provide vascularized tissue bulk, it must be combined with other procedures such as an omental flap that form the recipient bed when it is used for neovaginal reconstruction after exenteration or radical vaginal resections.[2]

After forming the vascularized recipient bed, the neovaginal skin graft must be immobilized with packing or a vaginal mold for 5-7 days to allow establishment of microcirculation. The split-thickness skin graft vagina will stricture during late healing. Patients should use a vaginal mold for at least 6 weeks and then use a vaginal dilator to prevent vaginal stricture. Sexually active patients will usually be able to maintain an adequate vaginal tube without long-term dilation, but late strictures may develop if the patient is not sexually active and does not continue mechanical dilatation.[1] If vaginal mucosa is retained at the introitus, the split-thickness skin graft will eventually assume characteristics of vaginal mucosa, including loss of keratin, response to estrogen and production of glycogen.

Flaps

Many different types of tissue flaps are available for pelvic reconstructions. The majority of flaps used in gynecologic reconstructions bring vascularized adipose tissue and skin with or without underlying fascia or muscle. Occasionally muscular flaps can be modified to bring in only vascularized soft tissue to obliterate a large tissue defect without the skin layer. Unlike skin grafts, tissue flaps generally do not develop contracture during the course of healing.[1] Clinically useful flaps have an arc of rotation that will allow the flap to be rotated into the tissue defect without jeopardizing the vascular supply to the flap. Because no single flap will be able to repair all pelvic defects in all situations, the practicing Gynecologic Oncologist should be familiar with a variety of flaps that are used for pelvic reconstructions.

Random skin flaps receive their blood supply through the dermal-subdermal plexus and are constructed of skin with a >1 cm margin of subcutaneous adipose tissue.[1] These are formed without intentionally including major vessels ("random" blood supply) and must be kept relatively short, with the length usually less than twice

20A

the width of the base of the flap. Random skin flaps should be used with caution in patients with smoking history, diabetes, peripheral vascular disease or prior radiation therapy to the operative site, as these comorbidities often indicate the presence of microvascular compromise which could lead to flap failure.[1] Wide mobilization of adjacent skin and fat to fill a tissue defect without tension is the most frequently performed random flap. Advancement flaps, lateral transposition or rhomboidal flaps, rotational flaps and Z-plasty procedures are other examples of specific random flaps that are used in pelvic reconstructions. In some locations, a specific subcutaneous vessel can be used to supply an axial or arterial flap, which can be relatively longer than a simple random skin flap.

In fasciocutaneous flaps the vessels course along the septa between the muscle belly and the superficial fascia, supplying a predictable territory of overlying skin. These flaps can usually be much longer than random flaps and are rotated into the defect with a relatively reliable vascular pedicle. Fasciocutaneous flaps from the labia and inner thigh can be used for vulvovaginal reconstructions or as V-Y transposition flaps.

Myocutaneous or musculocutaneous flaps comprise a unit of muscle and fascia with an overlying island of subcutaneous fat and skin supplied by a musculocutaneous perforator that runs through the belly of the muscle. The most versatile flaps have an arc of rotation that allows the flap to be employed for a variety of reconstructions. A major advantage of myocutaneous flaps lies in their provision of vascularized tissue to provide revascularization and obliteration of large defects after radical pelvic procedures.[1,2] Gracilis, rectus abdominis, bulbocavernosus, gluteus and tensor fascia lata flaps are examples of myocutaneous flaps that are often used for pelvic reconstructions. Free tissue flaps with microvascular reanastomosis can be used to transplant a myocutaneous unit from virtually any site in the body, but are only rarely employed in gynecologic procedures.

Random Skin Flaps

Advancement or axial translocation flaps (Fig. 20A.2) can be used to fill rectangular defects occurring on a relatively flat surface.[1] The entire flap and adjacent tissues must be mobilized to allow the flap to fill the defect. The flap length should be less than twice the width of the base.

Lateral transposition or rotational flaps can be used to fill a defect of virtually any configuration.[1] Again, the flap length should not exceed twice the width of

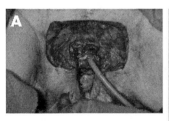

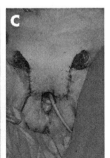

Figure 20A.2. Example of an advancement (axial translocation) flap used to repair an anterior vulvar defect. A) Flap and resection are planned prior to resection. B) Flap raised and widely mobilized. C) Flap advanced to fill the operative defect.

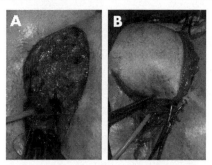

Figure 20A.3. Example of a mons rotational (lateral transposition) flap used to close an anterior vulvar defect. A) Extent of the vulvar defect. The flap will be developed from the adjacent mons to be rotated 80-90 degrees into the vulvar defect. B) The flap has been rotated into place and is secured with towel clips prior to suturing into position.

the base and adjacent tissues must be widely mobilized. Lateral transposition flaps should be rotated perpendicular to the skin wrinkle lines. Rotational skin flaps are often used to close round or triangular defects. The skin of the anterior abdominal wall, with its richly anastomotic blood supply, is able to support large rotational flaps to repair defects up to 10-15 cm in diameter (Fig. 20A.3). Rhomboidal rotational flaps (Fig. 20A.4) from the vulva and inner thigh are often used to repair rhomboidal vulvar defects.

Z-plasty flaps are used to substitute vertical and horizontal dimensions of a contracture or to break up a circular scar.[1] They are particularly useful for relieving contractures at the vaginal introitus or vaginal canal. Small Z-plasties can be used prophylactically at the introitus during neovaginal reconstruction to prevent an introital stenosis from developing during healing.

20A

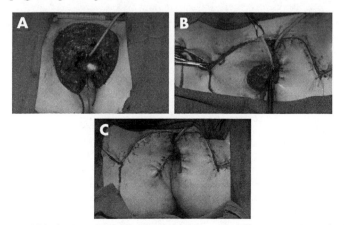

Figure 20A.4. Use of paired rhomboidal rotational flaps to repair radical vulvectomy defect (A). B) Paired flaps are raised from the inner thighs and rotated to fill the defect on each side of the vulva. C) The completed repair.

Myocutaneous Flaps

A variety of fasciocutaneous and myocutaneous flaps have been developed for pelvic reconstruction, including labiocrural (Singapore) fasciocutaneous flaps, gracilis muscle, rectus abdominis muscle, bulbocavernosus muscle, tensor fascia lata muscle and gluteus maximus muscle flaps. Although it is not technically a myocutaneous flap, the omentum can be developed into a vascularized pedicle that is a valuable adjunct to pelvic reconstructions.

The gracilis myocutaneous flap (Fig. 20A.5) has been used since the late 1970s for neovaginal and vulvar reconstructions after radical pelvic procedures.[2-4] Both a well-defined primary vascular pedicle derived from the medial femoral circumflex artery and a poorly-defined accessory blood supply derived from anastomotic terminal branches of the obturator and pudendal arteries close to the origin of the gracilis muscle supply this flap.[3,4] Because of the length of the flap and its wide arc of rotation, it is a versatile flap for pelvic reconstruction. Occasionally, it is used for reconstruction of vulvar or groin defects, or the gracilis muscle alone is used for reconstruction of vaginal fistulas or after disruption of the rectal sphincter.

The thin strap-like gracilis muscle is the most medial adductor of the thigh. It functions to stabilize the knee medially, but can be sacrificed without loss of function. The gracilis originates from the pubic tubercle and extends posterior to the adductor longus to insert into the medial tibial plateau. The classical gracilis flap derives its major blood supply from a vascular pedicle that enters the deep muscle 6-8 cm distal to the pubic tubercle after passing between the adductor longus and brevis muscles.[4] This supports a large skin territory that extends along the medial thigh to the distal one-third of the thigh. A shorter gracilis flap supplied from the accessory blood supply of the muscle from terminal branches of the obturator vessels has been successfully used for vulvovaginal reconstructions, without increasing flap loss.[3,4]

The most common use of gracilis flap is for neovaginal reconstruction after pelvic exenteration. The patient is placed in the modified Whittemore position using direct placement stirrups with the hips abducted 45 degrees and flexed slightly. A guideline is drawn from the pubic tubercle to the medial tibial plateau along the margin of the adductor longus. The skin island of the gracilis flap will be located posterior to this line (Fig. 20A.5A). For long flaps, a 14-20 cm long by 6-10 cm wide ellipsoid skin island will be developed with the proximal margin 4-6 cm distal to the crural fold.[4] Skin islands up to 24 cm can be used, but these run the risk of extending into the "watershed" region of the distal one-third of the medial thigh, where blood supply to the skin is derived from sartorius muscle perforators. A 10-14 × 5-8 cm ellipsoid with the proximal margin at the crural fold is used for the skin island of the short flap.[3,4] Because the muscular perforators fan out anteriorly and posteriorly from the gracilis muscle, a vertical, rather than transverse, skin island can be used.

A full-thickness incision is made along the anterior and distal margin of the planned skin island through the fascia lata. The skin is loosely sutured to the fascia to prevent shearing and disruption of the perforating vessels (Fig. 20A.5B). The belly of the gracilis muscle is identified distally, posterior to the adductor longus muscle, and divided. The gracilis is mobilized from its bed with sharp and blunt dissection (Fig. 20A.5C), working from the distal tip of the flap to its origin so that the dominant vascular pedicle can be identified. The skin island is developed as the muscle is elevated as traction on the muscle may aid in identifying its associated skin territory. The dominant vascular pedicle, comprising artery and paired venae commitantes, enters the deep belly of the gracilis muscle approximately 6-8 cm from the pubic tubercle, emerging from under the

20A

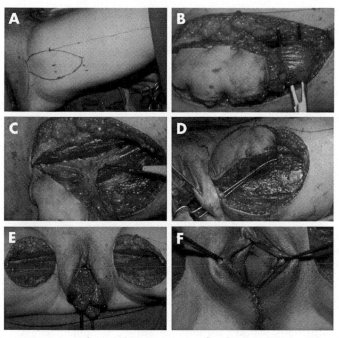

Figure 20A.5. Use of a gracilis myocutaneous flap for neovaginal reconstruction. A) Orientation of the skin flap. B) The skin flap has been developed, sutured to the fascia of the gracilis and the muscle is isolated prior to dividing it distally. C) The flap is mobilized off of the deep muscles. The neurovascular pedicle emerges from under the adductor longus approximately 6-8 cm from the pubic tubercle. The paired venae commitantes are clearly visible. D) The flap is being rotated posteriorly through the perineal tunnel to the vaginal introitus. E) The neovaginal tube will be formed from the paired flaps, which have been delivered through the perineal tunnels. F) The appearance of the introitus after the neovaginal tube has been rotated into the pelvis.

20A

belly of the adductor longus (Fig. 20A.5C). The nerve usually travels with the vascular bundle. If the long classical flap is used, the neurovascular pedicle is preserved, but it is deliberately divided when a short flap is employed. When a short flap is used, the proximal 2-3 cm of the muscle should not be aggressively skeletonized to prevent loss of the small accessory blood supply.

Subfascial tunnels under the vulva are developed, sufficient to allow passage of each flap without pressure (Fig. 20A.5D). The flaps are rotated posteriorly through the tunnels. The neovaginal tube is formed by approximating the skin edges with absorbable sutures, beginning at the distal tips of the flaps (Fig. 20A.5E). The proximal skin margins are left open to form the introitus. The neovaginal tube is then rotated posteriorly into the pelvic defect and anchored to the pelvis and/or the levator plate. The skin is approximated to the vaginal introitus with interrupted absorbable sutures (Fig. 20A.5F). The thigh incisions are closed over suction drains. Mechanical dilatation is not required to keep the gracilis neovagina patent.

Most series report some degree of flap loss in up to 20% of patients after gracilis neovaginal reconstructions, but major flap loss is encountered in only approximately 10-15% of patients.[3,4] Other flap-specific complications include donor site hematoma or infection, introital stenosis and flap prolapse. Introital stenosis is rare in the absence of major flap loss.[4] There is no significant increase in flap-specific complications with use of the short gracilis flap compared to the long flap.[4]

Rectus abdominis flaps (Fig. 20A.6) can be developed as myocutaneous flaps or without the skin island as muscular or myoperitoneal flaps for pelvic and neovaginal reconstructions.[5,6] The large and reliable vascular pedicle for this flap is from the deep inferior epigastric vessels. This flap has a very large arc of rotation and can be used for neovaginal, groin, perineal/presacral and anterior vulvar reconstructions.[1] Because of the widely anastomotic subcutaneous vessels derived from the muscular perforators, both transverse and vertical flaps can be developed, often supported by only a 4 × 4 cm strip of anterior rectus fascia.

The principles of rectus abdominis flaps are similar for both vertical and transverse flaps. The skin island is defined and incised along the superior border to the anterior rectus fascia. For vertical flaps the skin island usually measures 10-12 cm long × 5-8 cm wide, while slightly larger flaps ranging from 12-14 cm horizontally × 7-10 cm vertically are used for transverse rectus abdominis flaps (Fig. 20A.6A).[5,6] The muscle is divided at the superior border of the flap, and anastomotic vessels connecting to the superior epigastric vessels are ligated (Fig. 20A.6B). The flap is mobilized off of the rectus fascia, leaving a 4-5 × 4-5 cm strip of fascia that overlies the muscle under the skin island for its blood supply (Fig. 20A.6C). The inferior and lateral borders of the fascia are incised, and the skin is loosely anchored to the fascia with interrupted sutures. The rectus muscle is mobilized from attachments to the anterior and posterior rectus sheath to the level of the symphysis, working from medial to lateral border of the muscle and controlling subfascial anastomotic vessels. The deep inferior epigastric artery along with its venae commitantes enters the lateral border of the rectus muscle in the vicinity of the arcuate line and this must be preserved (Fig. 20A.6D).[5,6] The vascular pedicle can be mobilized, if needed, to allow rotation of the flap.

The rectus flap can be rotated posteriorly into the pelvis below the level of the arcuate line for neovaginal reconstruction. The flap is intubated with absorbable sutures to form a neovagina, often using a spiral closure and an open end is sutured to the introitus. If a partial vaginal defect is being repaired, it is used as a simple "patch" flap (Fig. 20A.6E) to fill the defect. If it is to be transferred to the groin or vulva, a subcutaneous tunnel is used.[1] During rotation of the flap, the vascular pedicle should be observed to ensure that there is no torsion of the vessels. The rectus fascia and skin defects can usually be closed primarily, but rarely synthetic mesh will be required to close the defect in the anterior rectus fascia.

Because of the extremely predictable vascular pedicle and richly anastomotic blood supply of the skin island, most series report a <5% incidence of major flap loss for rectus flaps.[5,6] The incidence of rectus flap loss appears to be increased in obese patients. There is no secondary blood supply for this flap; therefore, it should be used with caution in patients who have surgical incisions such as colostomy sites or transverse incisions over the rectus muscle as these might interrupt the inferior epigastric system. If there is a concern regarding the blood supply, this can be studied angiographically in the preoperative setting.

Similar to the gracilis flaps, rectus flaps provide a bulk of vascularized tissue that can readily obliterate dead space, while bringing a new blood supply into tissues that

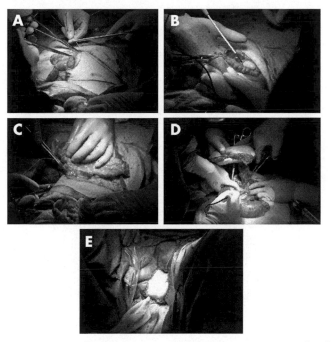

Figure 20A.6. Use of a transverse rectus abdominis myocutaneous flap for repair of a large posterior vaginal defect after posterior exenteration. A) Orientation of a transverse rectus abdominis skin island; suture is on the superior margin of the skin incision. B) After incising the superior rectus fascia to the lateral margin of the fascia, the muscle is divided. C) The skin island has been mobilized off of the lateral fascia and the rectus fascia incised laterally and inferiorly, preserving at least a 4 × 4 cm patch of fascia with perforating vessels to supply the skin island. D) The rectus muscle has been mobilized out of its sheath to its insertion on the pubis. The vascular pedicle is clearly visible, adjacent to the surgeon's clamp, entering from the posterior and lateral aspect of the muscle belly.

20A

may have been previously irradiated. Mechanical dilation is not required to maintain vaginal patency. Both types of neovaginas are satisfactory for vaginal coitus, with 40-60% of long-term survivors reporting vaginal coitus. Because keratinized skin will not convert into vaginal mucosa, however, most patients report poor lubrication and the development of a chronic watery vaginal discharge.

Myoperitoneal rectus flaps have also been used for neovaginal reconstructions. Unlike the myocutaneous flaps, there is a tendency for agglutination and vaginal stenosis because the peritoneal surface lines the vaginal canal and must become epithelialized via in-migration of epithelial cells from adjacent mucosa. While rectus myoperitoneal flaps are not suitable for total neovaginal reconstructions, they can be used to repair partial longitudinal defects of the vaginal wall.[1]

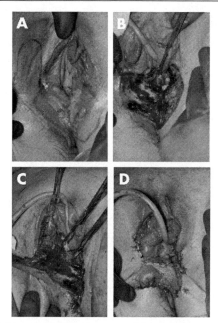

Figure 20A.7. Use of a bulbocavernosus (Martius) myocutaneous flap for perineal reconstruction. A) The patient had extensive vulvar carcinoma in situ involving the perineum. Planned resection and the approximate location/size of the flap are outlined. B) The operative defect involved the perineum, but did not extend laterally into the region of the posterior pudendal artery. C) The bulbocavernosus flap is mobilized. D) Appearance of the operative site after the bulbocavernosus has been rotated in to fill the defect and sutured into place.

The bulbocavernosus (Martius) flap has a long history of use as a myosubcutaneous or myocutaneous flap for repair of vesicovaginal and rectovaginal fistulas.[1] The bulbo-cavernosus muscle supports a stable skin island supplied from the perineal branches of the anterior or posterior pudendal arteries. Because this is a flap based on a reliable blood supply rather than a random flap, isolated skin islands and flaps exceeding the 2:1 length to width ratio can be developed. Using a bulbocavernosus flap based on the anterior pudendal artery, anterior vaginal defects can be repaired. Bulbocavernosus flaps based on the posterior pudendal artery are more often used to repair posterior vaginal or perineal defects (Fig. 20A.7). Paired posterior bulbocavernosus flaps can be used to create a partial vaginal tube at the time of pelvic exenteration but are usually not large enough for complete neovaginal reconstruction.[1]

The skin island can range from 4-7 cm wide, extending from the interlabial fold medially into the lateral labia majora and up to 8-10 cm in length from above the level of the clitoris anteriorly to approximately the level of the posterior fourchette posteri-orly. Pinching and elevating the labia majora essentially identifies the bulbocavernosus muscle and fat pad. After developing the skin island incision, the dissection is carried through the subcutaneous tissues between muscle and introitus to the level of the pubic arch. The muscle is elevated off of the deep tissues, proceeding from the tip and medial aspect of the flap. Introital skin and fat are dissected off of the pubic arch to create a short tunnel, and the flap is rotated medially into the vagina to fill the tissue defect. The bulbocavernosus flap tends to produce more hair and vaginal discharge than the gracilis or rectus flaps.

When used to repair posterior perineal defects, the skin island is rotated medially and sutured into place (Fig. 20A.7). Care must be taken to ensure that the perineal defect does not extend 2-3 cm lateral to the inferior pubic ramus, where the posterior pudendal artery supplying the skin island would be jeopardized.

The omentum can be developed into a pedicled flap that can act as a nonepithelialized subcutaneous flap for abdominal wall reconstructions, to provide vascularized tissue bulk in the pelvis after pelvic exenteration and to interpose a layer of vascularized tissue during repair of complicated gastrointestinal or urinary fistulas.[1] The omentum can also provide a recipient bed for partial or complete vaginal reconstructions with split-thickness skin grafts.

Blood supply to the omentum is richly anastomotic along the greater curvature of the stomach, with dual dominant vascular pedicles from the right and left gastroepiploic vessels.[1] The left gastroepiploic artery is a branch of the splenic artery, while the right is a branch of the gastric artery. The entire omentum can be vascularized from a single gastroepiploic artery pedicle.[1]

Either the entire omentum or the infracolic portion of the omentum can be used for an omental flap, depending on the size and location of the defect and the size and mobility of the omentum. Most frequently, the omentocolic ligament is divided, and the omentum is mobilized completely. One of the gastroepiploic pedicles is divided and the omentum is mobilized off of the greater curvature of the stomach. To fill in a pelvic defect after exenteration, the omentum is passed down the pericolic gutter and draped into the pelvis. This omental carpet is sutured into place with loose interrupted sutures over closed suction drains.

20A

Summary

The physician performing radical pelvic surgery should be aware of the large armamentarium of techniques that are useful for managing large pelvic surgical defects. It is essential to have an adequate knowledge of several options for reconstruction of any single type of defect so that the most appropriate procedure can be selected.

Suggested Reading

1. Soper JT. Grafts and flaps in gynecologic surgery. In: Mann WJ Jr, Stovall TG, eds. Gynecologic Surgery. New York: Churchill Livingstone, 1997:555-86.
2. Soper JT. Pelvic exenteration and pelvic reconstruction. In: Nichols DH, Clarke-Pearson DL, eds. Gynecologic, Obstetric and Related Surgery, 2nd Edition. St. Louis: Mosby, 2000:723-37.

References

1. Soper JT. Grafts and flaps in gynecologic surgery. In: Mann WJ Jr, Stovall TG, eds. Gynecologic Surgery. New York: Churchill Livingstone, 1997:555-86.
2. Soper JT. Pelvic exenteration and pelvic reconstruction. In: Nichols DH, Clarke-Pearson DL, eds. Gynecologic, Obstetric and Related Surgery, 2nd Edition. St. Louis: Mosby, 2000:723-37.
3. Soper JT, Larson DM, Hunter VJ et al. Short gracilis myocutaneous flaps for vulvovaginal reconstruction after radical pelvic surgery. Obstet Gynecol 1989; 74:823-7.

4. Soper JT, Rodriguez G, Berchuck A et al. Long and short gracilis myocutaenous flaps for vulvovaginal reconstruction after radical pelvic surgery: Comparison of flap-specific complications. Gynecol Oncol 1995; 56:271-5.

5. Soper JT, Secord AA, Havrilesky LJ et al. Rectus abdominis myocutaneous and myoperitoneal flaps for neovaginal reconstruction after radical pelvic surgery: comparison of flap-related morbidity. Gynecol Oncol 2005; 97:596-601.

6. Soper JT, Havrilesky LJ, Secord AA et al. Rectus abdominis myocutaneous flaps for neovaginal reconstruction after radical pelvic surgery. Int J Gynecol Cancer 2005; 15:542-8.

20A

Current Applications of Laparoscopy in Gynecologic Oncology

Lisa N. Abaid and John F. Boggess

Historical Perspective

As early as 1928, reports appeared in the European literature describing minimally invasive gynecologic procedures.[1] Over the next 30 years, minimally invasive diagnostic procedures evolved including hysteroscopy, pelviscopy and laparoscopy. Many improvements made in the 1960s, including a pressure-controlled carbon dioxide insufflator, an extra-abdominal light source and a uterine manipulator, helped transform minimally invasive diagnostic procedures.[2] Large European case series were reported in the late 1960s and 1970s describing the utility of diagnostic laparoscopy in the identification and management of benign gynecologic conditions. However, operative applications were limited to tubal ligation and small biopsies.[3] As technology improved and experience increased, operative laparoscopy flourished in the 1980s as a means to surgically manage a multitude of gynecologic conditions, including ectopic pregnancy, subserosal myomas and endometriosis.[4]

Applications of laparoscopy in gynecologic oncology were slower to occur, given the complexity of surgical procedures required to treat malignancy. However, by the late 1980s, case reports were emerging of minimally invasive techniques used by gynecologic oncologists. Assessment of pelvic lymph nodes was the first major application of laparoscopy in gynecologic oncology. In 1989, Dargent and Salvat described an extraperitoneal approach to pelvic lymph node sampling through a suprapubic incision.[5] Their technique evolved rapidly to a transperitoneal laparoscopic approach and was combined with radical vaginal surgery in the management of early cervical and endometrial carcinomas.[6,7]

Currently, there are laparoscopic techniques used to diagnose and treat patients with early and advanced cervical, endometrial and ovarian malignancies. Patients afflicted with gynecologic cancers tend to be older and in poorer health and thus are more likely to benefit from the reduced blood loss, shorter hospital stay and faster healing achieved by a minimally invasive approach. However, in addition to showing that laparoscopic procedures are feasible, it is also necessary to show that laparoscopic cancer management is equivalent from a recurrence and survival perspective. This section will review the most recent data on minimally invasive procedures used in the field of gynecologic oncology. Additionally, it will address concerns specific to the practice of laparoscopy in patients with cancer.

Gynecologic Oncology, edited by Paola Gehrig and Angeles Secord.
©2009 Landes Bioscience.

Cervical Cancer

Background

Since 1987, when Dargent first described a laparoscopic extraperitoneal pelvic lymphadenectomy in a patient with early cervical cancer, minimally invasive techniques for the treatment of cervical cancer have continued to evolve.[8] Most series have studied patients with early-stage disease, but some have also examined the utility of laparoscopy in patients with advanced-stage cervical cancer, typically defined as Stage IB2 or greater.

Initial treatment of cervical cancer is based on clinical staging criteria. A description of some common treatment options is shown in Table 20B.1. The majority of patients receive primary surgery, primary radiation, or a combined treatment regimen. Surgery should not be attempted unless there is a reasonable chance of complete tumor removal with an adequate disease-free tissue margin. Chemotherapy is not typically used as a primary treatment for cervical cancer due to a poor response rate but may be used as an adjuvant treatment or for recurrent disease.[9] Pelvic lymphadenectomy is typically performed in early-stage operable cervical cancer while para-aortic lymphatic metastases may be evaluated either radiologically or surgically in patients with advanced disease.

Laparoscopic techniques have been used to perform simple hysterectomy and both pelvic and para-aortic lymphadenectomies. Although more technically difficult, laparoscopic radical hysterectomy has been described in the literature using both a purely laparoscopic approach and a combined vaginal and laparoscopic approach. In the late 1990s, a robotic system was introduced with articulated instruments which allow the surgeon to mimic techniques used in open procedures. Additionally, it offers a three-dimensional view and superior fine-motor control, both of which are limited with conventional laparoscopy. Robotic-assisted surgery can overcome many of the difficulties encountered during a radical hysterectomy with conventional laparoscopic instruments and may prove to be a valuable adjunct in the treatment of cervical cancer.

20B

Table 20B.1. *General treatment plan for invasive cervical carcinoma*

Disease Stage	Treatment
Stage IA1	Cervical conization or total hysterectomy, abdominal or vaginal (Type I)
Stage IA2	Type II abdominal hysterectomy, bilateral pelvic lymphadenectomy
Stages IB1, IIA and good surgical risk	Type III abdominal hysterectomy, bilateral pelvic lymphadenectomy with postoperative radiation in selected high-risk patients
Stages IB2, IIB; IIIA, IIIB; IVA, IVB	Full external and intracavitary pelvic radiation

Adapted with permission from: Te Linde RW, Rock JA, Thompson JD. *Te Linde's Operative Gynecology*. 8th Ed. Philadelphia: Lippincott-Raven; 1997:1426.

Early-Stage Cervical Cancer

In 1991, Querleu et al performed transperitoneal laparoscopic pelvic lymph node dissection in 39 patients with early-stage cervical cancer. Thirty-two of these patients underwent immediate laparotomy and the remaining nodal tissue was negative for metastases.[7] In 1993, Fowler et al reported on 12 patients who underwent laparoscopic lymphadenectomy followed by laparotomy and removal of any remaining nodal tissue. The average number of nodes removed laparoscopically was 23.5, with no remaining positive nodes found at laparotomy. Additionally, the lymph node yield from the second six patients was 85%, compared to 63% from the first six patients suggesting that lymph node yield improves with experience.[10] These two studies prompted the Gynecologic Oncology Group (GOG) to perform a prospective trial. In this study, Schlaerth et al reported on 73 patients with Stage IA, IB and IIA cervical cancer who were scheduled to undergo radical hysterectomy with lymphadenectomy. Prior to surgery, laparoscopic pelvic and para-aortic lymphadenectomy was performed. Of the 40 evaluable patients, the para-aortic lymphadenectomy was considered to be adequate in all cases, but in 15% of cases, there was residual pelvic nodal tissue identified at laparotomy. Importantly, no metastatic disease was found in the retained lymph nodes.[11] Thus, it appears that while laparoscopic lymphadenectomy may increase the likelihood of retained nodal tissue, it does not appear to result in undiagnosed metastatic disease.

Five types of hysterectomy were described in 1974 by Piver: simple hysterectomy or extrafascial hysterectomy (Type I), modified radical hysterectomy (Type II), radical hysterectomy (Type III), extended radical hysterectomy (Type IV) and partial exenteration (Type V).[12] Type I, II and III hysterectomies have been described laparoscopically. Type II involves ligation of the uterine artery where it crosses the ureter and resection of the medial half of the cardinal ligaments, the proximal uterosacral ligaments and possible resection of the upper one-third of the vagina. Type III hysterectomy involves ligation of the uterine artery at its origin and removal of the entire cardinal ligament. In some cases, removal of the upper half of the vagina may be necessary (Fig. 20B.1).

Radical hysterectomy is the treatment of choice for early-stage cervical cancer, typically Stages IA2, IB1 and IIA. While laparoscopic lymphadenectomy has been combined with a variety of radical vaginal techniques, a purely laparoscopic radical hysterectomy, which avoids vulvar or perineal incisions, was first described by Nezhat et al in 1992.[13] Although initially limited to patients with Stage IA1 or IA2 disease, Spirtos et al described a technique (Type III radical hysterectomy) that included patients up to Stage IB.[14] Data abstracted from the most recent larger published series of patients receiving laparoscopic radical hysterectomy are summarized in Table 20B.2.

Radical hysterectomy is associated with more complications than simple hysterectomy due to the extensive dissection and presence of a malignancy. Complications seen with radical hysterectomy include large blood loss, fistula formation, bladder dysfunction, sexual dysfunction and lymphedema, as well as urinary tract infection, ileus, pulmonary embolism, wound infection and deep venous thrombosis. Laparoscopy is associated with a reduced blood loss and shorter hospital stay but has a longer operative time and may be associated with a longer time to return of normal bladder function and a higher intraoperative complication rate.[15]

Advanced-Stage Cervical Cancer

Patients with bulky cervical cancer (Stage ≥IB2 and/or tumor >4 cm) are currently offered chemoradiation therapy or neoadjuvant chemotherapy followed by radical surgery.[16] Identifying the presence of metastatic para-aortic disease assists in tailoring

20B

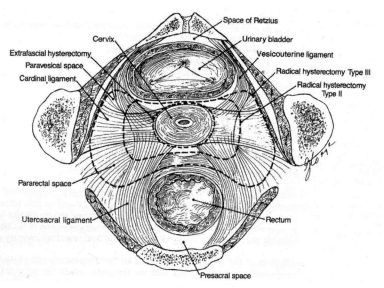

Figure 20B.1. The pelvic ligaments and spaces. Reprinted with permission from: Berek JS, Hacker NF. *Practical Gynecologic Oncology*. Fourth Ed. Philadelphia: Lippencott Williams & Wilkins, 2005:356.

therapy, including extending the field of radiation. Up to 38% of patients with Stage II disease and greater will have positive para-aortic nodes and only half of these will be detected by radiography alone.[17,18] Sonoda et al performed an extraperitoneal laparoscopic para-aortic lymphadenectomy on 111 patients with bulky or locally advanced cervical cancer, without intraoperative complications or laparotomies.[19] Mean nodes retrieved were 19 and 30 patients (27%) had metastatic disease. Average length of stay was 2 days and there were no blood transfusions. The node-positive group all received extended-field radiation and some also received chemotherapy. Median survival was 27 months, which is consistent with previously published reports.[20] It appears that minimally invasive techniques can offer valuable staging information, which can help tailor treatment regimens without conferring significant surgical risk.

Trachelectomy

Radical trachelectomy, which involves excision of the cervix, proximal parametrium and a cuff of vaginal mucosa, is a fertility-sparing procedure performed in young women with small tumors. It is preceded by a laparoscopic pelvic lymphadenectomy and only performed if the nodes are negative for metastatic disease. Pioneered and refined by Dargent and colleagues,[21] extensive work has been done by a Canadian group with this select patient population.

Plante, Roy and colleagues reported 82 patients with Stage IA, IB, or IIA cervical cancer who were scheduled to undergo vaginal radical trachelectomy, preceded by laparoscopic pelvic lymphadenectomy and parametrectomy.[22] The trachelectomy was aborted in 10 patients due to positive nodes or margins, or extensive tubal disease. Of the remaining 72 patients, there were five intraoperative complications (6%) resulting in

Table 20B.2. Summary of laparoscopic radical hysterectomy data in cervical cancer

Authors	n	PAN	PN	Lap	EBL	LOS	Comps	MFU	DFS
Gil-Moreno et al (46)	27	-	19.1	1 (3.7%)	400	5	7 (25.9%)	32	27 (100%)
Steed et al (15)	71	-	-	0	300	1	9 (13%)	17	67 (94%)
Abu-Rustum et al (47)	19	-	25.5	2 (10.5%)	301	4.5	10 (52.6%)	4-28 (range)	19 (100%)
Spirtos et al (48)	78	10.3	23.8	5 (93.6%)	225	2.9	12 (15.4%)	66.8 (mean)	70 (89.7%)

Abbreviations: n: number of patients; PAN: mean para-aortic lymph nodes; PN: mean pelvic nodes; Lap: conversion to laparotomy; EBL: estimated blood loss (milliliters); LOS: length of stay (days); Comps: intraoperative and postoperative complications; MFU: median follow-up (months); DFS: disease-free survival, number of patients without disease at the time of follow-up.

20B

three laparotomies, all for control of bleeding. There were no bladder or ureteral injuries and a low incidence of postoperative complications, with an average hospital stay of 3 days. During a mean follow-up of 60 months, there have been two recurrences (2.8%) and one death (1.4%), excluding a patient with a neuroendocrine tumor who rapidly recurred and died. In a subsequent study of these patients, 31 women conceived 50 pregnancies.[23] Thirty-six of the pregnancies (72%) were carried to the third trimester and 34 of those children are alive and well. This is a promising fertility-sparing option that is enhanced by the capacity to laparoscopically determine the presence of nodal metastases and thus patient suitability for this conservative treatment.

The laparoscopic approach appears feasible when treating Stage IA1 to IB1 cervical cancer with respect to lymph node retrieval and surgical margins. In general, operative time is significantly longer and while in some series hospital stay is shorter, most reported hospital stays appear comparable to traditional open surgery. It is difficult to determine from this data whether laparoscopic radical hysterectomy is sufficiently superior to open surgery to warrant the commitment to overcoming the learning curve. Robotic-assisted surgery may offer a promising option in the minimally invasive treatment of cervical cancer. Any prospective trials will need to focus on short-term and long-term quality of life postoperatively to highlight the most significant benefits.

Endometrial Cancer

The standard treatment of endometrial cancer begins with systematic surgical staging. Resection of the pelvic and para-aortic lymph nodes improves survival and is a vital component of staging.[24] Vaginal hysterectomy has been utilized in clinical Stage I disease; however, this method alone fails to detect the 20% of patients with clinical Stage I endometrial cancer who have extrauterine disease.[25] Laparoscopic techniques, either in conjunction with vaginal surgery or alone, have been gaining in popularity as a means of complete staging for endometrial cancer. The GOG performed a prospective randomized trial comparing the two modalities (LAP-2) and has enrolled over 2,000 patients. Preliminary data shows that intraoperative and postoperative complication rates appear to be consistent with previously published reports.

Laparoscopic hysterectomy was first described by Reich et al in 1989.[26] In 1992, Childers and Surwit described a combined laparoscopic and vaginal approach in two patients with Stage I endometrial cancer.[6] The development of the Koh ring and pneumo-occluder balloon, used with a uterine manipulator, helped to facilitate delineation of the vaginal fornices during colpotomy and to avoid ureteral injury.[27] With increasing experience and comfort with laparoscopic techniques, operating time and node retrieval have become comparable to that of open procedures.[28,29]

When comparing perioperative data over the last 5 years, it has been consistently demonstrated that despite slightly increased operative times, patients with laparoscopically treated endometrial cancer benefit from reduced blood loss and blood transfusions, shorter hospitalizations, fewer postoperative complications and equal or greater lymph node retrieval as compared to an open procedure. In addition, there has been no change in disease-free survival or recurrence among those undergoing laparoscopy (Table 20B.3). As a significant proportion of patients with endometrial cancer are obese and elderly and many have other comorbid conditions such as diabetes, hypertension and heart disease, this high-risk patient population stands to gain a significant benefit from a minimally invasive surgical approach.

An increased risk of positive peritoneal cytology is a concern specific to the laparoscopic treatment of endometrial cancer. In a retrospective study by Sonoda et al, of 131 "low-risk" patients undergoing LAVH, 10.3% were found to have positive peritoneal

20B

Table 20B.3. Summary of laparoscopic hysterectomy data in endometrial cancer

Authors	n	PAN	PN	Lap	EBL	LOS	Comps	MFU	DFS
Zapico et al[49]	38	-	15	2 (5.2%)	-	5.0	12 (31.2%)	36.3 (mean)	31 (81.6%)
Kuoppala et al[50]	40	-	11.1	1 (2.5%)	171	2.7	15 (37.5%)	38.3 (mean)	39 (97.5%)
Eltabbakh[51]	100	2.5	11.0	6 (6%)	200	2	11 (11%)	27	92 (92%)
Holub et al[52]	177	-	16.8	6 (3.4%)	211	3.9	-	33.6	166 (93.7%)

Abbreviations: n: number of patients; PAN: mean para-aortic lymph nodes; PN: mean pelvic nodes; Lap: conversion to laparotomy; EBL: estimated blood loss (milliliters); LOS: length of stay (days); Comps: intraoperative and postoperative complications; MFU: median follow-up (months); DFS: disease-free survival, number of patients without disease at the time of follow-up.

20B

cytology, compared to 2.8% of patients undergoing laparotomy.[30] This was thought to be secondary to the use of an intrauterine manipulator, as a second study by Vergote et al failed to show the same increase in a cohort of patients where uterine manipulators were not routinely used.[31] However, it does not appear that this increase in positive cytology confers a worse prognosis.[32] In an attempt to decrease this risk, some advocate cauterization of the Fallopian tubes in patients without a tubal ligation, prior to placing the intrauterine manipulator.

Although laparoscopic staging of endometrial cancer is gaining wider acceptance and studies published to date have been favorable, many are awaiting the results of the GOG LAP-2 trial to conclusively support this method as the preferred approach over conventional techniques. However, there is consistent and reproducible evidence that the minimally invasive approach is better for patients in the short-term, without apparent adverse long-term outcomes.

Ovarian Cancer

In 1990, Reich et al reported the first case of laparoscopic management of Stage I ovarian cancer.[33] As ovarian cancer most commonly presents in Stage III, requiring extensive surgical debulking, it has been less amenable to exclusively laparoscopic management. However, laparoscopy does have a valuable but limited role in the diagnosis and treatment of ovarian cancer. The most common indications are: (1) histologic confirmation of advanced or recurrent ovarian cancer, (2) second-look laparoscopy and (3) detection and staging or restaging of early ovarian cancer.

Advanced Ovarian Cancer

In patients presenting with stigmata of advanced ovarian cancer, an approach using neoadjuvant chemotherapy for presumed ovarian cancer has been advocated by some. As 10-20% of peritoneal carcinomatosis may be nonovarian in origin,[34] laparoscopy can be used to obtain an accurate histologic diagnosis and guide therapy. Additionally, laparoscopy can be used to determine if optimal debulking, defined as removing all tumor 1 cm in size or larger, is feasible.

Second-Look Laparoscopy

A second-look operation has typically been performed by laparotomy in order to evaluate the response to chemotherapy, with or without prior debulking. Husain et al reviewed records of 150 patients who underwent second-look laparoscopy after a previous open debulking and were clinically disease-free. Forty-six percent had negative second-look laparoscopies and their recurrence rate of 40% was equivalent to those who had second-look laparotomies. The rate of major complications was 2.7%.[35] Clough et al performed second-look laparoscopies followed by laparotomies in 20 patients and found the positive predictive value of laparoscopy for finding residual disease to be 100% (6/6). However, 2/14 patients with negative laparoscopics had disease found during laparotomy, for a negative predictive value of 86% (104).[36] Second-look laparoscopy may help patients with disease avoid a laparotomy, but the reliability of a negative laparoscopy remains in question.

Staging/Restaging Early Ovarian Cancer

Early-stage ovarian cancer is often difficult to reliably diagnose preoperatively, but once identified, lends itself to an exclusively laparoscopic approach. Common concerns regarding laparoscopic management of ovarian cancer include the risk of cyst rupture and abdominal dissemination.

In a study of 1545 patients with invasive epithelial ovarian cancer, Vergote et al found the rate of ovarian cyst rupture during laparotomy to be 45% overall. Rupture during surgery was associated with a reduction in disease-free survival, with a hazard ratio of 1.64 (1.07-2.51).[37] Inadvertent tumor rupture can occur during either laparotomy or laparoscopy and results in upstaging of the patient. One study found that 41.8% of ovarian masses ruptured during laparoscopy compared to 35.7% during laparotomy.[38] The authors concluded, "Ovarian cyst rupture was not related to the surgical route but to the frequency of cystectomy." Methods have been described to minimize spillage and its subsequent risks, including controlled drainage in an endobag, closure of all tissue layers and a shortened interval to the initiation of platinum-based chemotherapy.[39] Whenever ovarian masses are approached laparoscopically, care must be taken to remove the mass intact to prevent dissemination of potentially malignant cells.

The risk of abdominal dissemination, including both intra-abdominal spread and port-site metastases, has been widely debated. The theoretical mechanism is that the high humidity, CO_2-rich environment and blood act as trophic factors on neoplastic tissues. In addition, a "chimney" effect of the pneumoperitoneum is postulated to force malignant cells up and out through ports, seeding the tissues of the abdominal wall.

In 2004, Ramirez and colleagues reviewed all reported cases of port-site metastases in patients with gynecologic malignancies.[40] They found a total of 31 articles reporting 58 patients, 40 (69%) of who were diagnosed with low malignant potential or invasive ovarian cancer. Of these patients, 83% had Stage III or IV disease, 71% had ascites and 97% had carcinomatosis. These reports spanned 24 years and although they were unable to determine the incidence of port-site metastases, it is undoubtedly a rare occurrence. Childers et al calculated an incidence rate of 0.2% per puncture site in patients with gynecologic malignancies, after reviewing 105 procedures.[41] There was a frequency of 1.0% per procedure. This is similar to previously published data reporting abdominal wall metastases after a fine-needle aspiration biopsy in 0.1% of patients with abdominal malignancies.[42]

Although some still advocate for immediate laparotomy upon identification of an ovarian malignancy,[43] there have been two recent studies of patients with early stage disease managed laparoscopically (Table 20B.4). Leblanc et al performed laparoscopic restaging in 35 patients with early ovarian cancer diagnosed initially by either laparoscopy or laparotomy. Nine patients underwent fertility sparing procedures, but all patients underwent complete staging, which included pelvic and para-aortic lymphadenectomy.[44] Tozzi et al performed primary laparoscopic staging or restaging in 24 patients with early invasive ovarian cancer or low malignant potential tumors.[45] There were no laparotomies or intraoperative complications in either study. Recurrence rates for Stage IA were 6.4% and 6.3% respectively, and there were no port-site recurrences in either study during a median follow-up time of 54 and 46 months. Although both studies are small, they suggest acceptable outcomes for laparoscopic management of clinical Stage I ovarian cancer.

From conception to practical application, laparoscopic technique has evolved into a surgical discipline capable of replacing laparotomy for many gynecologic cancer treatments. Patient benefits of laparoscopic lymphadenectomy, Type I, II and III hysterectomy and disease status assessment of ovarian cancer have been well-documented and have proven the feasibility of these techniques. While survival data is limited, there does not appear to be a significant difference when laparoscopy is performed. Furthermore, early concerns regarding port-site metastasis, disease dissemination and inadequacy of laparoscopy to be a legitimate cancer operation seem to have been overestimated and are not likely to be significant.

Table 20B.4. Summary of laparoscopic staging of early invasive ovarian cancer

Authors	n	PAN	PN	Lap	EBL	LOS	Comps	MFU	DFS
Leblanc et al (44)	53*	20	14	1 (1.8%)	-	3.1	4/53 (7.5%)	54	32/35 (91%)
Tozzi et al (45)	24†	19.6	19.8	0	-	7	1/24 (4.1%)	46.4	22/24 (91.7%)

Abbreviations: n: number of patients; PAN: mean para-aortic lymph nodes; PN: mean pelvic nodes; Lap: conversion to laparotomy; EBL: estimated blood loss (milliliters); LOS: length of stay (days); Comps: intraoperative and postoperative complications; MFU: median follow-up (months); DFS: disease-free survival, number of patients without disease at the time of follow-up.

* = Seven patients had Fallopian tube carcinomas, 11 patients were undergoing a second-look laparoscopy and 35 patients were undergoing primary surgery for ovarian cancer

†= 11 patients under went completion laparoscopy after primary surgery for diagnosis, 13 patients underwent primary laparoscopic management.

Future Directions

As minimal access tools and technology continue to evolve, there does not appear to be a limit to the future indications. Technologies such as robotic surgical systems, which overcome many of the limitations of laparoscopy, will likely revolutionize what surgeons can achieve in the near future. While we are already improving survival with conventional therapy, minimally invasive approaches offer the same cancer outcomes with dramatic improvements in quality of life. Perhaps the most important issue currently is the need to standardize techniques to facilitate teaching, improve the reproducibility of procedures and to flatten the steep learning curve associated with laparoscopy that has been largely responsible for limiting widespread acceptance and application within Gynecologic Oncology.

References

1. Mikulicz-radecki FV, Freund A. A new hysteroscope and its practical use in gynecology. Z Geburtshilfe Gynakol 1928; 92:13-25.
2. Semm K. Laparoscopy in gynecology. Geburtshilfe Frauenheilkd 1967; 27(11):1029-42.
3. Kleissl HP, Christ F, Eberlein F. Report on 700 gynaecologic cases in diagnostic laparoscopy (author's transl). Geburtshilfe Frauenheilkd 1975; 35(5):354-9.
4. Semm K. Endoscopic intraabdominal surgery in gynecology. Wien Klin Wochenschr 1983; 95(11):353-67.
5. Dargent DSJ. Envahissement ganglionnaire pelvien: place de la pelviscopie retroperitoneale. Paris: McGraw-Hill, 1989.
6. Childers JM, Surwit EA. Combined laparoscopic and vaginal surgery for the management of two cases of stage I endometrial cancer. Gynecol Oncol 1992; 45(1):46-51.
7. Querleu D, Leblanc E, Castelain B. Laparoscopic pelvic lymphadenectomy in the staging of early carcinoma of the cervix. Am J Obstet Gynecol 1991; 164(2):579-81.
8. Dargent D. A new future for schauta's operation through presurgical retroperitoneal pelviscopy? Eur J Gynaecol Oncol 1987; 8:292.
9. Te Linde RW, Rock JA, Thompson JD. Te Linde's operative gynecology. 8th ed. Philadelphia: Lippincott-Raven, 1997.
10. Fowler JM, Carter JR, Carlson JW et al. Lymph node yield from laparoscopic lymphadenectomy in cervical cancer: a comparative study. Gynecol Oncol 1993; 51(2):187-92.
11. Schlaerth JB, Spirtos NM, Carson LF et al. Laparoscopic retroperitoneal lymphadenectomy followed by immediate laparotomy in women with cervical cancer: a gynecologic oncology group study. Gynecol Oncol 2002; 85(1):81-8.
12. Piver MS, Rutledge F, Smith JP. Five classes of extended hysterectomy for women with cervical cancer. Obstet Gynecol 1974; 44(2):265-72.
13. Nezhat CR, Burrell MO, Nezhat FR et al. Laparoscopic radical hysterectomy with paraaortic and pelvic node dissection. Am J Obstet Gynecol 1992; 166(3):864-5.
14. Spirtos NM, Schlaerth JB, Kimball RE et al. Laparoscopic radical hysterectomy (type III) with aortic and pelvic lymphadenectomy. Am J Obstet Gynecol 1996; 174(6):1763-7; discussion 1767-8.
15. Steed H, Rosen B, Murphy J et al. comparison of laparascopic-assisted radical vaginal hysterectomy and radical abdominal hysterectomy in the treatment of cervical cancer. Gynecol Oncol 2004; 93(3):588-93.
16. Morice P, Castaigne D. Advances in the surgical management of invasive cervical cancer. Curr Opin Obstet Gynecol 2005; 17(1):5-12.
17. Matsukuma K, Tsukamoto N, Matsuyama T et al. Preoperative CT study of lymph nodes in cervical cancer—its correlation with histological findings. Gynecol Oncol 1989; 33(2):168-71.

20B

18. Nelson JH Jr, Boyce J, Macasaet M et al. Incidence, significance and follow-up of para-aortic lymph node metastases in late invasive carcinoma of the cervix. Am J Obstet Gynecol 1977; 128(3):336-40.

19. Sonoda Y, Leblanc E, Querleu D et al. Prospective evaluation of surgical staging of advanced cervical cancer via a laparoscopic extraperitoneal approach. Gynecol Oncol 2003; 91(2):326-31.

20. Varia MA, Bundy BN, Deppe G et al. Cervical carcinoma metastatic to para-aortic nodes: extended field radiation therapy with concomitant 5-fluorouracil and cisplatin chemotherapy: a gynecologic oncology group study. Int J Radiat Oncol Biol Phys 1998; 42(5):1015-23.

21. Dargent D, Martin X, Sacchetoni A et al. Laparoscopic vaginal radical trachelectomy: a treatment to preserve the fertility of cervical carcinoma patients. Cancer 2000; 88(8):1877-82.

22. Plante M, Renaud MC, Francois H et al. Vaginal radical trachelectomy: an oncologically safe fertility-preserving surgery. An updated series of 72 cases and review of the literature. Gynecol Oncol 2004; 94(3):614-23.

23. Plante M, Renaud MC, Hoskins IA et al. Vaginal radical trachelectomy: a valuable fertility-preserving option in the management of early-stage cervical cancer. A series of 50 pregnancies and review of the literature. Gynecol Oncol 2005; 98(1):3-10.

24. ACOG practice bulletin, clinical management guidelines for obstetrician-gynecologists, number 65: management of endometrial cancer. Obstet Gynecol 2005; 106(2):413-25.

25. Creasman WT, Morrow CP, Bundy BN et al. Surgical pathologic spread patterns of endometrial cancer. A Gynecologic Oncology Group Study. Cancer 1987; 60(8 Suppl):2035-41.

26. Reich HDJ, McGlynn F. Laparoscopic hysterectomy. J Gynecol Surg 1989; 5:213-5.

27. Koh CH. A new technique and system for simplifying total laparoscopic hysterectomy. J Am Assoc Gynecol Laparosc 1998; 5(2):187-92.

28. Holub Z, Jabor A, Bartos P et al. Laparoscopic surgery in women with endometrial cancer: the learning curve. Eur J Obstet Gynecol Reprod Biol 2003; 107(2):195-200.

29. Eltabbakh GH. Effect of surgeon's experience on the surgical outcome of laparoscopic surgery for women with endometrial cancer. Gynecol Oncol 2000; 78(1):58-61.

30. Sonoda Y, Zerbe M, Smith A et al. High incidence of positive peritoneal cytology in low-risk endometrial cancer treated by laparoscopically assisted vaginal hysterectomy. Gynecol Oncol 2001; 80(3):378-82.

31. Vergote I, De Smet I, Amant F. Incidence of positive peritoneal cytology in low-risk endometrial cancer treated by laparoscopically assisted vaginal hysterectomy. Gynecol Oncol 2002; 84(3):537-8.

32. Sonoda YLD, Chi DS et al. The significance of positive peritoneal cytology in clinical early-stage endometrial cancer patients treated by laparoscopically-assisted vaginal hysterectomy. Abstract #108. San Diego: Society of Gynecologic Oncologists, 2004.

33. Reich H, McGlynn F, Wilkie W. Laparoscopic management of stage I ovarian cancer. A case report. J Reprod Med 1990; 35(6):601-4; discussion 604-5.

34. Dargent DF. Laparoscopic surgery in gynecologic oncology. Surg Clin North Am 2001; 81(4):949-64.

35. Husain A, Chi DS, Prasad M et al. The role of laparoscopy in second-look evaluations for ovarian cancer. Gynecol Oncol 2001; 80(1):44-7.

36. Clough KB, Ladonne JM, Nos C et al. Second look for ovarian cancer: laparoscopy or Laparotomy? A prospective comparative study. Gynecol Oncol 1999; 72(3):411-7.

20B

37. Vergote I, De Brabanter J, Fyles A et al. Prognostic importance of degree of differentiation and cyst rupture in stage I invasive epithelial ovarian carcinoma. Lancet 2001; 357(9251):176-82.

38. Fauvet R, Boccara J, Dufournet C et al. Laparoscopic management of borderline ovarian tumors: results of a French multicenter study. Ann Oncol 2005; 16(3):403-10.

39. van Dam PA, DeCloedt J, Tjalma WA et al. Trocar implantation metastasis after laparoscopy in patients with advanced ovarian cancer: can the risk be reduced? Am J Obstet Gynecol 1999; 181(3):536-41.

40. Ramirez PT, Frumovitz M, Wolf JK et al. Laparoscopic port-site metastases in patients with gynecological malignancies. Int J Gynecol Cancer 2004; 14(6):1070-7.

41. Childers JM, Aqua KA, Surwit EA et al. Abdominal-wall tumor implantation after laparoscopy for malignant conditions. Obstet Gynecol 1994; 84(5):765-9.

42. Lundstedt C, Stridbeck H, Andersson R et al. Tumor seeding occurring after fine-needle biopsy of abdominal malignancies. Acta Radiol 1991; 32(6):518-20.

43. Stelmachow J, Spiewankiewicz B. Possibilities and limitation of endoscopic procedures in oncological gynaecology. Eur J Gynaecol Oncol 2005; 26(1):21-3.

44. Leblanc E, Querleu D, Narducci F et al. Laparoscopic restaging of early stage invasive adnexal tumors: a 10-year experience. Gynecol Oncol 2004; 94(3):624-9.

45. Tozzi R, Kohler C, Ferrara A et al. Laparoscopic treatment of early ovarian cancer: surgical and survival outcomes. Gynecol Oncol 2004; 93(1):199-203.

46. Gil-Moreno A, Puig O, Perez-Benavente MA et al. Total laparoscopic radical hysterectomy (type II-III) with pelvic lymphadenectomy in early invasive cervical cancer. J Minim Invasive Gynecol 2005; 12(2):113-20.

47. Abu-Rustum NR, Gemignani ML, Moore K et al. Total laparoscopic radical hysterectomy with pelvic lymphadenectomy using the argon-beam coagulator: pilot data and comparison to laparotomy. Gynecol Oncol 2003; 91(2):402-9.

48. Spirtos NM, Eisenkop SM, Schlaerth JB et al. Laparoscopic radical hysterectomy (type III) with aortic and pelvic lymphadenectomy in patients with stage I cervical cancer: surgical morbidity and intermediate follow-up. Am J Obstet Gynecol 2002; 187(2):340-8.

49. Zapico A, Fuentes P, Grassa A et al. Laparoscopic-assisted vaginal hysterectomy versus abdominal hysterectomy in stages I and II endometrial cancer. Operating data, follow up and survival. Gynecol Oncol 2005; 98(2):222-7.

50. Kuoppala T, Tomas E, Heinonen PK. Clinical outcome and complications of laparoscopic surgery compared with traditional surgery in women with endometrial cancer. Arch Gynecol Obstet 2004; 270(1):25-30.

51. Eltabbakh GH. Analysis of survival after laparoscopy in women with endometrial carcinoma. Cancer 2002; 95(9):1894-901.

52. Holub Z, Jabor A, Bartos P et al. Laparoscopic surgery for endometrial cancer: long-term results of a multicentric study. Eur J Gynaecol Oncol 2002; 23(4):305-10.

20B

Surgical Procedures

Teresa Rutledge

Introduction

Treatment of gynecologic malignancies often involves a multimodality approach involving surgery, chemotherapy and radiation therapy in order to achieve cure. The optimal treatment is determined by the specific cancer type and the extent of disease. The goals of surgical management typically involve removing the primary tumor and assessing the extent of disease through staging techniques. Utilizing these principles, gynecologic oncologists hope to optimize cure while minimizing morbidity. In this chapter, the principles of surgical treatment for the specific gynecologic malignancies will be discussed. This chapter will provide the reader with an overview of the indications for certain surgeries, a description of the techniques and the most common potential complications.

Radical Hysterectomy

Indications

Stage IB—IIA cervical carcinoma can be treated with radical hysterectomy or radiation therapy with similar efficacy (please refer to Chapters 9 and 20B). The patient and physician therefore are faced with treatment options for these types of cancers. The criteria to select surgery versus radiation therapy is a controversial topic among both the surgeons and the radiation oncologist. Factors considered in this treatment decision include age, patient weight, tumor size, potential need for adjuvant radiotherapy/chemotherapy and presence of other comorbid conditions. See Figure 20C.1 as an example of a class III radical hysterectomy specimen—notice the additional vaginal margin and the inclusion of the parametrial tissue laterally.

Definitions

There is an array of types or classes of hysterectomy used in the treatment of benign and malignant gynecologic disease.[1]

Type I Extrafascial Hysterectomy
- Consists of removal of uterus, cervix and rim of vagina in a plane outside the pubocervical fascia
- Most common type of hysterectomy for benign indications
- Acceptable for microinvasive (<3 mm) cervical carcinoma if childbearing complete

Type II Modified Radical Hysterectomy (Figs. 20C.2)
- Consists of removal of uterus, cervix, medial one-half of the cardinal and uterosacral ligaments and proximal 1-2 cm of vagina
- Performed with or without pelvic lymphadenectomy
- Used to treat Stage IA2 (3-5 mm) cervical cancer or other small lesions

---

Gynecologic Oncology, edited by Paola Gehrig and Angeles Secord.
©2009 Landes Bioscience.

Figure 20C.1. Photograph of a radical hysterectomy specimen.

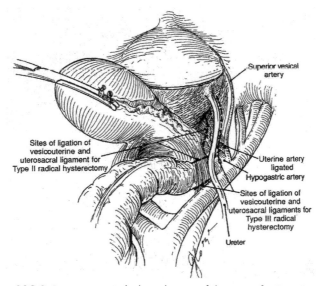

20C

Figure 20C.2. Demonstrating the lateral extent of dissection for Type II versus Type III hysterectomies. Reprinted with permission from: Berek JS, Hacker NF. *Practical Gynecologic Oncology*. Fourth Ed. Philadelphia: Lippencott Williams & Wilkins, 2005:354.

Type III Radical Hysterectomy
- Consists of removal of uterus, cervix, entire cardinal and uterosacral ligaments and the upper third of the vagina
- Includes a pelvic lymphadenectomy
- Commonly performed for Stage IA2-IIA cervical carcinoma

Type IV Hysterectomy
- Includes radical hysterectomy approach in addition removes periureteral tissue, superior vesicle artery and three-fourths of the vagina
- Rarely performed because radiation is more acceptable treatment approach

Type V Hysterectomy
- Includes removal of distal ureter and portion of the bladder
- Most commonly used to remove central recurrent disease

Technique

Radical hysterectomy can be performed through vertical midline, low transverse abdominal incisions, laparoscopically, or with robotic technology (refer to Chapter 20B). The peritoneal cavity is first inspected for any evidence of metastatic disease. It is important to rule out distant metastasis (chest X-ray, inspection liver surface and omentum) and local extension as this should cause the surgeon to abort the radical hysterectomy. A detailed description of radical hysterectomy technique is beyond the scope of this text but basically involves the development of key pelvic spaces and an extensive dissection of the ureter. The two most important pelvic spaces are the paravesical and pararectal areas. (Fig. 20C.3)

The borders of the paravesical space are:
- Medial—superior vesical artery and the bladder
- Lateral—obturator internus muscle along the pelvic sidewall
- Posterior—cardinal ligament
- Anterior—pubic symphysis

20C

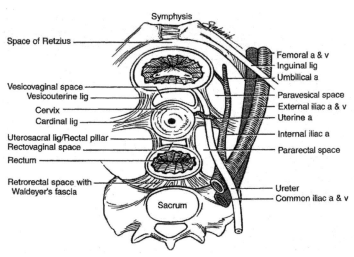

Figure 20C.3. The pelvic spaces. Reprinted with permission from: Morrow CP, Curtin JP. *Gynecologic Cancer Surgery*. Churchill Livingstone, 1996:111.

Table 20C.1. Common complications in women undergoing radical hysterectomy[2]

Operative/Postoperative Complications	Number of Patients (N = 195) (%)
Transfusion	19 (9)
Bowel/bladder injury	0 (0)
Ureteral injury	5 (2)
Venous thrombosis	3 (1.5)
Fistula formation	3 (1.5)
Bowel obstruction	5 (2)
Fever	50 (25)
Lymphedema	7 (3.5)

The borders of the pararectal space are:
- Medial—ureter along the broad ligament and the rectum
- Lateral—internal iliac artery and vein
- Posterior—sacrum
- Anterior—cardinal ligament

Complications

Radical hysterectomy has all the standard operative risk as any other major surgical procedure including infection, need for transfusion, venous thrombosis, cardiac injury and death. The average blood loss is between 500 and 1500 mL so transfusion is not uncommon. Bladder dysfunction can be a long-term unique consequence of radical pelvic surgery. This is related to partial dennervation of the detrusor muscle causing inability to empty the bladder and the need for prolonged use of bladder catheter or self-catheterization. Table 20C.1 lists common complication rates after radical surgergy seen in approximately 200 patients.[2]

Pelvic and Para-Aortic Lymphadenectomy

20C

Pelvic and para-aortic lymphadenectomy is a common procedure in gynecologic oncology surgery based on the known lymphatic drainage of the major gynecologic organs. Figure 20C.4 depicts the anatomical location of the pelvic and para-aortic lymph nodes and their relationship to the major retroperitoneal vasculature. The pelvic lymph node dissection is carried out by identifying the bifurcation of the common iliac artery, the external and internal iliac vessels, and the ureter. The paravescial and pararectal spaces are created as noted above. The boundaries of the pelvic lymph node dissection include the distal portion of the common iliac artery proximally, the deep circumflex iliac vein distally, the genitofemoral nerve laterally, the superior vescial artery medially and the obturator nerve inferiorly. All lymph node-bearing tissue within these boundaries is removed. The para-aortic lymph node dissection requires identification of the aortic bifurcation, the inferior vena cava, the inferior mesenteric artery and the ureter. The nodal tissue over the distal vena cava from the inferior mesenteric artery to the mid-portion of the right common iliac artery is removed as the right para-aortic lymph nodes. The nodal tissue between the aorta and the left ureter from the inferior mesenteric artery to the middle of the common iliac artery is removed as the left para-aortic lymph nodes.

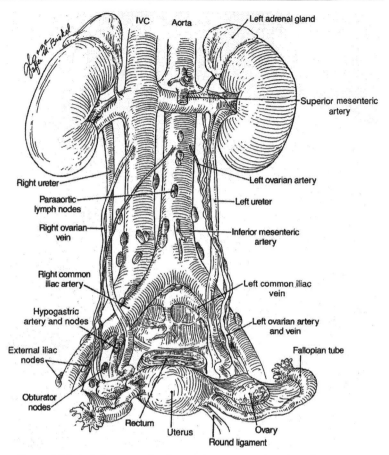

Figure 20C.4. Schematic of pelvic and para-aortic lymph node anatomy. Reprinted with permission from: Berek JS, Hacker NF. *Practical Gynecologic Oncology*. Fourth Ed. Philadelphia: Lippencott Williams & Wilkins, 2005:357.

Vulvectomy and Inguinofemoral Lymphadenectomy

Vulvar cancer surgery has evolved over the last few decades in order to decrease operative morbidity and sexual dysfunction while maintaining curative outcomes. Classically operative management of vulvar cancer would have consisted of the traditional radical vulvectomy with en bloc inguinofemoral lymphadenectomy performed through the classic butterfly incision (Fig. 20C.5). This procedure was associated with a wound separation in approximately 50% of patients.[1] The procedure has been modified to a radical wide excision with lymphadenectomy via a separate incision technique (Fig. 20C.6). The modified procedure is associated with similar cure rates, but significantly less morbidity and a reduction in wound dehiscence to approximately 15%.[1] Today the surgical procedure is often tailored based on tumor location, size and depth of

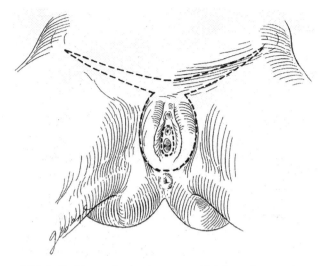

Figure 20C.5. En bloc radical vulvectomy with bilateral groin dissection. Reprinted with permission from: Berek JS, Hacker NF. *Practical Gynecologic Oncology*. Fourth Ed. Philadelphia: Lippencott Williams & Wilkins, 2005:559.

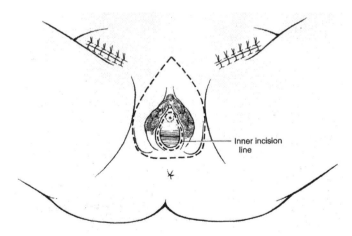

20C

Figure 20C.6. Radical vulvectomy with separate groin incisions. Reprinted with permission from: Morrow CP, Curtin JP. *Gynecologic Cancer Surgery*. Churchill Livingstone, 1996:415.

invasion. Based on these tumor characteristics, most vulvar cancers can be treated with modified radical vulvectomy (i.e., radical wide excision) which involves an elliptical incision with a 1-2 cm margin around the lesion. The excision should extend to the depth of the deep perineal fascia. Most excisions can be closed primarily; however flaps

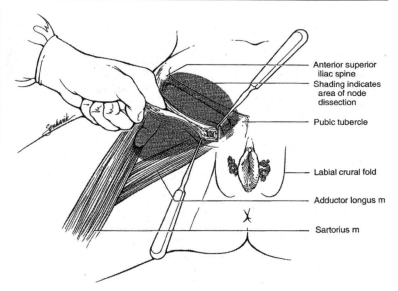

Anterior superior
iliac spine

Shading indicates
area of node
dissection

Pubic tubercle

Labial crural fold

Adductor longus m

Sartorius m

Figure 20.7. Figure 8 Inguinal lymph node dissection incision. Reprinted with permission from: Morrow CP, Curtin JP. *Gynecologic Cancer Surgery.* Churchill Livingstone, 1996:427.

can be used to facilitate closure in some cases (refer to Chapter 20A). Access to the inguinal lymphadenectomy is achieved through a separate incision made parallel to the inguinal ligament (Fig. 20C.7). The groin lymph nodes are divided into superficial and deep groups by the cribiform fascia. The superficial nodes are located in the femoral triangle formed by the inguinal ligament superiorly, the sartorius muscle laterally and the adductor longus muscle medially (Fig. 20C.8). The saphenous vein can be identified in the medial inferior aspect of the dissection and should be preserved to decrease postoperative lymphedema rates. The deep nodes are located medial to the femoral vein. The cribiform fascia medial to the vein is entered allowing removal of the nodes adjacent to the vein. Depending on the extent of deep dissection, the surgeon can cover the exposed femoral vessel with sartorius muscle transposition. Drains are usually placed in the groin after lymph node dissection to avoid lymphocyst formation and are left until the drainage is <30 mL/day. Prophylactic antibiotics are sometimes used while the drain is in place to reduce the risk of superficial cellulitis. The role of sentinel lymph node biopsy is currently under evaluation in clinical trials (refer to Chapter 16).[3]

Pelvic Exenteration

Pelvic exenteration is most commonly performed for centrally recurrent or advanced cervical cancer. Total pelvic exenteration (removal of uterus, bladder and rectum) is the only curative option for recurrent or persistent cervical cancer after radiation therapy (Figs. 20C.9). The procedure obviously involves massive reconstruction efforts with either continent or incontinent urinary diversion, colostomy and/or low rectal anastomosis and neovagina formation. The surgeon must ensure that the disease is confined to the central pelvis prior to proceeding with the procedure. This involves preoperative

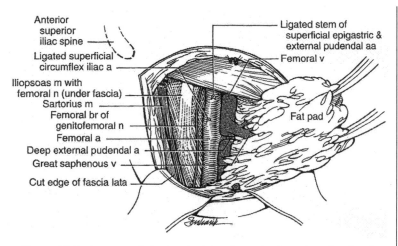

Figure 20.8. Anatomy of the inguinal lymph node dissection. Reprinted with permission from: Morrow CP, Curtin JP. *Gynecologic Cancer Surgery*. Churchill Livingstone, 1996:430.

imaging to rule out distant metastasis and intraoperative abdominal exploration and lymph node dissection. The morbidity and mortality related to total pelvic exenteration is high mostly related to large blood loss and length of surgery.

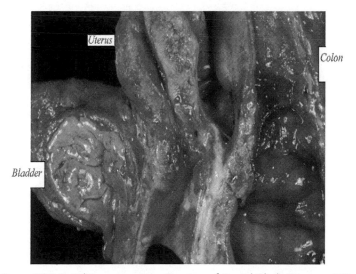

Figure 20C.9. Pelvic exenteration. Courtesy of Daniel Clarke-Pearson, MD, Division of Gynecologic Oncology, University of North Carolina at Chapel Hill, Chapel Hill, North Carolina.

Conclusion and Future Directions

The cornerstone of treatment for many gynecologic malignancies is surgery. As we strive to improve outcomes, the balance between cure and morbidity is a constant battle. Advances in surgical technology allow gynecologic oncologists to achieve similar if not better outcomes with decreased morbidity and improved quality of life for the patient. Probably the best example of this impact is the growing use and acceptance of laparoscopic approaches to gynecologic cancer surgeries. In the future, robotics can be expected to further these improvements. Also, through better understanding of the specific cancers and their spread patterns we can offer less radical techniques. The application of sentinel node techniques could further tailor our surgical approaches allowing for better outcomes.

Suggested Reading

1. Berek et al. Practical Gynecologic Oncology 4th edition.
2. Hoskins et al. Prinicples and Practice of Gynecologic Oncology 4th edition, Surgical Principles in Gynecologic Oncology, 311-31.

References

1. Hoskins et al. Prinicples and Practice of Gynecologic Oncology, 4th edition, Surgical Principles in Gynecologic Oncology, 311-31.
2. Rutledge TL et al. A Comparison of stages IB1 and IB2 cervical cancers treated with radical hysterectomy. Is size the real difference? Gyn Oncol 2004; 95:70-6.
3. Coleman RL. Santoso JT. Vulvar carcinoma. Current Treatment Options in Oncology 2000; 1(2):177-90.

20C

CHAPTER 21

Palliative Care

Cecelia H. Boardman

Introduction

Many women with gynecologic malignancies will ultimately succumb to their disease. Optimal patient management in women with progressive gynecologic malignancies involves the recognition and active management of symptoms associated with these diseases. Increased recognition of the importance of the quality of life in the treatment of patients with cancer has led to the emergence of the field of palliative care medicine. The recognition that active management of symptoms in cancer patients significantly improves the quality of life underscores the importance of diagnosing and treating these symptoms. Patients with progressive malignancy experience a multitude of symptoms and are often afraid to bring these symptoms to their physicians' attention. Patients often do not want to voice their concerns regarding the negative effects of chemotherapy and their disease process for fear that their physician will say "there is nothing more to be done" and will abandon them or that important decisions may not be made as time was spent on less "important" issues. On the contrary, when treatments can no longer be effective to cure the cancer, there are many treatments that can be prescribed to improve the patient's quality of life while recognizing that quantity is not infinite. Adequate and aggressive management of pain, depression, constitutional symptoms and gastrointestinal and pulmonary dysfunction allows for improved quality of life. Home hospice allows for the patient to spend quality time with their families in a home environment, treating death as a truly natural process, while actively managing symptoms.

Gynecologic malignancies account for 77,000 new cancer cases annually in the United States, and 28,000 women will succumb to their disease. Principles that are applicable to end-of-life care for a patient with any type of progressive malignancy can be applied to women with progressive gynecologic cancers, as many of the issues and symptoms are similar. These include adequately addressing symptoms such as pain, depression, sleeplessness and anorexia. Unique features of progressive gynecologic cancers include fistulas, functional bowel obstructions secondary to carcinomatosis, pelvic hemorrhage, groin and perineal recurrences, which require specific understanding of appropriate interventions. Terminal physiology is usually referable to the specific disease. Progressive ovarian cancer often involves the peritoneal surfaces with miliary seeding of tumor which inhibits intestinal motility, leading to an inability to absorb nutrients and move intestinal contents downstream. This phenomenon is known as a carcinomatous ileus, or a functional malignant bowel obstruction, and leads to death through dehydration and malnutrition. Ovarian cancer that is progressive in the pulmonary or lymphatic system may lead to death through anorexia and cachexia. Endometrial cancer can recur in the abdominal or pelvic cavity, the retroperitoneal lymph nodes and/or

Gynecologic Oncology, edited by Paola Gehrig and Angeles Secord.
©2009 Landes Bioscience.

the lungs. Endometrial cancer also has a propensity to metastasize to bone and brain. Bony metastases can be quite painful and put the patient at risk for pathologic fractures. Progressive endometrial cancer often results in death from similar mechanisms as seen in ovarian cancer. Progressive cervical cancer generally leads to death from two distinct pathways—localized disease and distant disease. Localized cervical cancer can invade into the soft tissues of the pelvis, resulting in significant pain. Fistulization into the bowel or bladder presents significant hygiene issues and can lead to sepsis. Locally progressive cervical cancer can lead to bilateral ureteral obstruction with death due to hyperkalemia with cardiac dysrhythmia (less commonly) or uremia secondary to renal failure (more commonly). Additionally, locally progressive disease is often accompanied by cachexia. Metastatic cervical cancer often involves the retroperitoneal lymph nodes and the lungs, resulting again in cachexia and respiratory distress, hypoxemia and resultant respiratory failure, respectively. Vulvar cancer is similar to cervical cancer in terminal physiology. Locally progressive disease in the vulva or groin can cause significant issues in terms of hygiene, pain control and hemorrhage. Metastatic disease to the lungs is the other common terminal event for women who succumb to vulvar cancer.

Goals of optimal palliative care in gynecologic malignancies include effective symptom management, making a graceful transition from active anticancer management to end-of-life care, meeting both the patient's and her family's needs, neither hastening nor postponing death and remembering that death is a natural process (Fig. 21.1). Familiarity with a variety of pharmacologic therapies aimed at symptom management is imperative (Table 21.1). It is important to remember that patients often have multiple symptoms. Open communication with the patient and family and simple explanations are critical to the success of palliative care.

Figure 21.1, viewed on following page. Algorithm for effective palliative care. Reproduced with permission from the NCCN (1.2008) Palliative Care Guidelines, Clinical Practice Guidelines in Oncology. ©National Comprehensive Cancer Network, 2008. Available at: http://www.nccn.org.

21

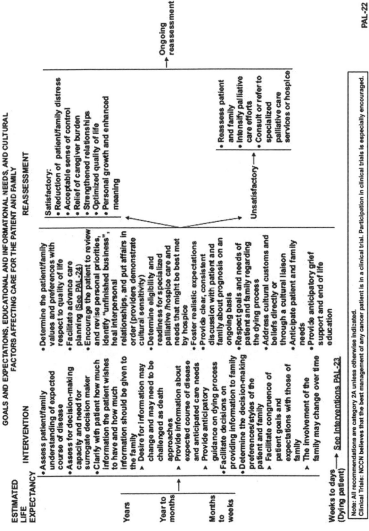

Figure 21.1. Please see legend on previous page.

Table 21.1. Medications commonly used in palliative care

Drug	Dose	Usage
Albuterol	0.5 mL in 2.5 mL saline nebulized over 5-25 min	Bronchospasm
Amytryptiline	10-25 mg qhs initially, increase by 10-25 mg prn; max dose 75-100 mg	Pain, depression, bladder spasm
Baclofen	5 mg tid; dose increase q3d by 5 mg as needed to max 20 mg tid	Muscle spasm, neuropathic pain
Belladonna and opium suppository	1 sup vaginally or rectally qd to qid	Bladder spasm
Bisacodyl	1-2 tab po qhs to bid or 1 sup pr qd	Constipation
Carbamazepine	200 mg qhs; dose increase q3d to bid-tid-qid as needed to max 1200 mg/d	Neuropathic pain
Clonazepam	0.5 mg tid; increase q3d by 0.5 mg to max dose 20 mg/d	Neuropathic pain, myoclonus
Dexamethasone	4 mg qAM	Nausea, anorexia, pain
Docusate	100 mg qd to tid	Constipation
Dronabinol	2.5 mg bid; 5 mg/m^2 po q 4-6 hr	Anorexia; nausea
Fentanyl nebulized	25 μg in 3 mL NSS q 1-3 hr prn	Dyspnea
Fluoxetine	20-60 mg po qAM	Depression
Furosemide	40-80 mg qd	Edema, ascites

continued on next page

Table 21.1. Continued

Drug	Dose	Usage
Gabapentin	Titrate to effect from 300 qd to 1200 mg tid	Neuropathic pain
Granisetron	1 mg po bid or 2 mg po qd	Nausea
Haloperidol	0.5 mg tid; 3 to 5 mg to start, increase by 3 to 5 mg to effect, can repeat q 1 hr	Nausea; agitation
Hydroxyzine	25 mg qhs	Pruritus
Lactulose	15-30 mL qd to tid	Constipation
Lidocaine	2% jelly apply to affected area up to 5x/d	Pain
Lorazepam	0.5-2 mg po q 6 hr prn	Anxiety, nausea
Megestrol acetate	400-800 mg/d	Anorexia
Methylphenidate	5 mg at 8 AM and at 10 AM; increase to 10 +10, up to a max of 60 mg/d	Narcotic induce sedation, depression
Metaclopramide	10 mg qid	Nausea, ileus, anorexia
Octreotide	150 μg sq bid, max dose 600 μcg/24 hr	Carcinomatous ileus
Ondansetron	8 mg bid; can increase 2-3x if needed	Nausea
Phenazopyridine	100-200 mg po tid	Dysuria
Scopolamine	1 patch behind ear q 3 d	Nausea
Senna	1 tab po qd to 4 tab bid, titrate to effect	Constipation
Tegaserod	6 mg po bid	Carcinomatous ileus, constipation

21

Pain Management

Pain is a very common symptom in patients with progressive malignancy. Approximately 30% of patients on treatment and 60-90% of patients with progressive disease experience pain. Adequate analgesia may often create other symptoms from the side effects of the medicines prescribed. Pain is often multifactorial; visceral pain may have a constant ache, muscular pain is described as a spasm, bony pain is often worse with movement and neuropathic pain (as a result of chemotherapy or direct invasion of tumor into neural structures) has a burning or tingling quality. Different types of medications are used to address these types of pain and therefore adequate pain relief may require combinations of medications. Pain control is more effective if dosing of analgesics is scheduled rather than taken in response to symptoms; the goal is prevention. It is important to be familiar with the dosing intervals, common side effects, toxicities and maximal doses of the medications provided. For the patient with a dysfunctional intestinal tract, sublingual, rectal, transdermal, subcutaneous and intravenous routes of administration can be employed. For the patient who can tolerate oral medication, this is the preferred route of administration from both a logistical and a cost standpoint.

The World Health Organization has designed a three-step ladder to help clinicians adequately assess and manage pain. The lowest step is mild pain, which is best managed with nonnarcotic analgesics such as acetaminophen or nonsteroidal anti-inflammatories. These drugs are characterized by a ceiling effect, above which increasing the dose will not achieve increased analgesia. High doses of acetomenophen can result in hepatic toxicity and chronic doses of nonsteroidal anti-inflammatories can result in gastrointestinal bleeding as well as renal toxicity, especially in the elderly or those patients with multiple comorbidities. Pain that is persistent despite nonnarcotic management will be best addressed by a step up the ladder to low potency opioids, such as codeine or propoxyphene. These drugs are not characterized by a ceiling effect and are best used for short-term pain management, such as immediately postoperatively or in the acute management of a bony metastasis that is also being treated with radiation. Pain that is still persistent or increases despite low to medium potency opioids mandates a step up to high potency narcotics (Fig. 21.2). These are the mainstay of the management of patients with progressive disease and include morphine, fentanyl and methadone. Long-acting forms are used for scheduled dosing and short-acting forms are used for breakthrough pain. Fentanyl comes in a transdermal form, which is often helpful in patients with gastrointestinal dysfunction. Adjuvant measures can actually be added at any level and are intended to address specific types of pain. The goal is pain control, restful sleep, daytime alertness to maximize and minimize narcotic-induced side effects.

Narcotic-induced side effects include nausea, sedation and constipation. Sedation can be managed by the addition of methylphenidate or dextroamphetamine. Haloperidol, lorazepam, dexamethasone and the serotonin antagonists can be used to treat nausea. Itching can be controlled with hydroxyzine or diphenhydramine, though these may increase sedation. Myoclonus can be managed with clonazepam. Constipation can be a significant problem. Effective bowel regimens in patients on chronic narcotics are critical. Optimal management of constipation can take a variety of approaches and is discussed further in the section on gastrointestinal dysfunction. Switching the long-acting narcotic (e.g., from morphine to methadone) may provide significant relief from side effects without polypharmacy. Decreasing the dose and adding an adjuvant may also help minimize side effects. For patients with intractable symptoms or side effects from narcotics, regional anesthetic blockade via an epidural or intrathecal approach can provide significant relief and allow for significant dose reduction in systemic narcotic use and related side effects.

21

OPIOID PRINCIPLES, PRESCRIBING, TITRATION, AND MAINTENANCE

III. ORAL AND PARENTERAL OPIOID EQUIVALENCES AND RELATIVE POTENCY OF DRUGS AS COMPARED WITH MORPHINE BASED ON SINGLE DOSE STUDIES

Opioid Agonists	Parenteral Dose	Oral Dose	Factor (IV to PO)	Duration of Action[1]	Not Recommended
Codeine	130 mg	200 mg	1.5	3-4 h	Meperidine[8]
Fentanyl[2]	100 µg	--	--	1-3 h	Propoxyphene[8]
Hydrocodone[3]	--	30-200 mg	--	3-5 h	Partial agonist (buprenorphine)[9]
Hydromorphone	1.5 mg	7.5 mg	5	2-3 h	Mixed agonist-antagonist (pentazacine, nalbuphine, butorphanol, dezocine)[9]
Levorphanol[4]	2 mg	4 mg	2	3-6 h	
Methadone[4]	10 mg	3-20 mg[5]	2	4-8 h	
Morphine[6]	10 mg	30 mg	3	3-4 h	
Oxycodone	--	15-20 mg	--	3-5 h	
Oxymorphone	1 mg	10 mg	10	3-6 h	
Tramadol[7]	--	50-100 mg	--	3-7 h	

Special Note: Partial agonists and mixed agonists-antagonists have limited usefulness in cancer pain. They should NOT be used in combination with opioid agonist drugs. Converting from an agonist to an agonist-antagonist could precipitate a withdrawal crisis in the narcotic dependent patient.

1. Shorter time generally refers to parenterally administered opioids (except: for controlled-release products which have more variability); longer time generally applies to oral dosing.
2. Available in transdermal system for sustained dosing (see instructions on PAIN-E 3 of 3) and oral transmucosal or buccal systems for breakthrough pain.
3. Equivalence data not substantiated. Clinical experience suggests use as a mild, initial use opioid but effective dose may vary. Usually combined with ASA or acetaminophen in doses from 325 to 750 mg. Dosage must be monitored for safe limits of ASA or acetaminophen. Dose listed refers only to opioid portion.
4. Long half-life, observe for drug accumulation and side effects after 2-5 days. May need to be dosed every 4 h initially then changed to every 6-8 h after steady state achieved (1-2 wks).
5. With higher doses of morphine, the oral conversion ratio of morphine to methadone may be closer to 10 to 1 rather that 3 to 2. PRACTITIONERS ARE ADVISED TO CONSULT WITH A PAIN SPECIALIST IF THEY ARE UNFAMILIAR WITH METHADONE PRESCRIBING.
6. Conversion factor listed for chronic dosing. Avoid using morphine in renal failure due to accumulation of morphine-6-glucuronide metabolite.
7. Weak opioid receptor agonist with some antidepressant activity. For mild to moderate pain. Recommended dose of 100 mg 4 times a day (maximum daily dose 400 mg) to avoid CNS toxicity. Even at maximum dose 100 mg four times a day, tramadol is less potent than other opioid analgesics such as morphine.
8. Not recommended for long term or high dose use because of CNS toxic metabolites (normeperidine, norpropoxyphene).
9. Partial agonists and mixed agonist-antagonists may produce withdrawal in opioid-dependent patients.

Note: All recommendations are category 2A unless otherwise indicated.
Clinical Trials: NCCN believes that the best management of any cancer patient is in a clinical trial. Participation in clinical trials is especially encouraged.

21

Figure 21.2. Opioid prescribing, titration and maintenance. Reproduced with permission from the NCCN (1.2008) Adult Cancer Pain Guidelines, Clinical Practice Guidelines in Oncology. ©National Comprehensive Cancer Network, 2008. Available at: http://www.nccn.org. Please see the NCCN copyright statement in the legend for Figure 21.1.

Adjuvant medications are intended to address specific components of the pain complex. For bony pain, nonsteroidal anitinflammatories, bisphosphonates and radiation therapy act as adjuvants. For neuropathic pain, dexamethasone, some of the older antidepressants such as amitriptyline and nortriptyline and even clonazepam or carbamazepine may be beneficial. Some patients find relief with a TENS unit. For chemotherapy-induced neuropathy, gabapentin may be beneficial. Muscle spasms can be managed with diazepam, baclofen, or a TENS unit. Pain associated with a malignant ulcer may be due to bacterial superinfection and will respond to treatment with oral metronidazole or clindamycin. Topically, lidocaine jelly 2%, morphine gel 3%, or transdermal lidocaine patches may also provide some relief.

Mood Disorders

Neuropsychiatric issues for patients with progressive malignancy include anxiety, depression, weakness, fatigue and sleep disturbance. Benzodiazepenes provide adequate anxiolysis, and alprazolam has the added benefit of having some antidepressant properties. Depression is underappreciated and undertreated in patients with progressive malignancy. Aggressive treatment of depression can significantly improve quality of life. While the selective serotonin reuptake inhibitors (SSRIs) provide a wonderful class of effective drugs for treating depression, the delayed onset of action may not make them the first choice for patients with a short life expectancy. Fluoxetine can be especially problematic as it can cause anorexia, weight loss, insomnia and nausea and therefore can exacerbate already existing symptoms. Methylphenidate is an excellent choice in patients with depression, sedation from narcotics and psychomotor retardation. It has an almost immediate onset of action and enhances the analgesic effects of narcotics. Methylphenidate should be dosed only in the morning as it can disrupt restful sleep at night if taken after noon. Weakness is often attributable to an underlying cause such as hypokalemia, hypercalcemia, anemia, or narcotics. Management is therefore geared to treating the underlying cause. For patients with anorexia and cachexia in addition to weakness, dexamethasone or megestrol acetate may be helpful.

Gastrointestinal Disorders

Gastrointestinal dysfunction represents the most challenging aspect of caring for women with progressive gynecologic malignancy. It is also the most common pathophysiology contributing to death in women with gynecologic cancers. Recurrent ascites can contribute to pain and poor intestinal motility. As noted previously, carcinomatous ileus results from recurrent disease along the intestinal serosa which inhibits motility, presenting clinically as either a high grade ileus or a bowel obstruction. Radiographic imaging, such as a CT scan of the abdomen and pelvis performed with oral contrast, an upper GI with small bowel follow-through, or a barium enema may reveal diffuse bowel dilation with slow transit time. A carcinomatous ileus is best managed in a palliative fashion with gastric decompression. If radiographic imaging reveals a specific point of obstruction, there are times that this can be managed surgically, even in a palliative care setting.

Anorexia

Anorexia may cause more distress for the family than the patient. If it is not distressing to the patient, it may not require any specific intervention, other than reassurance. The dying patient does not need to eat and, at times, the physician may need to give the patient permission not to eat. Simple dietary modifications—small meals with foods that have high caloric value—may prove beneficial. The patient should be instructed to eat what appeals to her and not worry about "proper" nutrition (yes, it is okay to have ice cream for breakfast). Anorexia may be due to delayed gastric emptying and may respond to prokinetics such as metoclopramide or tegaserod. Corticosteroids and progestational agents (e.g., megesterol acetate) can be employed to stimulate the appetite. If these are unsuccessful, dronabinol may be used.

Nausea

Nausea is often multifactorial and has a complex physiology and hence, may require multiple medications for optimal management. Nausea may contribute to anorexia. Physiologically, nausea can be related to gastrointestinal dysfunction, neuropsychiatric problems, treatment itself and metabolic derangements. Gastrointestinal dysfunction can include oral thrush, delayed gastric emptying (gastroparesis), carcinomatous ileus, ascites, or constipation. If ascites is present-paracentesis may be beneficial for management of nausea. Neuropsychiatric problems, such as anxiety or brain metastasis, may also be contributory or causative. Treatment itself, such as chemotherapy, radiation, or medications, can contribute to the sensation of nausea. As noted previously, changing the type of long-acting narcotic can improve symptoms. An attempt to decrease or discontinue all unnecessary medications should be made. Metabolic disturbances including uremia, hyperkalemia, hyponatremia and hypomagnesemia can also contribute to nausea.

Effective treatment of nausea is often best directed at identifying and treating the underlying problems. Simple interventions, such as avoiding strong or unpleasant odors, small frequent meals and acupuncture can be beneficial. Pharmacologic management of nausea may require multiple medications. As noted previously, promotility agents may help with ileus and the sensation of nausea. Metoclopromide is particularly effective for chemotherapy-induced nausea, but side effects include sedation, akesthesia and dystonic reactions. Octreotide can be used to decrease gastrointestinal secretions. Central causes may respond to corticosteroids. Haloperidol also has an extremely effective central action to treat nausea. Antihistamines such as hydroxyzine are effective for metabolic related nausea as well as treating the extrapyradimal side effects of other medications. Benzodiazepenes are effective for anticipatory emesis. Phenothiazines are effective in radiation-induced nausea, but again, extrapyramidal side effects may limit usefulness. Anticholinergics are inferior in efficacy to the phenothiazines, but scopolamine is useful in patients with carcinomatous ileus due to its transdermal delivery system and its efficacy in decreasing retrograde gastrointestinal motility. Cannabinoids such as dronabinol are also effective antinausea agents in addition to their appetite stimulating properties. Certainly, management of nausea was revolutionized when the selective 5-hydroxytriptamine 3 (5-HT3) receptor antagonists became available in the 1990s. This class of drugs includes granisetron and ondansetron. These are extremely effective medications but are also the most expensive. Given their cost, in a palliative care setting, these drugs are best utilized only if all other pharmacologic interventions are ineffective. One approach to antinausea management is to start with metoclopromide, then add haloperidol or dexamethasone; if nausea persists, then add an antihistamine or dexamethasone if not already initiated. If all of these measures are ineffective, then try a 5-HT3 receptor antagonist.

Constipation

Constipation is another common gastrointestinal problem with a multitude of causes, including medications, dehydration, diet, physical inactivity and electrolyte abnormalities. History and physical examination, including digital rectal exam, helps to guide intervention. Often, simple measures such as increasing fluid intake, adding additional fiber to the diet and increasing physical activity will improve bowel function. For the patient with an abdomen filled with soft stool but minimal stool in the rectum, an osmotic cathartic such as lactulose or a stimulant laxative is effective. For the patient with soft stool filling the rectum, a peristalsis-stimulating laxative such as senna or bisacodyl will lead to evacuation. In the patient with painful defecation due to hemorrhoids or an anal fissure, bulk-forming agents such as psyllium in addition to stool softeners can improve function. Impacted stool should be manually disimpacted. Acute constipation can be managed with enemas such as soap suds or milk of molasses.

Pharmacologic management of constipation is again complex and geared towards addressing root causes. Agents that directly stimulate peristalsis include senna and bisacodyl. Lubricants, such as mineral oil, allow for easier passage of the stool but should not be combined with docusate. Osmotic cathartics, such as lactulose and sorbitol, draw fluid into the bowel lumen and stimulate peristalsis. Lactulose is effective for chronic use and is simple to titrate to efficacy. Stool softeners like docusate add fluid to the stool but do not have much effect on peristalsis. Bulk-forming soluble fiber increase stool bulk and soften stool consistency. For patients with narcotic-induced constipation, changing the narcotic may improve symptomatology. An alternative approach is naloxone given orally. This opioid antagonist acts directly on opioid receptors in the gut wall to counteract the effect of narcotic locally. This is not systemically absorbed and therefore will not precipitate pain crisis or narcotic withdrawl. One stepwise approach is to start with lifestyle and dietary modifications. It is important to be as proactive as possible. If this is unsuccessful, addition of a stimulant and a stool softener may be effective. If not, adding an osmotic cathartic and, if further intervention is required, periodic enemas may prove beneficial.

Carcinomatosis

The most challenging aspect of managing gynecologic cancer patients at end of life is the bowel dysfunction that often accompanies progressive peritoneal malignancy, known as carcinomatous ileus. Clinically these patients present with intractable nausea, vomiting and obstipation. Medical management can include scopolamine to decrease retrograde peristalsis and treat nausea, octreotide to decrease gastrointestinal secretions, stimulant cathartics, promotility agents, antinausea medications, adequate analgesia and periodic enemas. Intravenous hydration should be avoided, as it only prolongs the dying process. If the patient experiences colicky pain, promotility agents and senna should be discontinued and an antispasmodic such as loperamide or scopolamine prescribed. In the patient with significant large volume ascites, paracentesis may be beneficial. Aggressive surgery should be avoided unless radiographic imaging demonstrates one specific point of obstruction. For the patient with persistent nausea or large volume secretions, gastric decompression is beneficial. Initially, nasogastric suction is used to acutely decompress the gastrointestinal tract. For persistent significant drainage, consideration should be made for placement of a gastrostomy tube. This is not intended for feeding but for venting the intestinal tract via gravity. This simplifies home care and improves patient appearance and comfort. As the tube allows for decompression, the patient can then take liquids orally, which simply pass out through the tube. Large particulate solid food should be avoided, as it can lead to obstruction of the tube and hence exacerbation of symptoms, although soft foods are acceptable. The largest bore tube possible should be placed. Placement can be performed endoscopically in the GI suite, percutaneously by interventional radiology, or via an open approach in the operating room. The most difficult aspect of caring for patients with a carcinomatous ileus is the recognition that this is a terminal event and efforts should be intensely focused on improving patient comfort.

Ascites

Ascites is often a problem for women with progressive gynecologic malignancy; it effects 60% of terminal ovarian cancer patients. Small volume ascites may respond to diuretics such as spironolactone. Large volume, tense, symptomatic ascites may be managed with periodic paracentesis. Complications of this intervention include the introduction of bacteria into the peritoneal cavity, subacute bacterial peritonitis and further depletion of already low albumin stores with worsening of the ascites. Rarely, peritoneo-venous shunts are placed. Chronic indwelling catheters placed for frequent paracentesis have recently become available and are utilized by some hospice companies. Avoidance of hypotonic intravenous fluids will help minimize this problem.

Gastrointestinal disorders represent a significant percentage of the end-organ dysfunction seen in women with progressive gynecologic malignancies. It is important not to be overwhelmed by the choices of pharmacologic interventions available. A systematic and simplified approach works well. Not every medication will be effective for every patient. Therefore, flexibility and willingness to try several different approaches to achieve the desired result is important. Above all, open and honest communication with the patient about her wishes and the goals of care is critical.

Respiratory Disorders

The respiratory system is adversely affected by progressive gynecologic malignancy. Most commonly, dysfunction presents as shortness of breath or dyspnea on exertion. The sensation of air hunger may be due to anemia, anxiety, parenchymal metastasis, radiation pneumonitis or pulmonary fibrosis related to prior therapy, lymphangitic spread of tumor, or recurrent malignant pleural effusions. Exacerbation of underlying medical conditions, such as chronic obstructive pulmonary disease or congestive heart failure may initially be confused with progressive malignancy. Reversible causes should be sought. The addition of an anxiolytic even in the patient who denies anxiety is often beneficial. Simple measures such as upright positioning, avoidance of strong odors, a well-ventilated and cooled room and relaxation techniques can be very effective for symptom management. A short course of steroids may significantly improve symptoms related to radiation pneumonitis and may have some effect on lymphangitic spread. Supplemental oxygen may be required to improve patient comfort and should be provided. Nebulizer treatments are the mainstay of managing patients with shortness of breath. Albuterol is effective for wheezing. Narcotic nebulizer treatments such as fentanyl or morphine mixed with dexamethasone can be very beneficial. Nebulized furosemide may also provide some symptomatic relief. Respiratory panic can be managed with midazolam, benzodiazapenes and intravenous morphine. For the patient with large volume ascites where shortness of breath is a result of poor diaphragmatic excursion, paracentesis may be beneficial.

Recurrent pleural effusions may present with cough, dyspnea, tachypnea, or shortness of breath. Thoracentesis may provide symptomatic relief. Removal of greater than one liter at a time may result in re-expansion pulmonary edema with resultant worsening of respiratory compromise. If pleural fluid reaccumulates, pleurodesis with talc or bleomycin should be considered. Video assisted thorascopic sclerosis (VATS) has been shown to have higher efficacy and lower reaccumulation rates associated with a shorter hospitalization and has become the preferred approach. Additionally, similar to the chronic indwelling paracenetesis catheters that some patients benefit from, indwelling catheters can be placed in the pleural cavity to allow the patient to drain her effusions for palliation of symptoms on an as needed basis in the home setting.

21

Bone Metastases

Bony metastases can cause significant pain. Hypercalcemia is seen in 50% and pathologic fractures in 10% of patients with bony metastasis. Bone scans are more sensitive than plain radiographs for diagnosing bony metastasis. Pain management follows along the basic principles of the WHO three step ladder discussed previously, although acetomenophen and nonsteroidal antiinflamatories are often effective. Radiation is extremely effective at providing relief from pain and symptoms. Additionally, bisphosphonates and intralesional local anesthetics may also provide relief. Lytic metastasis greater than 2.5 cm in diameter or that result in more than 50% cortical bone loss are at high risk for pathologic fracture. Pathologic fracture can cause a significant detriment to quality of life and is best managed preventatively. Patients with a life expectancy of more than 2 months and a performance status of 2 or better are good candidates for primary internal fixation to prevent pathologic fracture.

Hypercalcemia may be secondary due to bony metastasis or a primary paraneoplastic process due to production of a parathyroid hormone-like substance by the tumor. Symptoms include pain, polyuria, dehydration, anorexia, nausea, vomiting, weakness and confusion. Hydration is the first step in management. Corticosteroids, calcitonin and bisphosphonates may then provide additional benefit.

Genitourinary Dysfunction

Patients with progressive gynecologic malignancy may also have site-specific problems related to the genitourinary system. Dysuria can be managed with oral phenazopyridine or lidocaine bladder washes. Bladder spasms may be due to reversible causes that should be treated directly, such as infection or fecal impaction. If an indwelling Foley is present, changing the catheter may relieve symptoms. Pharmacologically, oxybutinin, imipramine or amytriptyline may be beneficial. B and O (belladonna and opium) suppositories placed rectally or intravaginally are also very effective, particularly in the patient who has difficulty taking oral medications.

Hematuria may be mild, without anemia, clotting and urinary retention; or it may be severe. Mild cases of hematuria are often precipitated by a urinary tract infection and will improve with initiation of antibiotic therapy. For patients with significant hematuria associated with clotting and urinary retention, a large bore three way Foley catheter should be placed and the bladder irrigated to remove clots. If this is not effective in clearing the hematuria, cystoscopy with fulguration is usually effective. Rarely, installation of 1% alum solution or formaldehyde is required to control hematuria. Palliative radiation can be considered, but this is usually not necessary.

Ureteral obstruction often requires no intervention. If hydronephrosis results in pain or infection, stent placement or percutaneous nephrostomy may improve symptoms. If bilateral ureteral obstruction is noted, no intervention should be undertaken. Renal failure is one of the most painless and comfortable ways to expire. If renal failure is acute, electrolyte abnormalities such as hyperkalemia may lead to cardiac dysrhythmias and sudden death. Subacute or chronic ureteral obstruction slowly leads to the accumulation of urates in the blood stream with progressive somnolence and ultimately, death without pain over a 2-6 week period. Urate crystals may be noted on the tongue. For patients with progressive malignancy, intervention to relieve bilateral ureteral obstruction merely prolongs the dying process, leading to a more painful and lingering death and should be avoided. Again, open communication with the patient and her family is critical in helping them understand these processes.

Patients with advanced gynecologic malignancies involving the vulva, groin, or central pelvis may be concerned about and embarrassed by odor from necrotic tumor. Often odor comes from anaerobic bacterial overgrowth, which can be improved significantly through the use of oral antibiotics such as metronidazole or clindamycin or topically applied metronidazole gel. The area should be cleansed daily with a disinfectant solution such as Dakin's (sodium hypochlorite) or chlorhexidine. Activated charcoal slurry or charcoal impregnated dressings are also effective for absorbing odors and baby diapers act as excellent overdressings for areas of necrotic tumor with significant drainage that make personal hygiene difficult. Raw areas of tumor involving the groin or perineum that are irritated with dressing changes can be dressed with xeroform gauze (petrolatum impregnated dressings) to improve local comfort. Pain from skin involvement with tumor can be managed with topically applied aluminum with magnesium hydroxide antacid, morphine gel, or lidocaine gel. Locally directed radiation may also be used for palliation of symptoms.

Hospice Care

End of life care should be family centered, respectful of the patient's wishes and realistic in terms of goals (Figs. 21.3 and 21.4). Many resources exist to assist patients and their families at end of life. Inpatient palliative care units, home hospice and inpatient hospice facilities all provide resources to support patients at end of life. Palliative care medicine is a rapidly growing field and these specialists can provide excellent guidance for difficult issues such as intractable pain. Hospice certification requires that the treating physician certify that the patient has 6 months or less to live. If patients exceed this time span, they are recertified for hospice. Patients and families derive the greatest benefit from hospice if they are enrolled prior to the last few days of life. Therefore, a hospice discussion is best initiated at the time of evidence of progression of disease that prompts a change in or discontinuation of therapy rather than when death is imminent. Home hospice benefits the patient and family by allowing for symptoms to be addressed in the home environment by skilled care givers who have immediate access to medications without the patient needing to come to the physician's office. In most cases enrollment in hospice does require a signed do not resuscitate order.

A do not resuscitate (DNR) order should be discussed with the patient and her family well before end of life. This allows for the patient to clearly express her wishes regarding life sustaining therapies. A popular construct is Five Wishes, which allows the patient to specify their desires regarding end of life care. Upon initiating this discussion, it is important for the physician to reassure the patient that signing a DNR order does not mean that she will not receive further care. Care is still provided but the goals of that care shift from active anticancer management to aggressive symptom management. Many patients have an unconscious or conscious fear of abandonment by their physician, particularly if they elect to not pursue further anticancer therapy. The physician must address this concern openly in a caring fashion. Once the DNR order is signed, a copy should be placed in the patient's medical record and another copy should be kept in a prominent location in the home. This allows for respect of the patient's wishes in the event that emergency response personnel become involved in the patient's care.

Palliative care is about improving quality of life at end of life without prolonging life. Optimally delivered palliative care should include aggressive pain management, assessment for the need for antidepressants and anxiolytics and a careful determination of other symptoms that require intervention. Family and friends should be involved as

21

21

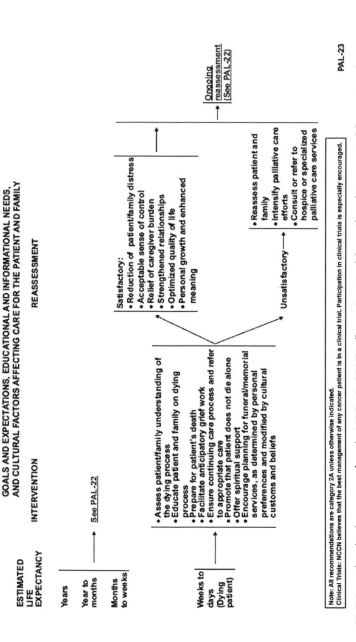

Figure 21.3. Reproduced with permission from the NCCN (1.2009) Palliative Care Guidelines, Clinical Practice Guidelines in Oncology. ©National Comprehensive Cancer Network, 2009. Available at: http://www.nccn.org. Please see the NCCN copyright statement in the legend for Figure 21.1.

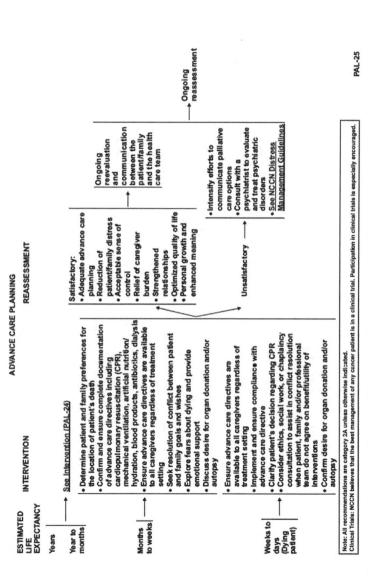

ADVANCE CARE PLANNING

ESTIMATED LIFE EXPECTANCY | INTERVENTION | REASSESSMENT

Years

Year to months → See Intervention (PAL-24)

- Determine patient and family preferences for the location of patient's death
- Confirm and ensure complete documentation of advance care directives including cardiopulmonary resuscitation (CPR), mechanical ventilation, artificial nutrition/hydration, blood products, antibiotics, dialysis

Months to weeks →

- Ensure advance care directives are available to all caregivers regardless of treatment setting
- Seek resolution of conflict between patient and family goals and wishes
- Explore fears about dying and provide emotional support
- Discuss desire for organ donation and/or autopsy

Weeks to days (Dying patient) →

- Ensure advance care directives are available to all caregivers regardless of treatment setting
- Implement and ensure compliance with advance care directive
- Clarify patient's decision regarding CPR
- Consider ethics, social work, or chaplaincy consultation to assist in conflict resolution when patient, family and/or professional team do not agree on benefit/utility of interventions
- Confirm desire for organ donation and/or autopsy

Satisfactory:
- Adequate advance care planning
- Reduction of patient/family distress
- Acceptable sense of control
- Relief of caregiver burden
- Strengthened relationships
- Optimized quality of life
- Personal growth and enhanced meaning

Ongoing reevaluation and communication between the patient/family and the health care team

→ Ongoing reassessment

Unsatisfactory →
- Intensify efforts to communicate palliative care options
- Consult with a psychiatrist to evaluate and treat psychiatric disorders
- See NCCN Distress Management Guidelines

Note: All recommendations are category 2A unless otherwise indicated.
Clinical Trials: NCCN believes that the best management of any cancer patient is in a clinical trial. Participation in clinical trials is especially encouraged.

PAL-25

Figure 21.4, Reproduced with permission from the NCCN (1.2009) Palliative Care Guidelines, Clinical Practice Guidelines in Oncology. ©National Comprehensive Cancer Network, 2009. Available at: http://www.nccn.org. Please see the NCCN copyright statement in the legend for Figure 21.1.

much as possible, and care should be home centered whenever possible, minimizing costs and treating death truly as a natural process.

The past 10 years has seen a dramatic shift from death in the inpatient setting to home hospice. However, there is still not a large body of literature addressing palliative care management in gynecologic oncology. With increasing physician emphasis on quality of life and improvement in research methods, there is more research to be done in this arena looking at issues of palliative chemotherapy versus supportive care, malignant bowel obstruction and pain management. Other areas of future research include cost analysis of end of life care and symptom management. Obviously, determining the appropriate timing for hospice referral in a patient with progressive malignancy remains a difficult question, and research aimed at determining objective parameters to use as an indicator for hospice referral is also an area for further development.

Future Directions

The past 10 years has seen a dramatic shift from death in the inpatient setting to home hospice. However, there is still not a large body of literature addressing palliative care management in gynecologic oncology. With increasing physician emphasis on quality of life and improvement in research methods, there is more research to be done in this arena looking at issues of palliative chemotherapy versus supportive care, malignant bowel obstruction and pain management. Other areas of future research include cost analysis of end of life care and symptom management. Obviously, determining the appropriate timing for hospice referral in a patient with progressive malignancy remains a difficult question, and research aimed at determining objective parameters to use as an indicator for hospice referral is also an area for further development.

Suggested Reading

1. WallerA, Caroline NL. Handbook of Palliative Care in Cancer, 2nd Ed. Butterworth Heinemann, 2000. An excellent practical guide to symptom management in cancer patients.
2. ASCO Curriculum: Optimizing Cancer Care—The Importance of Symptom Management. Kendall/Hunt Publishing Company, 2001. Practical management guide and slide set. Good for self study as well as for staff education.
3. Berger AM, Portenoy RK, Weissman DE, eds. Principles and Practice of Palliative Care and Supportive Oncology, Second Edition. Lippincott Williams and Wilkins, 2002. Definitive textbook for more advanced study.

21

Updated FIGO Staging (2008)

Table A1. Carcinoma of the vulva

Stage	Definition
I	Tumor confined to the vulva
IA	Lesions ≤2 cm in size, confined to the vulva or perineum and with stromal invasion ≤1.0 mm*, no nodal metastasis
IB	Lesions >2 cm in size or with stromal invasion >1.0 mm*, confined to the vulva or perineum, with negative nodes
II	Tumor of any size with extension to adjacent perineal structures (1/3 lower urethra, 1/3 lower vagina, anus) with negative nodes
III	Tumor of any size with or without extension to adjacent perineal structures (1/3 lower urethra, 1/3 lower vagina, anus) with positive nodes
IIIA	(i) With 1 lymphnodal metastasis (≥5 mm), or (ii) 1-2 lymphnodal metastasis(es) (<5 mm)
IIIB	(i) With 2 or more lymphnodal metastases (≥5 mm), or (ii) 3 or more lymphnodal metastases (<5 mm)
IIIC	With positive nodes with extracapsular spread
IV	Tumor invades other regional (2/3 upper urethra, 2/3 upper vagina), or distant structures
IVA	Tumor invades any of the following: (i) Upper urethral and/or vaginal mucosa, bladder mucosa, rectal mucosa, or fixed to pelvic bone, or (ii) Fixed or ulcerated femoral-inguinal lymph nodes
IVB	Any distant metastasis including pelvic lymph nodes

*The depth of invasion is defined as the measurement of the tumor from the epithelial-stromal junction of the adjacent most superficial dermal papilla to the deepest point of invasion.

Reprinted with permission granted by the International Federation of Gynecology and Obstetrics. FIGO Committee on Gynecologic Oncology. Revised FIGO staging for carcinoma of the vulva, cervix, and endometrium. Int J Gynecol Obstet 2009; 105(2): in press.

Gynecologic Oncology, edited by Paola Gehrig and Angeles Secord.
©2009 Landes Bioscience.

Table A2. Carcinoma of the cervix uteri

Stage	Definition
I	The carcinoma is strictly confined to the cervix (extension to the corpus would be disregarded)
IA	Invasive carcinoma which can be diagnosed only by microscopy, with deepest invasion ≤5 mm and largest extension ≥7 mm
IA1	Measured stromal invasion of ≤3.0 mm in depth and extension of ≤7.0 mm
IA2	Measured stromal invasion of >3.0 mm and not >5.0 mm with an extension of not >7.0 mm
IB	Clinically visible lesions limited to the cervix uteri or preclinical cancers greater than stage IA*
IB1	Clinically visible lesion ≤4.0 cm in greatest dimension
IB2	Clinically visible lesion >4.0 cm in greatest dimension
II	Cervical carcinoma invades beyond the uterus, but not to the pelvic wall or to the lower third of the vagina
IIA	Without parametrial invasion
IIA1	Clinically visible lesion ≤4.0 cm in greatest dimension
IIA2	Clinically visible lesion >4 cm in greatest dimension
IIB	With obvious parametrial invasion
III	The tumor extends to the pelvic wall and/or involves lower third of the vagina and/or causes hydronephrosis or nonfunctioning kidney**
IIIA	Tumor involves lower third of the vagina, with no extension to the pelvic wall
IIIB	Extension to the pelvic wall and/or hydronephrosis or non-functioning kidney
IV	The carcinoma has extended beyond the true pelvis or has involved (biopsy proven) the mucosa of the bladder or rectum. A bullous edema, as such, does not permit a case to be allotted to Stage IV
IVA	Spread of the growth to adjacent organs
IVB	Spread to distant organs

*All macroscopically visible lesions—even with superficial invasion—are allotted to stage IB carcinomas. Invasion is limited to a measured stromal invasion with a maximal depth of 5.00 mm and a horizontal extension of not >7.00 mm. Depth of invasion should not be >5.00 mm taken from the base of the epithelium of the original tissue—superficial or glandular. The depth of invasion should always be reported in mm, even in those cases with "early (minimal) stromal invasion" (~1 mm). The involvement of vascular/lymphatic spaces should not change the stage allotment.
**On rectal examination, there is no cancer-free space between the tumor and the pelvic wall. All cases with hydronephrosis or non-functioning kidney are included, unless they are known to be due to another cause.

Reprinted with permission granted by the International Federation of Gynecology and Obstetrics. FIGO Committee on Gynecologic Oncology. Revised FIGO staging for carcinoma of the vulva, cervix, and endometrium. Int J Gynecol Obstet 2009; 105(2): in press.

Table A3. Carcinoma of the endometrium

Stage	Definition
I*	Tumor confined to the corpus uteri
IA*	No or less than half myometrial invasion
IB*	Invasion equal to or more than half of the myometrium
II*	Tumor invades cervical stroma, but does not extend beyond the uterus**
III*	Local and/or regional spread of the tumor
IIIA*	Tumor invades the serosa of the corpus uteri and/or adnexae#
IIIB*	Vaginal and/or parametrial involvement#
IIIC*	Metastases to pelvic and/or para-aortic lymph nodes#
IIIC1*	Positive pelvic nodes
IIIC2*	Positive para-aortic lymph nodes with or without positive pelvic lymph nodes
IV*	Tumor invades bladder and/or bowel mucosa, and/or distant metastases
IVA*	Tumor invasion of bladder and/or bowel mucosa
IVB*	Distant metastases, including intra-abdominal metastases and/or inguinal lymph nodes

*Either G1, G2, or G3.
**Endocervical glandular involvement only should be considered as Stage I and no longer as Stage II.
#Positive cytology has to be reported separately without changing the stage.
Reprinted with permission granted by the International Federation of Gynecology and Obstetrics. FIGO Committee on Gynecologic Oncology. Revised FIGO staging for carcinoma of the vulva, cervix, and endometrium. Int J Gynecol Obstet 2009; 105(2): in press.

Table A4. Staging for uterine sarcomas (leiomyosarcomas, endometrial stromal sarcomas, adenosarcomas, and carcinosarcomas)

(1) Leiomyosarcomas

Stage	Definition
I	Tumor limited to uterus
IA	<5 cm
IB	>5 cm
II	Tumor extends to the pelvis
IIA	Adnexal involvement
IIB	Tumor extends to extrauterine pelvic tissue
III	Tumor invades abdominal tissues (not just protruding into the abdomen)
IIIA	One site
IIIB	> one site
IIIC	Metastasis to pelvic and/or para-aortic lymph nodes
IV	
IVA	Tumor invades bladder and/or rectum
IVB	Distant metastasis

(2) Endometrial stromal sarcomas (ESS) and adenosarcomas*

Stage	Definition
I	Tumor limited to uterus
IA	Tumor limited to endometrium/endocervix with no myometrial invasion
IB	Less than or equal to half myometrial invasion
IC	More than half myometrial invasion
II	Tumor extends to the pelvis
IIA	Adnexal involvement
IIB	Tumor extends to extrauterine pelvic tissue
III	Tumor invades abdominal tissues (not just protruding into the abdomen)
IIIA	One site
IIIB	> one site
IIIC	Metastasis to pelvic and/or para-aortic lymph nodes
IV	
IVA	Tumor invades bladder and/or rectum
IVB	Distant metastasis

(3) Carcinosarcomas

Carcinosarcomas should be staged as carcinomas of the endometrium.

*Note: Simultaneous tumors of the uterine corpus and ovary/pelvis in association with ovarian/pelvic endometriosis should be classified as independent primary tumors. Reprinted with permission granted by the International Federation of Gynecology and Obstetrics. FIGO Committee on Gynecologic Oncology. FIGO staging for uterine sarcomas. Int J Gynecol Obstet 2009; 104(3):179.

Index